GAUGE THEORIES AND MODERN FIELD THEORY

The MIT Press

Cambridge, Massachusetts, and London, England

GAUGE THEORIES AND MODERN FIELD THEORY

Proceedings of a Conference
held at
Northeastern University, Boston
September 26 and 27, 1975

Edited by
Richard Arnowitt and Pran Nath

PUBLISHER'S NOTE

This format is intended to reduce the cost of publishing certain works in book form and to shorten the gap between editorial preparation and final publication. The time and expense of detailed editing and composition in print have been avoided by photographing the text of this book directly from the authors' typescripts.

Printed in the United States of America

Second printing, 1977

Library of Congress Cataloging in Publication Data

Conference on Gauge Theories and Modern Field
Theory, Northeastern University, 1975.
Gauge theories and modern field theory.

1. Gauge fields (Physics)--Congresses.
2. Field theory (Physics)--Congresses.
I. Arnowitt, Richard. II. Nath, Pran, 1939-
III. Title.
QC793.3.F5C66 1975 530.1'4 76-5836
ISBN 0-262-01046-1

CONTENTS

PREFACE

The Conference on Gauge Theories and Modern Field Theory was held at Northeastern University on September 26 and 27, 1975. Over 190 high energy theorists and graduate students attended. A majority of the institutions in the United States carrying on research in this area were represented, as well as a number of institutions from outside the United States. During the past few years, there has been a large increase in the use of field theory as a framework for understanding high energy phenomena. This includes work on the structure of gauge theories, unified theories of interactions, theories of quark confinement, supersymmetry and coherent state phenomena. Several of these approaches involve innovative methods of applying field theory, and perhaps some have the possibility of developing into fundamental theories. Research has been progressing at a rapid pace and whole new areas have recently sprung up. This volume contains the talks given at the Conference. We hope that the book will be a useful reference for high energy theorists already working in this area, as well as a helpful introduction to other theorists and experimentalists who wish to learn the present status of the field.

The Program Committee of the Conference consisted of J. D. Bjorken (SLAC-Stanford), L. D. Faddeev (Steclov), D. Gross (Princeton), R. Jackiw (MIT), B. W. Lee (FNAL), P. Nath (Northeastern), and S. Weinberg (Harvard). We should like to express our thanks to the Program Committee for their many months of work involving many meetings, telephone conferences, and mail correspondence. Their work in deciding which topics were most important to be presented, and who could best present the material, was crucial in making the Conference a success.

The Organizing Committee of the Conference consisted of R. Aaron, R. Arnowitt, H. Goldberg, P. Nath, G. Pancheri-Srivastava, Y. Srivastava, M. Vaughn, E. von Goeler and R. Weinstein, all of Northeastern. In addition, P. Argyres,

D. Garelick, M. Gettner, M. Glaubman, E. Saletan, and C. Shiffman also helped with organizing the Conference. A number of Northeastern graduate students contributed large amounts of time and effort to the running of the Conference. Some of these were G. Blanar, C. Boyer, H. Buhay, D. Cardarelli, N. Chase, D. Chiaverini, S. Kanavi, and J. Kwiecien. We should like to take this opportunity to thank all these people for the work that they did. Particular thanks are due to E. von Goeler and G. Blanar for the many hours they spent in arranging the audio visual aspects of the Conference, and to the special efforts of R. Aaron and H. Goldberg. We should also like to thank the secretarial staff, B. Cairns, H. Gettner, and G. Tang, who labored unstintingly before, during and after the Conference. Certainly it was the dedicated efforts of the Conference Secretary, G. Tang, which allowed the Conference to run as smoothly as it did, and it is a special pleasure to thank her for her time and effort. Many members of the Northeastern University community aided in the organization of the Conference. We should like to express our appreciation to President K. Ryder and Deans K. Ballou, M. Essigmann, and S. Herman for the support they gave us.

The financial aid received from the United States Energy Research and Development Administration, the United States National Science Foundation, and Northeastern University are gratefully acknowledged. Their sponsorship made the success of the Conference possible. The Organizing Committee felt that some space should be reserved for graduate student attendees since students are almost completely excluded from the large international high energy conferences. We are pleased to say that about 40 graduate students attended this Conference (without registration fee).

R. Arnowitt
P. Nath
Organizing Committee
Chairmen and Editors

CONTRIBUTORS TO THE CONFERENCE PROCEEDINGS

S. ADLER, The Institute for Advanced Study

P. W. ANDERSON, Bell Laboratories

J. M. CORNWALL, University of California - Los Angeles

R. DASHEN, The Institute for Advanced Study

A. DE RUJULA, Harvard University

S. DESER, Brandeis University

S. L. GLASHOW, Harvard University

F. GURSEY, Yale University

R. JACKIW, Massachusetts Institute of Technology

B. W. LEE, Fermi National Accelerator Laboratory

P. NATH, Northeastern University

A. PAIS, Rockefeller University

J. C. PATI, University of Maryland

J. SCHWINGER, University of California - Los Angeles

S. WEINBERG, Harvard University

K. WILSON, Cornell University

B. ZUMINO, CERN

GAUGE THEORIES AND MODERN FIELD THEORY

GAUGE SYMMETRY BREAKING

Steven Weinberg*

*Lyman Laboratory of Physics, Harvard University, Cambridge, Mass. 02138.
Work supported in part by the National Science Foundation, Grant MPS75-20427.

1. Varieties of Spontaneous Symmetry Breaking

It is widely believed that the weak, electromagnetic, and perhaps the strong interactions are governed by exact gauge symmetries, some of which are spontaneously broken.[1] Right now there is intense interest in the problem of how to put the newly discovered particles, which to some extent were anticipated in gauge theories, into detailed gauge models. Today I would like to talk about a different problem, which however underlies all attempts at model building. It is the problem of how the gauge symmetries of the weak interactions get broken.

Let us begin by recalling the various possibilities. For every spontaneously broken symmetry there is a Goldstone boson, which may or may not be eliminated by the Higgs mechanism. The nature of the symmetry breaking can therefore be categorized according to the nature and properties of the Goldstone bosons. First, the Goldstone bosons may be elementary or composite. In the former case, the Lagrangian will contain elementary scalar fields, whose vacuum expectation values determine the Fermi couplings:

$$\langle\phi\rangle \approx G_F^{-\frac{1}{2}} \approx 300 \text{ GeV} \tag{1.1}$$

Also, the vector bosons will have masses of order

$$m_W \approx e\langle\phi\rangle \approx 50 \text{ GeV} \tag{1.2}$$

However, we have a free choice of the magnitudes of the coupling constants in both the Yukawa interaction $\overline{\psi}\Gamma\psi\phi$ and the scalar self-interaction $\lambda\phi^4$. Three cases may be distinguished:

I. $\underline{\Gamma<<1,\ \lambda<<1}$: Gauge theories of this type are the most familiar, and have by far the greatest predictive power. The small value of the Yukawa coupling constant Γ implies that the observed weak interactions are predominately produced by vector boson exchange, and the small value of the self-coupling λ implies that the ratios of vector boson masses (e.g. m_Z/m_W) are determined by weak mixing angles in the usual way. Also, these theories necessarily involve Higgs bosons with masses of order

$$M_{HIGGS} \approx \lambda^{\frac{1}{2}} \langle\phi\rangle << 300 \text{ GeV} \tag{1.3}$$

This mass plays the role of an effective ultraviolet cut-off in weak interaction processes, so the integrals for weak radiative corrections will generally converge at momenta much less than 300 GeV. (There are indications[2] that this argument can be turned around: it is the low cut-off that implies all the relations among couplings and masses usually associated with gauge theories.)

II. $\underline{\Gamma<<1,\ \lambda\approx 1}$: The weak interactions here are still predominately due to exchange of single vector bosons, and the couplings of the vector bosons to quarks and leptons obey the usual gauge-theoretic constraints. However, in this case we would know nothing about the ratios of vector boson masses. The Higgs boson mass would be of order 300 GeV, which is another way of saying that we really know nothing about the properties or even the existence of the Higgs bosons in such theories. Similarly, the effective ultraviolet cut-off would be at the "unitarity" value of 300 GeV. As in Case I, the quark and lepton masses would be of order

$$M_{FERMION} \approx \Gamma \langle \phi \rangle << 300 \text{ GeV} \qquad (1.4)$$

suggesting that Γ might be of order 10^{-2} to 10^{-3}.

III. $\underline{\Gamma \approx 1,\ \lambda \approx 1}$: Unlike Cases I and II, the weak interactions here could be due <u>both</u> to multiple scalar boson exchange and to exchange of single vector bosons; the two would give contributions of about the same order of magnitude. (Even though strongly coupled, the scalars here would have such large masses that their exchange would generate Fermi interactions of order G_F.) Also unlike Cases I and II, the fermion masses $\Gamma \langle \phi \rangle$ would at first glance appear to be of order 300 GeV. This is much too high, so one is led to suppose that the Yukawa couplings and the vacuum expectation value $\langle \phi \rangle$ must respect some zeroth-order symmetry, which keeps some of the fermion masses zero to zeroth order in e. If (as is often the case) this symmetry is <u>not</u> respected by the gauge couplings, then fermion masses will arise, of order

$$M_{FERMION} \approx e^2 \times 300 \text{ GeV} \qquad (1.5)$$

which is not unacceptable. That part of the observed weak interactions produced by multiple exchanges of scalar bosons will respect the zeroth order symmetry of $\langle \phi \rangle$ and Γ. This symmetry is supposed to keep the ordinary quarks and leptons massless to zeroth order in e, so it is likely to include chiral transformations on the fermions; the contribution of multiple scalar exchanges to the effective Fermi interaction would then be of the vector/axial-vector type, just like the vector boson exchange contribution. However, the two parts of the weak interaction might be distinguished by other symmetry properties, and certainly could eventually be separated by their space-time structure at energies above m_W.

Finally, we have the possibility that the symmetry breaking is dynamical,[3] by which is meant that the Goldstone bosons are composite particles. Any composite particle must have strong coupling to its constituents, so theories with dynamical symmetry breaking are similar to the above theories of type III. There are however important differences, to be discussed at the end of this talk.

(Incidentally, I have not included here the possibility that $\lambda << 1$ and $\Gamma \approx 1$. This is because strong Yukawa couplings will always generate strong ϕ^4 couplings, while strong ϕ^4 couplings cannot themselves generate strong Yukawa couplings.)

The models of type I are certainly the most attractive, especially so because, despite their greater predictive power, they have done well so far in comparison with observation. Why then should we torture ourselves with these other possibilities? The reason is that no one has yet been able to find a theory of type I in which some of the gauge symmetries are much more strongly broken than others.

In the balance of this talk I will discuss some old and some new reasons why such a hierarchy of gauge symmetry breaking is needed; then I will indicate why gauge hierarchies do not seem to arise in theories of type I, or in a related but somewhat different class of theories with elementary scalars; and then I will discuss the implications of one kind of theory which can exhibit a gauge hierarchy, the theories with dynamical symmetry breaking.

2. Gauge Hierarchies

There is a familiar old argument why some gauge symmetries must be broken much more strongly than others.[4] The gauge group that describes the ordinary weak and electromagnetic interactions appears to be something like $SU(2)\otimes U(1)$. This is not simple, so there is more than one free gauge coupling constant, and we have no a priori way to determine the mixing angle. (The problem is even more acute if we want to unify the strong with the weak and electromagnetic interactions.[5]) In order to have a more satisfying theory, we therefore suppose that $SU(2)\otimes U(1)$ is a subgroup of a larger simple gauge group G. But we do not observe weak interactions corresponding to the generators of G outside the $SU(2)\otimes U(1)$ subalgebra, so since there is supposed to be only one gauge coupling constant, we must assume that the corresponding vector bosons are much heavier than the W and Z. This can only happen if the breaking of G down to $SU(2)\otimes U(1)$ is much stronger than the breakdown of $SU(2)\otimes U(1)$ to electromagnetic gauge invariance.

In addition to our prejudice in favor of simple groups, there is another reason for wanting to put the weak and electromagnetic interactions into a larger gauge group, a reason which also provides indications of how big this group must be. The requirement is that the symmetries of the strong interactions, and in particular isospin conservation, arise naturally in zeroth order — that is, that they do not depend on any precise choice of parameters in the Lagrangian.

In a gauge theory of strong interactions, without strongly interacting scalar fields, the symmetries of the strong interactions may be identified with the symmetries of the quark mass matrix m. In particular, isospin conservation holds if and only if two of the eigenvalues of m, corresponding to states differing in charge by one unit, are equal. In the class of theories we are considering here, the zeroth order quark mass matrix takes the form

$$m = \Gamma \cdot \langle\phi\rangle \tag{2.1}$$

where Γ represents the matrices appearing in the Yukawa interaction

$$\bar{\psi}\Gamma\psi\cdot\phi \tag{2.2}$$

Isospin conservation is natural if the equality of the pair of eigenvalues of m is stable against variations of the Yukawa coupling constants in Γ and of the parameters in the polynomial part of the Lagrangian which determine $\langle\phi\rangle$.

One necessary condition for naturalness can be derived solely from considerations of the effect of changing the Yukawa coupling constants. We may decompose the bilinear products of quark fields into irreducible representations of the symmetry group G of the theory

$$\bar{\psi}_p \psi_q = \sum_{\ell m} O_{\ell m} C^{\ell m}_{pq} \tag{2.3}$$

where $O_{\ell m}$ is an operator transforming as the m-th vector of the ℓ-th irreducible representation of G, and C is an appropriate Clebsch-Gordan coefficient. Hence, without knowing anything about the nature of the scalar fields, we may write a corresponding decomposition for the quark mass matrix

$$m_{pq} = \sum_{\ell m} M_{\ell m} C^{pq}_{\ell m} \tag{2.4}$$

where $C^{pq}_{\ell m}$ is the reciprocal Clebsch-Gordan coefficient. The coefficient $M_{\ell m}$ is proportional to an arbitrary Yukawa coupling constant g_ℓ, so if isospin conservation is to be natural, it must be preserved when we multiply $M_{\ell m}$ with an arbitrary ℓ-dependent factor a_ℓ. The quark mass matrix then becomes

$$m_{pq}(a) \equiv \sum_{\ell m} M_{\ell m} C^{pq}_{\ell m} a_\ell \tag{2.5}$$

Our requirement is that the two equal eigenvalues of m(a) remain equal for all a_ℓ in at least some finite range.

The above condition is necessary, but not sufficient. To get closer to a sufficient condition, we must ask how the zeroth order symmetry arises. In a large class of cases[6] (known in the trade as models of "type 1" and "type 2") the zeroth order symmetries of the quark mass matrix arise because the vacuum expectation values of those scalar fields ϕ that enter into Yukawa couplings preserve some subgroup of the symmetry group G of the theory. [This zeroth order symmetry subgroup is supposed to be broken by the vacuum expectation values of other scalar fields χ, which belong to some irreducible representations of G which cannot have G-invariant Yukawa couplings; the symmetries of the quark mass matrix will then be broken in order α by weak and electromagnetic effects. Such a situation can occur naturally, either because the ϕ do not include enough irreducible representations to break G entirely, or because there is some "unlocking" in the quartic polynomial $P(\phi,\chi)$, so that it is naturally separately invariant under transformations of ϕ and χ.] But if the isospin symmetry of m is to be <u>required</u> by symmetries of the theory which are unbroken by $\langle\phi\rangle$, then these symmetries must include at least one element of the isospin group, other than a mere rotation around the three-axis.

Let us see how this works in a simple more-or-less realistic example. Suppose there are just four quark flavors: $\mathfrak{p}$, $\mathfrak{n}$, λ, $\mathfrak{p}'$. Suppose also that the symmetry group G includes at least the familiar SU(2) ⊗ U(1) gauge group, with the left-handed Cabibbo-rotated quarks forming SU(2) doublets; the right-handed quarks forming SU(2) singlets; and U(1) chosen so that electric charge is a gauge generator. It is straightforward to show that a necessary and sufficient condition for the stability of the relation $m_{\mathfrak{p}} = m_{\mathfrak{n}}$ against a finite variation (2.5) in Yukawa couplings, is that the various quark bilinears must form at most three irreducible representations D_1, D_2, D_3 of G:

$$\left.\begin{matrix} \bar{\mathfrak{p}}_L\mathfrak{p}_R & \bar{\mathfrak{n}}_L\mathfrak{p}_R & \bar{\mathfrak{p}}'_L\mathfrak{p}_R & \bar{\lambda}_L\mathfrak{p}_R \\ \bar{\mathfrak{p}}_L\mathfrak{n}_R & \bar{\mathfrak{n}}_L\mathfrak{n}_R & \bar{\mathfrak{p}}'_L\mathfrak{n}_R & \bar{\lambda}_L\mathfrak{n}_R \end{matrix}\right\} \text{ belong to } D_1$$

$$\left.\begin{matrix} \bar{\mathfrak{p}}_L\mathfrak{p}'_R & \bar{\mathfrak{n}}_L\mathfrak{p}'_R & \bar{\mathfrak{p}}'_L\mathfrak{p}'_R & \bar{\lambda}_L\mathfrak{p}'_R \end{matrix}\right\} \text{ belong to } D_2$$

$$\left.\begin{matrix} \bar{\mathfrak{p}}_L\lambda_R & \bar{\mathfrak{n}}_L\lambda_R & \bar{\mathfrak{p}}'_L\lambda_R & \bar{\lambda}_L\lambda_R \end{matrix}\right\} \text{ belong to } D_3 \tag{2.6}$$

This definitely requires G to be larger than just SU(2) ⊗ U(1), but there are several possibilities among which we can choose. However, if we also require that G contains at least one element of the isospin group (other than a rotation around the 3-axis) then we can show that G must contain at least the gauge group

$$SO(5) \otimes SU(2) \otimes U(1) \tag{2.7}$$

with the SO(5) ⊗ SU(2) transformation properties of the quarks given by

Left: (4,1)

Right: $(1,2) \oplus (1,1) \oplus (1,1)$

This is a remarkably large group; the superunified gauge theories[5] that have been proposed do not contain a weak and electromagnetic gauge group anywhere near so large.

There are of course many ways of modifying these conclusions. If isospin is what is called a "type 3" zeroth order symmetry,[7] then G might not contain any isospin rotations, though quark bilinears would still have to fall into irreducible representations according to (2.6). Our conclusions could also be modified if there were more than four quark flavors, especially if there were degeneracies among the new quarks, or if the right-handed quarks were not singlets under the SU(2) gauge group. It might be interesting to explore these various possibilities in the light of this analysis.

Another alternative is that the approximate isospin invariance of the quark mass matrix is a mere accident.[8] The close degeneracy of the neutron and proton masses may well give a misleading impression of the conservation of isospin in the strong interactions, because most of the nucleon mass comes from the spontaneous breakdown of chiral $SU(2) \otimes SU(2)$, not from the quark masses. A better measure of the isospin violation in the quark mass matrix is given by the ratio

$$\frac{m_n - m_p}{m_n + m_p} = \frac{m_{K^0}^2 - m_{K^+}^2}{m_\pi^2} = 0.21 \qquad (2.8)$$

This may be compared with the weak gauge coupling constant,

which for a weak mixing angle $\theta \approx 35°$ has the value

$$\frac{g^2}{4\pi} = \frac{\alpha}{\sin^2\theta} = 0.022 \tag{2.9}$$

It is an open question whether the order-of-magnitude discrepancy between (2.8) and (2.9) should be taken as a serious problem.

Finally, the conclusions reached here depend on the assumption that the Cabibbo angle θ_c is non-zero in zeroth order. However, $\sin\theta_c$ is 0.25, not very different from (2.8). Georgi[9] and Zee[10] have recently proposed different models in which θ_c and $m_n - m_p$ both arise from "radiative" corrections. In such theories G need not be as large as (2.7).

3. Theories with Elementary Scalars

At first sight, it might seem that a hierarchy of gauge symmetry breaking could occur naturally in a wide variety of theories. After all, we know of mechanisms[6,7] whereby a global symmetry like isotopic spin conservation can be preserved in zeroth order but broken in higher orders. However, in these cases the hierarchy of symmetry breaking occurs precisely because the symmetry which is preserved in zeroth order is only a symmetry of part of the Lagrangian. A gauge symmetry must be a symmetry of the whole Lagrangian, so these mechanisms are not available to explain a hierarchy of gauge symmetry breaking.

It is possible to be more precise about this.[11] Suppose we find a minimum of the potential at which a certain subgroup of the symmetry group of the theory is left invariant. If the potential at this point is very flat, then a small perturbation, even if it preserves the symmetry of the potential, can shift the minimum to a point where the unbroken subgroup is slightly broken. On the other hand, if the potential at the minimum has a quadratic curvature, then a symmetry-preserving perturbation cannot change the symmetry group of the minimum. The requirement that the second derivative of the potential in some direction vanish at the minimum requires us to impose conditions in the theory which are usually quite unnatural.

In the last few months I have been looking into the question of whether this situation would be improved if the minimum were of the Coleman - E. Weinberg type,[12] produced by a balance between the quartic terms in the polynomial and the one-loop corrections, rather than between the quartic and the quadratic terms in a polynomial. It would of course be unnatural simply to set the quadratic terms equal to zero. However, Eldad Gildener, a student of Sydney Coleman, has shown in his recent Ph.D. thesis[13] that even in theories with non-zero scalar masses, it is natural to find minima of the potential at field strengths much larger than the boson masses, where these masses can be regarded as small perturbations. I would like to show you a generalized version of Gildener's analysis, because it reveals some intriguing possibilities for the ways that symmetries can be spontaneously broken, even though it does not seem to lead to hierarchies of gauge symmetry breaking.

We want to explore a region of the potential where the fields ϕ are so large that the scalar boson masses (and other super-renormalizable couplings, like ϕ^3 interactions) are negligible. The tree approximation to the potential takes the form of a homogeneous polynomial of fourth order

$$V_0(\phi) = \frac{1}{24} f_{ijk\ell}\phi_i\phi_j\phi_k\phi_\ell \qquad (3.1)$$

We assume, as usual, that the f's are of order e^2, where e is a typical gauge coupling of the theory. The minima (if any) of this polynomial lie along lines

$$\phi_i = n_i\rho \qquad (3.2)$$

where n_i is a direction vector such that

$$f_{ijk\ell}\, n_j n_k n_\ell = 0 \qquad (3.3)$$

The demand that this have a solution for some non-zero n imposes <u>one</u> non-trivial condition on the f's, because for N field variables there are N equations to be satisfied and only N - 1 variables available in the direction vector n. For instance, for the adjoint representation of SU(N) the ϕ form a traceless Hermitian tensor Φ, and in general we would have

$$V_0(\Phi) = f_1(\mathrm{Tr}\,\Phi^2)^2 + f_2\,\mathrm{Tr}\,\Phi^4$$

The condition that this have a stationary point will be satisfied for instance if $-f_2/f_1$ is an even integer between 1 and N.

This sort of condition appears highly unnatural. However, we also have one inessential variable here, the scale M_R of the renormalization point at which the f's are defined. Gildener shows that in a variety of special cases the trajectories of the f's generated by the renormalization-group equations do cross values which satisfy these constraints. The question thus shifts to whether the scale of the fields at the actual minimum is close enough to the scale of the renormalization point to justify the use of a perturbation expansion in which the zeroth order field is in the direction given by (3.3).

To answer this, we include one-loop terms in the potential. These are of the form[14]

$$V_1(\phi) = \frac{1}{64\pi^2}\,\mathrm{Tr}\left[M_\phi^4 \ln(M_\phi^2/M_R^2)\right] + \frac{3}{64\pi^2}\,\mathrm{Tr}\left[\mu_\phi^4 \ln(\mu_\phi^2/M_R^2)\right] - \frac{1}{16\pi^2}\,\mathrm{Tr}\left[m_\phi^4 \ln(m_\phi^2/M_R^2)\right] \tag{3.4}$$

where M_ϕ, μ_ϕ, and m_ϕ are the mass matrices of the particles of spin 0, 1, and 1/2 at a given value ϕ of the vacuum expectation value of the scalar fields. Substituting (3.2) in (3.1) and (3.4) gives a potential of the form

$$V(n\rho) \simeq V_0(n\rho) + V_1(n\rho) = A_n\rho^4 + B_n\rho^4 \ln(\rho^2/M_R^2) \tag{3.5}$$

Because of Eq. (3.3), both A_n and B_n are of order e^4, with B_n in particular given by

$$B_n = \frac{1}{64\pi^2}\,\mathrm{Tr}\,M_n^4 + \frac{3}{64\pi^2}\,\mathrm{Tr}\,\mu_n^4 - \frac{1}{16\pi^2}\,\mathrm{Tr}\,m_n^4 \tag{3.6}$$

This is always positive if there are no fermions in the theory, and it remains positive even if fermions are present, provided the Yukawa coupling constants are not too large. For $B > 0$, the function (3.5) necessarily has a <u>minimum</u> at a point ρ_0:

$$\ln(\rho_0^2/M_R^2) = -\frac{1}{2} - \frac{A_n}{B_n} \tag{3.7}$$

$$V_{min} = \frac{-B_n}{2}\rho_0^4 < 0 \tag{3.8}$$

Note that if it were not for the condition (3.3), A_n would be of order e^2 while B_n is of order e^4, so $\ell n(\rho_0^2/M_R^2)$ would be of order $1/e^2$, and perturbation theory would be invalid. The choice of a renormalization point at which (3.3) can be satisfied is what allows us to use perturbation theory to locate the minimum.

Unfortunately, it does not seem that the constraints on f imposed by Eq. (3.3) are of the type required to produce hierarchies of symmetry breaking. What we need is that the potential be flat in some transverse direction, so that small perturbations can change the symmetry of the minimum. (The minimum would still be near a direction satisfying (3.3), but our calculation of the effect of one-loop graphs on its true position would of course need revision in this case.) Instead, (3.3) just says that the potential is flat in a radial direction, so that one-loop corrections only determine the scale of the minimum, not its direction.

For instance, let us consider a theory with an O(n) gauge group and two n-vector scalar fields χ_i and η_i. If it turns out that at the minimum of the potential $\vec{\chi}$ and $\vec{\eta}$ are not parallel, and one of them is much larger than the other, then we have a hierarchy of symmetry breaking: O(n) is broken strongly down to O(n-1) and then more weakly down to O(n-2). For large field strengths we may neglect boson masses and trilinear couplings, so the tree approximation to the potential takes the form

$$V_0(\vec{\chi},\vec{\eta}) = g_\chi(\vec{\chi}^2)^2 + g_\eta(\vec{\eta}^2)^2 + f\,\vec{\chi}^2\,\vec{\eta}^2 + h(\vec{\chi}\cdot\vec{\eta})^2$$

There are four distinct kinds of stationary point here, each associated with a corresponding constraint on the couplings, which determines the choice of renormalization point:

STATIONARY POINT	CONSTRAINT
(a) $\vec{\chi} \perp \vec{\eta}$, $\vec{\chi}^2/\vec{\eta}^2 = \sqrt{g_\eta/g_\chi}$	$f = -2\sqrt{g_\chi g_\eta}$
(b) $\vec{\chi} \parallel \vec{\eta}$, $\vec{\chi}^2/\vec{\eta}^2 = \sqrt{g_\eta/g_\chi}$	$f+h = -2\sqrt{g_\chi g_\eta}$
(c) $\vec{\chi} = 0$ $\vec{\eta} \neq 0$	$g_\eta = 0$
(d) $\vec{\eta} = 0$ $\vec{\chi} \neq 0$	$g_\chi = 0$

In case (a), O(n) is broken down to O(n-2) in zeroth order, and there is no particular reason why any O(n-1) subgroup should be less strongly broken than O(n) itself. In cases (b)-(d), O(n) is only broken down to O(n-1) in zeroth order, but the tree-approximation potential does not have vanishing second derivatives in any direction which could lead to a breakdown of O(n-1) in higher order. Thus this model offers no possibility of a gauge hierarchy.

4. Dynamical Symmetry Breaking

There is one kind of theory which does naturally lead to a hierarchical breakdown of gauge symmetries.[15] Consider a conventional gauge model, perhaps based on a simple gauge group, with gauge and Yukawa couplings of order e and scalar self-coupling of order e^2, but with an enormous scalar mass $\overline{M}$ in the Lagrangian. Spontaneous symmetry breaking will produce a super-large scalar vacuum expectation value $\langle\phi\rangle \approx \overline{M}/e$, and vector, spinor, and scalar masses of order $\overline{M}$. Our laboratory observations at energies $E \ll \overline{M}$ will be described by an effective field theory involving only those particles that did not get masses of order $\overline{M}$; these are the gauge bosons associated with symmetries unbroken by $\langle\Phi\rangle$, plus those fermions which are kept massless by unbroken chiral symmetries, but no scalars. We may suppose that this effective field theory is just the gauge field theory we think we observe, based on a non-simple gauge group more or less like

$$SU(2) \otimes U(1) \otimes SU(3)_{color} \otimes ?$$

with the massless fermions of the theory including the "observed" quarks and leptons. Any simple subgroup of this gauge group which happens to satisfy the condition for asymptotic freedom will have a gauge coupling g^2_{eff} at energy E of order $be^2 \ln \overline{M}/E$ (where b is a pure number of order unity) and will therefore produce strong interactions at energies of order $\overline{M} \exp(-1/be^2)$. These strong forces may produce Goldstone bosons as bound states, thus generating a further spontaneous symmetry breakdown. It is this <u>dynamical</u> symmetry breaking which in this sort of theory would have to be responsible for the masses of the W and Z and the quarks and leptons.

Whether or not one believes this scenario, it is worth asking what predictive power a theory would have if the spontaneous symmetry breaking has to be produced by strong forces. Surprisingly, quite a lot. If we suppose that the theory has N "color" multiplets, each with the same strong interactions, then in the absence of the weak and electromagnetic interactions, this theory would have an accidental global chiral $U(N) \otimes U(N)$

symmetry. The dynamical spontaneous symmetry breakdown is characterized by a parameter M [perhaps of order $\overline{M}\exp(-1/be^2)$] in the sense that the couplings of Goldstone bosons to the corresponding currents as well as the fermion masses produced by the spontaneous symmetry breaking are of order M. In addition, there is an intrinsic $U(N) \otimes U(N)$ breakdown, caused by the weak and electromagnetic interactions. Despite the fact that the spontaneous symmetry breaking is due to strong forces, one can still do perturbative calculations at energies $E \ll M$, using for this purpose a new effective Lagrangian which incorporates all the soft-Goldstone-boson theorems implied by the broken $U(N) \otimes U(N)$ symmetry.

I have gone into all this in great detail in a recent long preprint.[16] For the present, I just want to mention a few of the noteworthy aspects of this sort of theory:

(i) For every solution of the dynamical symmetry breaking problem, there will, in the absence of the weak and electromagnetic interactions, be an infinite number of other equally good solutions, related to the first by arbitrary $U(N) \otimes U(N)$ transformations. This ambiguity is removed by the weak and electromagnetic interactions; a theorem of Dashen[17] can be used to show that the correct solution is the one for which the effective Lagrangian contains no terms of first order in the Goldstone boson fields.

(ii) It is possible to define a "unitarity gauge" in which those Goldstone bosons that correspond to broken elements of the weak and electromagnetic gauge group are eliminated. The masses of the intermediate vector bosons can be determined by inspection of the effective Lagrangian in the unitarity gauge. It turns out that these masses are of order eM, so that M must be of order 300 GeV. As in the "type III" theories discussed in the first section, the quarks and leptons in this kind of theory must be identified as the fermions which do not get masses of order M from the spontaneous breakdown of $U(N) \otimes U(N)$; their masses are then of order e^2M.

(iii) The global symmetry group $U(N) \otimes U(N)$ will in general contain some exact symmetries of the weak and electromagnetic interactions, which are not elements of the weak and electromagnetic <u>gauge</u> group. If such symmetries are spontaneously broken, the theory will include "true" Goldstone bosons, which are not eliminated by the Higgs mechanism, and which remain massless to all orders in e. The occurrence of such particles presents great difficulties for realistic theories based on dynamical symmetry breaking.

(iv) There are usually also broken elements of $U(N) \otimes U(N)$ which are not symmetries of the weak and electromagnetic interactions. To each of these there corresponds a "pseudo"-Goldstone boson, with mass of order eM. The interactions of the pseudo-Goldstone bosons are weak at ordinary energies, and may be read off from the effective Lagrangian.

The occurrence of pseudo-Goldstone bosons is the most strik- consequence of the general class of theories in which the masses of the intermediate vector bosons arise from dynamical symmetry breaking. Looking forward to the day when ISABELLE goes on the air, we can anticipate that the experiments designed to find intermediate vector bosons will find scalar bosons as well, with roughly similar masses. These will be Higgs bosons if the symmetry breaking is of type I, which we hope is the case. But if despite our hopes the symmetry breaking is dynamical, then in place of Higgs bosons we must expect to find pseudo-Goldstone bosons.

References

1. For reviews, see E. S. Abers and B. W. Lee, Phys. Rep. 9C, 1 (1973); S. Weinberg, Rev. Mod. Phys. 46, 255 (1974).

2. J. M. Cornwall, D. N. Levin, and G. Tiktopoulos, Phys. Rev. Lett. 30, 1268 (1973); 31, 572 (E)(1973); C. H. Llewellyn-Smith, Phys. Lett. B 46, 233 (1973); J. Sucher and C. H. Woo, Phys. Rev. D 8, 2721 (1973); D. N. Levin and G. Tiktopoulos, Phys. Rev. 12, 415 (1975); S. Sakakibara and J. Sucher, Univ. of Maryland Technical Report No. 76-047, to be published.

3. R. Jackiw and K. Johnson, Phys. Rev. D 8, 2386 (1973); J. M. Cornwall and R. E. Norton, Phys. Rev. D 8, 3338 (1973). Other references are given in Ref. 16.

4. S. Weinberg, Phys. Rev. D 5, 1962 (1972).

5. J. C. Pati and A. Salam, Phys. Rev. D 8, 1240 (1973); H. Georgi and S. L. Glashow, Phys. Rev. Lett. 32, 438 (1974); H. Fritsch and P. Minkowski, to be published. Also see Ref. 15.

6. S. Weinberg, Phys. Rev. Lett. 29, 388, 1698 (1972).

7. H. Georgi and S. L. Glashow, Phys. Rev. D 6, 2977 (1972); *ibid.* D 7, 2457 (1973). Also see H. Georgi and A. Pais, Phys. Rev. D 12, 508 (1975).

8. I thank H. Fritsch for a conversation on this possibility.

9. H. Georgi, to be published.

10. A. Zee, Phys. Rev. D 9, 1772 (1974), and to be published.

11. H. Georgi and S. L. Glashow, Phys. Rev. D 6, 2977 (1972); S. Weinberg, in *Proceedings of the XVII International Conference on High Energy Physics*, London, 1974, edited by J. R. Smith (Rutherford Laboratory, Chilton, Didcot, Berks., England, 1974), p. III-59.

12. S. Coleman and E. Weinberg, Phys. Rev. D 7, 1888 (1973).

13. E. Gildener, "Radiatively-Induced Spontaneous Symmetry Breaking for Asymptotically Free Gauge Theories", Harvard University Ph.D. thesis, May 1975; and to be published.

14. See Ref. 12, or S. Weinberg, Phys. Rev. D 7, 2887 (1973).

15. H. Georgi, H. Quinn, and S. Weinberg, Phys. Rev. Lett. 33, 451 (1974).

16. S. Weinberg, to be published.

17. R. Dashen, Phys. Rev. D 3, 1879 (1971).

DISCUSSION

J. PATI (Maryland)

If spontaneous symmetry breaking is dynamical, which I consider to be an attractive possibility, the problem of "naturalness" of isospin, or SU(3) - symmetry would take a completely different perspective. Thus the criterion for choosing extended symmetries would not apply.

WEINBERG

Yes, and I can comment on that, but I certainly have not seen any model that's really very attractive. If, as in the case of theory of type III where it's not dynamical, the natural mass of the fermions which is produced by the dynamical symmetry breaking is of the order of that mass scale 300 GeV because until you turn on the weak and electromagnetic interactions they don't know about any other mass scale so that the fermions that we see, the quarks and the leptons in that kind of theory or in a theory of type III would have to be fermions whose mass is zero in lowest order because of the nature of this spontaneously unbroken subgroup here that then has to get broken in higher order in the weak and electromagnetic interactions. You can draw pictures where there's a very heavy fermion here, and an ordinary fermion here, and the gauge boson mixes them and produces a mass of the light fermion of the order of α x 300 GeV which is a perfectly nice mass scale for quarks to be, or at least the heavier of the observed quarks. The heaviest observed quarks would still be light quarks from this point of view. Now it's quite possible that this unbroken subgroup H has in it something like isospin and that the light quarks could inherit that. That is, for example, imagine a theory where you have - just take the simplest thing that comes to mind - $np\lambda p'$ (just to be perverse about the ordering), and then partners $NP\Lambda P'$ which are very heavy. If the weak interactions which connect these act equally on all of these, that is, if the group for instance is SU2 x SU2 where the second SU2 interchanges light and heavies, then whatever degeneracies occur here will be inherited by these. They will be inherited to first order in α, but then to next order in α you will get all the weak interactions that connect these producing perturbations which will then break isospin. So one can see in a sort of qualitative way how symmetries like isospin could naturally arise. The arguments I mentioned in my talk would certainly not apply to such a case if that's what your question is and one would have to analyze a complete model. I don't know of any model of that type which is really satisfying. I don't hope that dynamical symmetry breaking is the case. It seems to me that it takes away so much of the predictive power of these theories that I just hope it's wrong but, of course, it would be sort of nice if these pseudo-Goldstone bosons were found. That would be an interesting confirmation.

H. GOLDBERG (Northeastern)

Just a comment - that if these masses occur like 300 and they are coupled in a maximally CP violating manner you could explain the CP violation in the K mass matrix in a natural way.

WEINBERG

I'm not sure. The fact that the ordinary fermion have zero mass means that lots of things are constrained. The unbroken symmetry group has to be big enough to keep them zero mass. That's why, by the way, the Fermi interactions which are not due to gauge boson exchange are of the VA type in such theories because the unbroken symmetry has to preserve the masslessness of the ordinary quarks. The heavy quarks wouldn't have VA type Fermi interactions. In fact, they have any interactions you like.

B. F. L. WARD (Purdue)

Why are you not concerned that the pattern of symmetry breaking in the first case depends so crucially on the normalization point?

WEINBERG

Well, I don't think it does - that's a good question because of the way I said it makes it sound as if it does, but I don't think it does. The thing is, you have a theory - let's say the theory is of scalar fields with $M^2\varphi^2 + \lambda\varphi^4$. Now you say, let's say there are gauge interactions also. You ask, where is the minimum of the potential in this theory? You start doing perturbation theory and say that the minimum is here at a point $\frac{M}{\sqrt{2}}$. Then you ask are there any other minima which are for very very much larger fields. Where are they? Now, you cannot find the answer to that question if you choose a renormalization point at random because perturbation theory simply doesn't work. There is only one minimum. The minimum doesn't depend on the choice of the renormalization point, it's just how you find it that does. The perturbation theory has to be based on starting from a tree approximation which is defined with an appropriate renormalization point if you're going to do perturbation theory. If you want to use some other renormalization point, the minimum is still there but you're not going to find it in perturbation theory. What I'm saying is that if you choose the renormalization point where one of these weird looking constraints like f_2/f_1 is a negative integer - if you choose the renormalization point so that that constraint is satisfied then the minimum is so close that you can use perturbation theory. If you don't you just can't answer any question.

J. P. HSU (Texas)

I would like to comment on an unexpected problem in the hope that some interested people will check our conclusions: Despite the general formal proof of unitarity in the usual formalism for non-Abelian gauge theories, we find that in the Weinberg theory unitarity is violated in the 4th order W^{+}-photon scattering, if we use bilinear gauge conditions, e.g., $(\partial_{\mu} - ieA_{\mu})W^{+\mu} - iMS^{+}/\xi = a^{+}(x)$, $\partial_{\mu}A^{\mu} = a_{A}(x)$, $\partial_{\mu}Z^{\mu} + M_{Z}\chi/\eta = a_{Z}(x)$. We note that this happens because $\partial_{\mu}A^{\mu}$ fulfill the free equation in this case and the gauge condition $\partial_{\mu}A^{\mu} = a_{A}(x)$ is stable, yet the usual formalism treats it as if it were unstable and needs gauge compensating terms.

(Comment submitted after session had ended.)

THE WORLD OF BASIC ATTRIBUTES: VALENCY AND COLOR

Jogesh C. Pati*

Within the gauge theory approach to color, I discuss the compatibility of the hypotheses of integer-charge quarks and physical color with lepto-production experiments. Two important consequences of this hypothesis for deep inelastic phenomena are: (a) color-gluons contribute in a scale-invariant manner to structure functions; (b) (σ_L/σ_T) is nonvanishing and a function of x only. It is suggested that the anomalous $(\bar{\mu}e)$-events observed at SPEAR are progenies of integer-charge quark-pairs produced by e^-e^+-annihilation decaying into leptons (a possibility which was suggested by the quark-lepton unification hypothesis). It is remarked that the dimuons produced by neutrinos with their observed rate can be attributed to production of charged color-gluons, while the prompt leptons observed in hadronic collisions may be attributed to the production of quarks and/or color gluons in pairs. Distinguishing features of these explanations and also distinctive signatures for color are noted.

*Center for Theoretical Physics, Department of Physics and Astronomy, University of Maryland, College Park, Md. 20742.
This paper is based upon work done in collaboration with Professor Abdus Salam.
Work supported in part by the National Science Foundation, Grant GP 43662X.

I. INTRODUCTION

There is every reason to believe that the known hadrons are made out of quarks and that quarks possess twin attributes: valency and color. Valency (also called "flavor") is at least of four kinds (p,n, λ,χ = charm), while quark-colors are of three varieties: (red, yellow, blue) = (r,y,b). The quark-iepton unification hypothesis[1] suggests that lepton number is the fourth color, so that the leptons have the same valency as the quarks, but just a different color. Within this hypothesis, the basic set of fermions is a sixteen fold (of four component objects) $F_{L,R}$ consisting of 12 quarks and 4 leptons, whose universal interactions (weak, electromagnetic and strong) might be generated by a left↔right symmetric minimal gauge structure[2] ($SU(2)_L \times SU(2)_R \times SU(4)'_{L+R}$). Imbedding of this minimal symmetry within its "natural" extensions[3] allows a description of the basic forces (weak, electromagnetic and strong) in terms of one generating universal coupling constant.

There are a few characteristic features of the unification hypothesis as outlined above, which distinguish it from other alternative approaches:[4]

(1) Quarks and leptons in one multiplet:

In this case lepton number L (or baryon number B) is part of a nonabelian symmetry like I_3 or Y.

(2) Spontaneous Violations of Baryon Number, Lepton Number and Fermion Number:

Baryon number, lepton number and fermion number F = (B+L) are preserved in the basic lagrangian. Violations[1] of these quantum numbers in general arise <u>only</u> through spontaneous symmetry breaking (for example via mixing of gauge mesons carrying different quantum numbers); this leads in general to unstable short lived quarks (if they are integer charged) and to unstable proton (regardless[5] of quark-charges). One important consequence of this manner of symmetry violation, however, is that no ultra-heavy gauge masses[4] are ever needed to account for the known proton stability. Furthermore, if quarks are integer-charged, the theory would have no need for the hypothesis of quark-confinement, since integer-charged quarks decaying[1] rapidly ($\tau_q \lesssim 10^{-10}$ sec.) into leptons ($q \to \ell + \ell + \bar{\ell}$, $q \to \ell$ + pions, etc.) would not have been detected by present particle searches, even if their production cross-section was not insignificant ($\sigma_q \gtrsim 10^{-32}$ cm^2 (say) with $m_q \sim 2$ to 3 GeV).

(3) Spontaneous Breakdown of Parity and CP:

Because of the left-right symmetric gauge-structure, the minimal symmetry as well as its natural extensions allow parity violation to be spontaneous in origin;[6] simultaneously they provide the basis for a desirable milliweak theory of CP violation[7] (through spontaneous symmetry breaking), which automatically yields $\eta_{+-} = \eta_{oo}$ and[8] $|\eta_{+-}| \approx (m_{W_L}/m_{W_R})^2 (\sin 2\theta_R/\sin 2\theta_L) |\sin(\delta_R - \delta_L)|$; the smallness of $|\eta_{+-}|$ may be attributed in this case naturally to $m_{W_R} \gg m_{W_L}$.

(4) Co-existence of V+A Interactions and Right-Handed Neutrino:

Being left-right symmetric, such a theory must introduce[1] a right-handed counterpart of the two-component left-handed neutrino (for the electron and muon-neutrinos separately). Eventually, whether the two two-component objects (ν_L and ν_R) combine[9] (perhaps through radiative corrections) to form a light four-component neutrino, or whether they remain as massless two two-component objects depends upon further details of the theory and the nature of spontaneous symmetry breaking. The experimental circumstance that in the familiar weak interactions only the left-handed neutrino (ν_L) appears (or equivalently the V-A chiral projection dominates over V+A) is attributed within such a theory to the fact that the gauge mesons coupled to V+A-currents (combining known fermions) acquire a mass heavier than those coupled to V-A currents of the familiar fermions. There is the necessary consequence, however, that V+A-interactions and right-handed neutrino ν_R must appear at a level $\gtrsim 10^{-3}$ of (V-A)-interactions.[10]

(5) Possible Visible Effects of Anomalous Lepton-Hadron Interactions at Present or Moderately High Energies:

Since proton-stability does not constrain the masses of exotic gauge bosons X coupled to lepton-hadron currents ($\bar{\ell}q$), within the approach outlined above, it becomes possible[11] for a whole new class of lepton-hadron (semi-leptonic) interactions to manifest at moderately high energies.[12] For example, the mass of X-particles, it turns out, need be no heavier than 10^5 GeV in the minimal symmetry model and it may be much lighter ($\simeq$ 100 GeV) if we enlarge the symmetry. With $m_X \simeq 300$ GeV, we estimate that the anomalous-interaction-contribution (with its different energy dependence compared to one photon-contribution) should be visible in $e^-e^+ \rightarrow$ hadrons at centre of mass energies $\gtrsim 10$ GeV, something, which should be available in the near future. Such anomalous interactions may also manifest in neutral-current neutrino interactions[12] (at a level $\lesssim G_{Fermi}$) and in lepton-pair production by hadronic collisions ($p+p \rightarrow \ell^+\ell^-$+hadrons).

The theory introduces a new class of neutral current neutrino-interactions (in addition to those of the $SU(2)_L \times U(1)$-theory), which could already be a major-contributor to the observed neutral current effects. These arise through gauge mesons (called S° for the minimal symmetry model and[13] S°_V and S°_A for the extended symmetries), which are coupled to diagonal currents of the form $(\Sigma\bar{q}q - 3\Sigma\bar{\ell}\ell)$, which are pure iso-scalar. (S° is coupled to pure vector; while S°_V and S°_A are coupled to vector and axial-vector-currents respectively). These gauge mesons need be no heavier than $\simeq$ 300 GeV, the effective strengths of their interactions at low energies being $\lesssim G_{Fermi}$.

(6) Mirror-Partners of the Basic Fermions:

If the unification is to be carried to a stage where all basic forces (weak, electromagnetic and strong) are governed by one generating coupling constant, the minimal symmetry $(SU(2)_L \times SU(2)_R \times SU(4)'_{L+R})$ must be imbedded within a higher symmetry. Consistent with the approach outlined above, such extended symmetries are, for example, $SU(4)_L \times SU(4)_R \times SU(4)'_L \times SU(4)'_R$; or $SU(16)_L \times SU(16)_R$, or $SU(32)$, all of which would possess Adler-Bell-Jackiw-anomalies unless we postulate[14] that the basic set of fermions $F_{L,R}$ are accompanied by a mirror set F'_{L+R}; the two sets of F and F' being coupled with opposite chiralities to the same set of gauge bosons. The theory would thus be symmetric under the mirror symmetry transformation

$$F_{L,R} \leftrightarrow F'_{R,L} \tag{1}$$

At a stage when the fermions are massless, such a theory is vector-like[15] for the theory outlined above, where the basic set of fermions F is a sixteen-fold, the mirror set F' will contain four new valency quantum numbers (p',n',λ',χ') with the "familiar" four colors (r,y,b,ℓ); the first three colors (r,y,b) would thus give 12 heavier quarks (p' being the mirror of p), while the fourth color (ℓ) would give four heavy leptons (E°, E^-, M^-, M°).

The existence of mirror provides altogether eight valency quantum numbers $\{(p,n,\lambda,\chi) + (p',n',\lambda',\chi')\}$, the strong interactions of the theory being invariant under $U(8)_L \times U(8)_R \times SU(3)'_{L+R}$; the chiral $U(8)_L \times U(8)_R$-symmetry is broken primarily by quark-mass-terms. Quite clearly, the existence of the mirror-quantum numbers would provide new possibilities for the interpretation of J/ψ-particles especially in accounting for their extreme narrowness (this I discuss briefly later). Furthermore, allowing for Cabibbo-like mixings [i.e., mixings of (n,λ), (p,χ), (n',λ') and (p',χ')] as well as F-F'-mixings [i.e., mixings of (p,p'), (n,n'), (λ,λ') and (χ,χ') etc.] in the Fermi-mass matrix would lead to a rich variety of new complexions[16] in the weak interactions.

The above discussion summarizes some of the salient features of our unification hypothesis. In this talk, I discuss in more detail mainly two topics:

(i) First a new result regarding the problem of color-brightening for integer-charge quark theories -- we show[17,18] that within the gauge theory approach to color, insofar as lepto-production experiments are concerned, integer charge quarks behave asymptotically just the same way as fractionally charged quarks and that charged spin-1-color-gluon contributions scale, both contrary to common expectation. This takes away the familiar objections (stemming from lepto-production experiments) against the twin hypotheses that quarks carry integer-charges and that color-threshold is relatively low (< 5 GeV) and thereby provides new impetus for the concept of physical color as an alternative to the hypothesis of confined (unphysical) color.

(ii) Second, I stress that if quarks are integer-charged; (a) the dimuon production[19] by neutrino-interactions receive a simple explanation in terms of production of color-gluons (or similar color-octet states); (b) the $(\bar{\mu}e)$-events[20] seen in e^-e^+-annihilation may be attributed to production of integer-charge-quark-pairs followed by their rapid decays into leptons and (c) the anomalous direct leptons[21] observed in hadronic collisions (especially at higher energies $\gtrsim$ 10 GeV) receive a simple explanation in terms of pair production of color gluons and/or quarks. Experimental tests to distinguish these explanations from alternative hypotheses are discussed.

More specifically the topics discussed are:

II. The Minimal Symmetry: A Brief Review

III. Quark-Charges, Color-Brightening:

(i) Unphysical versus Physical Color

(ii) Familiar Arguments on Color Brightening

(iii) A Flaw in the Familiar Arguments

(iv) Lepto-Production of Color through Quark and Gluon Partons

(v) Distinctive Signatures for Color.

IV. The New Phenomena:

(i) Dimuons

(ii) $(\bar{\mu}e)$-Events: Pair Production of Quarks and Color-Gluons

(iii) Direct Leptons

(iv) The J/ψ-Particles

V. Summary and Concluding Remarks

II. THE MINIMAL SYMMETRY: A BRIEF REVIEW

1. To set the notations, I first present briefly the main features of the basic model[2] built upon a sixteen-fold of four component fermions $F_{L,R}$:

$$F_{L,R} = \begin{bmatrix} p_r & p_y & p_b & p_\ell = \nu_e \\ n_r & n_y & n_b & n_\ell = e^- \\ \lambda_r & \lambda_y & \lambda_b & \lambda_\ell = \mu^- \\ \chi_r & \chi_y & \chi_b & \chi_\ell = \nu_\mu \\ \text{red} & \text{yellow} & \text{blue} & \text{lilac} \end{bmatrix}_{L,R} \tag{2}$$

Here χ denotes charm. Note the necessity of color <u>and</u> charm for putting baryonic and leptonic matter together. All "low-energy" phenomena may be described by gauging a minimal non-abelian local symmetry:

$$G = SU(2)_L \times SU(2)_R \times SU(4)'_{L+R} \tag{3}$$

where $SU(2)_{L,R}$ gauges valency-indices $\{(p,n) + (\lambda,\chi)\}_{L,R}$, while $SU(4)'$ gauges the four colors (red, yellow, blue and lilac) = (r,y,b,ℓ); the fourth color "ℓ" for lilac denotes lepton-number. This is the <u>minimal symmetry</u> capable of uniting baryons and leptons and providing a unified description of weak, electromagnetic and strong interactions. It satisfies the left $\leftrightarrow$ right discrete symmetry and thus incorporates the desirable milliweak-theory of CP-violation[7] referred to before. In addition, it is anomaly-free. The gauge fields of the theory $W_{L,R}$ and V generated by $SU(2)_{L,R}$ and $SU(4)'$ respectively are:

$$W_{L,R} = \frac{1}{2}\begin{bmatrix} \vec{\tau}\cdot\vec{W} & 0 \\ 0 & \tau_1(\vec{\tau}\cdot\vec{W})\tau_1 \end{bmatrix}_{L,R}, \quad V = \begin{bmatrix} V_{11} & V_\rho^- & V_{K*}^- & \bar{X}^\circ \\ V_\rho^+ & V_{22} & \bar{V}_{K*}^\circ & X^+ \\ V_{K*}^+ & V_{K*}^\circ & V_{33} & X'^+ \\ X^\circ & X^- & X'^- & \sqrt{3/4}S^\circ \end{bmatrix}$$

where $V_{11} = 1/\sqrt{2}\ (V_3 + V_8/\sqrt{3} - S^\circ/\sqrt{6}\ )$; $V_{22} = 1/\sqrt{2}\ (-V_3 + V_8/\sqrt{3} - S^\circ/\sqrt{6}\ )$, $V_{33} = 1/\sqrt{2}\ (-2V_8/\sqrt{3} - S^\circ/\sqrt{6}\ )$ and $S^\circ \equiv V_{15}$. The color-octet of gauge-particles (gluons) V(8) given by ($V_\rho^\pm$, $V_{K*}^\pm$, V_{K*}°, $\bar{V}_{K*}^\circ$, V_3 and V_8) appearing in the topleft 3 x 3 block of V are coupled to (red, yellow and blue)-colors only (i.e. only to quarks). [The charges of V-particles shown above correspond to the integer-charge quark-model (see below). For the fractionally-charged quark model, V(8)-gluons are electrically neutral and the X's carry charges ± 2/3 e]. The coupling parameters[22] $g_{L,R}$ and f associated with the gauge groups $SU(2)_{L,R}$ and SU(4)' are:

$$g^2_{L/4\pi} \simeq g^2_{R/4\pi} \simeq 2\alpha;\ f^2/4\pi \simeq (1\text{-}10) \tag{4}$$

The model described above leads to two notable possibilities for quark-charges with a unique prediction for the lepton-charges (0,-1,-1,0):

$$[Q(F)] = \begin{bmatrix} 0 & +1 & +1 & 0 \\ -1 & 0 & 0 & -1 \\ -1 & 0 & 0 & -1 \\ 0 & +1 & +1 & 0 \\ \text{red} & \text{yellow} & \text{blue} & \text{lilac} \end{bmatrix}, \text{ or} \begin{bmatrix} \frac{2}{3} & \frac{2}{3} & \frac{2}{3} & 0 \\ -\frac{1}{3} & -\frac{1}{3} & -\frac{1}{3} & -1 \\ -\frac{1}{3} & -\frac{1}{3} & -\frac{1}{3} & -1 \\ \frac{2}{3} & \frac{2}{3} & \frac{2}{3} & 0 \\ \text{red} & \text{yellow} & \text{blue} & \text{lilac} \end{bmatrix} \tag{5}$$

The electric charge formula (in terms of generators of SU(4) x SU(4)') for the two cases are:

$$Q = (F_3 + F_8/\sqrt{3} - \sqrt{2/3}\,F_{15}) + (F'_3 + F'_8/\sqrt{3} - \sqrt{2/3}\,F'_{15}) \quad \text{(Integer-charge quarks)}$$

$$= (F_3 + F_8/\sqrt{3} - \sqrt{2/3}\,F_{15}) - \sqrt{2/3}\,F'_{15} \quad \text{(Fractionally charged quarks)} \qquad (6)$$

For the integer-charge-quark-model photon distinguishes between red-color on the one hand and yellow and blue on the other; while for the fractionally charged quark-model, photon does not distinguish between the first three colors. For the integer-charge case, electric charge thus contains a SU(3)'-color-octet piece $(F'_3 + F'_8/\sqrt{3})$, while for the fractionally charged case, electric charge Q is entirely a singlet of SU(3)'-color. Defining $Q_{col} \equiv (F'_3 + F'_8/\sqrt{3})$ and grouping the SU(3)'-singlet part together under $Q_{val} \equiv (F_3 + F_8/\sqrt{3} - \sqrt{2/3}\,F_{15}) - \sqrt{2/3}\,F'_{15}$; we may write

$$Q = Q_{val} + Q_{col} \quad \text{(Integer-charge-quarks)}$$

$$= Q_{val} \quad \text{(Fractionally charged quarks)} \qquad (7)$$

$$Q_{val} = (2/3, -1/3, -1/3, 2/3) \text{ For } (p,n,\lambda,\chi)\text{-quarks}$$

$$= (0,-1,-1,0) \text{ For leptons}$$

$$Q_{col} = (-2/3, 1/3, 1/3) \text{ For (red, yellow, blue)-quarks}$$

$$= 0 \quad \text{(For leptons)} \qquad (8)$$

The experimental distinctions between the two hypotheses especially in lepto-production experiments are discussed in detail in the next section. Here, we first note some qualitative <u>similarities and dissimilarities</u> between the cases:

(i) SU(3)'-Color as a Classification Symmetry and the Masses of the Gluon-octet V(8):

Integer as well as fractional possibilities for quark-charges originate under the same local symmetry[1,2] (e.g. $G = SU(2)_L \times SU(2)_R \times SU(4)'_{L+R}$). Prior to spontaneous symmetry breaking, photon is not distinguished, since all gauge mesons are massless. Spontaneous symmetry breaking "decides" the composition of the distinguished gauge particle -- The massless photon. Either charge-assignments for quarks can be realized under allowed patterns of spontaneous symmetry breaking. However, one notable distinction emerges between the two cases, assuming that the theory is non-abelian (as is the case for G) -- this concerns the masses of the color-octet of gauge mesons V(8). For the integer-charge-case, color as a local symmetry is broken, the color-octet of gluons acquire masses -- yet a global color-symmetry is maintained[23] approximately to $O(\alpha)$. For the case of fractionally charged quarks, on the other hand, simply in order to maintain color as a global-classification symmetry, it appears[23] that one must leave the local-SU(3)'-color-symmetry unbroken, and thus the V(8) octet of gluons massless (in addition to the photon). This necessitates that for the fractionally charged case one must solve the infrared problem associated with the octet of massless gluons. Correspondingly since no fractionally charged (stable) objects and massless gluons have appeared experimentally, one must prove (for this case) that the theory can confine[24] quarks and all color. No such hypothesis is needed (as mentioned before) for the case of integer-charge-quarks.

(ii) Asymptotic Freedom:

The gauge-structure $SU(3)'_{col}$ (indeed also the full symmetry G or its extensions) being nonabelian, the field theory of the gauge-interactions is asymptotically free[25] for either quark-charge-assignments.

For the integer-charge-hypothesis, since Higgs-Kibble particles are used to assign masses to the octet of color-gluons, asymptotic freedom is lost in the Higgs-sector (due to the quartic couplings of such fields). Such a loss of asymptotic freedom is, however, not serious, as expressed and hoped for by several authors, if spontaneous symmetry breaking turns out to be dynamical (and Higgs-Kibble-fields arise effectively as $(\bar{\psi}\psi)$-type-composites through non-perturbative solutions). (Alternatively, even with elementary scalars, as long as the effective quartic couplings of the scalar fields at present energies are sufficiently small ($\lesssim e$), the theory would be "temporarily" asymptotically free[26] in the present energy regime anyhow). Thus the <u>bonus</u> of asymptotic-freedom would apply equally well for either quark-charge assignment at least in the present energy regime. For the theory to be "free" at truely asymptotic energies (exceeding the masses of W and X-particles), one may have to rely upon the possibility of dynamical symmetry breaking for <u>either</u> quark-charge-assignment.

(iii) <u>"Naturalness" of Parity and Strangeness</u>:

Since the weak symmetry group (e.g. $SU(2)_L \times SU(2)_R$) as well as F'_{15}, <u>commute</u> with the symmetry group SU(3)'-color (which generate effective strong interactions) <u>both charge assignments share the following common advantages</u>: (a) parity and strangeness violations remain[27] "natural" ($\mathcal{O}(\alpha/m_W^2)$). (b) The saturation properties[28] of hadrons treated as composites of qqq and $q\bar{q}$ together with the lowest lying hadrons being color-singlets receives a simple explanation.

2. Masses of Gauge Particles, Eigenstates:

The masses of gauge particles for the minimal symmetry-model (and the relevant physical processes, which restrict these masses) are listed below (using $f^2/4\pi = 10$):

Mass	Process
$m(V(8)) \lesssim (2\text{-}5)$ GeV	(q-q strong force)
$m(W_L) \simeq (50 - 100)$ GeV	V-A Int.
$m(W_R) \gtrsim 300$ GeV	V+A Int.
$m(S°) \gtrsim 1000$ GeV	$\nu + p \to \nu +$ Hadrons
	$\nu_\mu + e \to \nu_\mu + e$
$m(X) \gtrsim 10^5$ GeV	$K_L \to \bar{\mu} + e$

(Note that the restriction on X-particle masses within the minimal symmetry model (arising from the absence of $K_L \to \bar{\mu}e$-decays) does not, in general, apply within extended symmetries[12] such as $SU(16)_L \times SU(16)_R$. The masses shown above can be obtained in the integer-charge quark-model with simple representations of Higgs-Kibble-multiplets and an allowed pattern of spontaneous symmetry breaking.[2] (For example, the multiplets A = (2+2, 2+2, 1), B = (1, 2+2, $\bar{4}$), C = (2+2, 1, $\bar{4}$), D = (1, 1, 15) and E = (1, 3, 1) are adequate to generate the desired mass-pattern for the gauge particles.[29] For the fractionally charged quark-model, on the other hand, simple representations of Higgs-Kibble-multiplets cannot be utilized to give masses to the V(8) octet of color-gluons, though they can furnish masses to the W's, X's and S°. (One may (if one wishes to) provide masses to V(8)-gluons in this case through multiplets such as (1, 1, 4x4x4x4); however such multiplets, while breaking local SU(3)'-symmetry, do not permit any global SU(3)'-symmetry to be preserved. This is the reason, why one ought to leave the color-gluons V(8) massless for the fractionally charged model, as mentioned before).

Spontaneous symmetry breaking, while giving masses to the gauge bosons induces mixing between several of them, so that the physical gauge-particles are in general mixtures of the canonical gauge particles: The nature of such mixing is <u>determined</u> to a large extent by the composition of the photon and the values of the effective gauge coupling constants g and f in the W and V-sectors. I list below the five neutral and six charged eigenstates for the <u>integer-charge-model</u> for the simplest case of spontaneous symmetry breaking (ignoring correction terms of order (g^2/f^2) or smaller and setting $G_L = g_R = g$ (for convenience of writing)):

$$\begin{aligned}
A &= (e/fg)\,[f\,W_{valency} + (2/\sqrt{3})g\,U^\circ]\,; && m_A = 0 \\
\tilde{U} &= (3f^2 + 2g^2)^{-1/2}\,[\sqrt{3}\,f\,U^\circ - g\,W_{valency} \\
&\quad + O(\varepsilon^2)]\,; && m_U \lesssim 2 \text{ to } 5 \text{ GeV} \\
V^\circ &= 1/2\,[\sqrt{3}\,V_8 - V_3]\,; && m_{V^\circ} \lesssim 2 \text{ to } 5 \text{ GeV} \\
Z^\circ &\simeq 1/\sqrt{2}\,[(W_R^3 - W_L^3) - \sqrt{2/3}\,(g/f)\,S^\circ]\,; && m_{Z^\circ} \simeq 100 \text{ GeV} \\
\tilde{S} &\simeq [S^\circ + \sqrt{2/3}\,gW_R^3]\,; && m_S \gtrsim 1000 \text{ GeV}
\end{aligned} \tag{9}$$

where,

$$\begin{aligned}
U^\circ &= (1/2)\,(\sqrt{3}\,V_3 + V_8) \\
W_{valency} &= W_{3L} + W_{3R} - \sqrt{2/3}\,(g/f)S^\circ \\
2e^2 &= g^2f^2/(g^2 + f^2) \simeq g^2 \\
O(\varepsilon^2) &= O[(m_U/m_{W_L})^2] \ll 1
\end{aligned} \tag{10}$$

The charged particle eigenstates[30] (ignoring W-X mixing[31] for this purpose and corrections of $O(\delta^2)$; $\delta \simeq 10^{-4}$) are given by:

$$\tilde{V}^{\pm}_{\rho} \simeq \cos\beta\, V^{\pm}_{\rho} + \sin\beta\, W^{\pm}_{L}$$
$$\tilde{V}^{\pm}_{K*} \simeq \cos\alpha\, V^{\pm}_{K*} + \sin\alpha\, W^{\pm}_{L} \qquad (11)$$
$$\tilde{W}^{\pm}_{L} \simeq W^{\pm}_{L} - V^{\pm}_{\rho}\sin\beta - V^{\pm}_{K*}\sin\alpha$$

where,

$$\sin\alpha = -\sin(\theta_L + \phi_L)\,(m_V/m_{W_L})^2\,(g/f)$$
$$\sin\beta = -\cos(\theta_L + \phi_L)\,(m_V/m_{W_L})^2\,(g/f) \qquad (12)$$

Here θ_L and ϕ_L are the Cabibbo rotations in $(n,\lambda)_L$ and $(p,\chi)_L$-spaces respectively, the observed Cabibbo-angle being $\Theta_c = \Theta_L - \phi_L$. Note the mixings of the color-gluons with the weak valency gauge mesons W's both in the neutral and in the charged sectors. These mixings are inevitable, if the photon is to contain a mixture of valency and color-gauge mesons (Eq. (9)), i.e. if quarks are integer-charged. Such mixings have profound physical implications on the one hand on lepto-production of color (i.e. $e + p \to e + X_{col}$; $\nu_\mu + p \to \mu^- + X_{col}$); on the other hand on the decays of lowest lying color-octet states (which, without such mixings, would contain stable members by conservation of color I'_3 and Y'-quantum numbers). It should be stressed at the same time that the above mixings being small do not lead to any undesirable effect either as regards parity or strangeness violation in hadronic processes or as regards q.e.d.-processes such as (g-2) for the electron and the muon.[32] We return to a detailed investigation of the consequences of such mixing in the next section.

III. QUARK-CHARGES; COLOR BRIGHTENING

1. Unphysical Versus Physical Color:

The unification hypothesis outlined in the previous section uniquely fixes the charges of the leptons; but it does not apriori fix the quark-charges. It allows, in general (as also other alternative approaches) two very distinct possibilities:

(i) Quarks are fractionally charged; the octet of color-gluons are massless. In this case, from experimental considerations, one must assume that quarks and all color are confined. In short by hypothesis color is unphysical (although there exists no theoretical proof for such a hypothesis at present).

(ii) Alternatively; quarks are integer-charged. The octet of color-gluons (possessing charged members) acquire mass ($\simeq$ 2 to 5 GeV). Local color-symmetry is broken, yet a global color-symmetry is preserved up to order α corrections. As mentioned before, there is no need to assume confinement in this case, since integer charge quarks, following the quark-lepton-unification hypothesis[2], may decay rapidly ($\tau_q \lesssim 10^{-10}$ sec.) into leptons ($q \to \ell + \ell + \bar{\ell}$, $q \to \ell$ + pions) and such short lived quarks are likely to have been missed by standard particle and quark-searches, even if their production cross section was not insignificant ($\gtrsim 10^{-32}$ cm^2 say). In short, with integer-charge quarks color is as physical as valency (or flavor); there lies a rich world of color to be discovered.

It is quite possible in fact that color has already appeared. We point out that the anomalous ($\bar{\mu}e$)-events[20] seen at SPEAR may owe their origin to pair production of quarks in e^-e^+-annihilation or quark-like objects (see later)-(rather than heavy leptons), followed by quark and antiquark decays into leptons. In addition, of course, one or several of the multitude of J/ψ-particles may represent color. Familiar arguments[33] against the color interpretations for the J/ψ-particles do not apply to the cases[34] where some of them represent color, others representing new valency quantum numbers. We return later to a discussion of these matters and to the question of how to distinguish experimentally between heavy lepton versus quark-origin of the $\bar{\mu}e$ events on the one hand, and independently between color versus new valency quantum number-origin of the J/ψ particles on the other.

2. Familiar Arguments Against Physical Color:

The hypothesis of decaying integer-charge-quarks (physical color) is an attractive alternative to the more commonly held hypothesis of fractionally charged quarks (unphysical color), since it does not need to hide the basic particles of the theory and yet it provides a reason for the "missing quark". Despite these possible advantages, there have been a set of standard objections[35] to the twin hypothesis that (i) quarks carry integer-charges (color is physical) and (ii) color-threshold is not very high (i.e. it is within reach of lepto-production experiments at least at Fermilab). We first list below these familiar objections and then point out a flaw in them; which invalidate these objections:

(i) If quarks carry integer-charges, sufficiently above color-threshold, the color-part of quark-charges (Q_{col}) would be expected to contribute to electro (or muon)-production structure functions on par with the valency part of quark-charges (Q_{val}). If color-threshold was as low as 3 to 5 GeV, then color ought to have brightened in present electro or muon-production experiments with a large rise (~ 100%) in the structure functions, contrary to observations.

By the same token, sufficiently above color-threshold, familiar parton model based <u>sum rules</u> should have exhibited integer-character for quark-charges rather than fractional. For example, (ignoring gluon-contributions) the famous ratio

$$r \equiv \frac{4G_F^2 M_N E_\nu}{2\pi(\sigma_\nu + \sigma_{\bar{\nu}})} \int F_2^{\gamma N} \, dx \tag{13}$$

of electro and neutrino-production cross-sections should have acquired a value 0.5, if quarks are integer-charged, rather than[35,36] a value ≃ 5/18 appropriate for fractional charges.

(ii) It is known that the gluon-content of the nucleon in the deep inelastic regime is non-neglible since nearly 50% of the nucleon-momentum resides within gluons. If quarks are integer-charged, then sufficiently above color-threshold, the Yang-Mills coupling of the charged members ($V_\rho^\pm$, $V_{K*}^\pm$) of the spin-1 gluons to the photon would be expected to lead to a non-vanishing (large contribution) to σ_L/σ_T, which should be growing[37] with $|q^2|$. This too is qualitatively not supported by the data.

Accepting the above arguments, one would be inclined to conclude that either quarks are fractionally charged, or that color-threshold is sufficiently above 10 GeV beyond the reach of present experiments. This has been the standard point of view over the years. If this were true, none of the observed new phenomena (such as J/ψ-particles, or dimuons, or ($\bar{\mu}$e)-events) could have been attributed to excitations of color. We show[17,18] below that the above arguments, while they might have been sound within a phenomenological approach to color, do not hold within the gauge theory approach.

3. A Flaw in the Familiar Arguments:

First, we argue that even though photon carries color, color cannot brighten on par with valency for all (spontaneously broken) guage theories, satisfying the following two criteria: (1) valency and color are gauged independently with weak interactions associated with valency gauging, strong with color-gauging and (2) leptons are treated as singlets of SU(3)'-color.

The argument is simple: express the gauging pattern above symbolically in the form:

$$L_{int} = g\, W_\mu\, (J_\mu^{valency} + J_\mu^{leptons}) + f\, V_\mu\, J_\mu^{color} \tag{14}$$

where W stands for the weak and V for the strong gauges. Notice now that before spontaneous symmetry breaking (when all gauge fields are massless), leptons interact with the quark-valency-current $J_\mu^{valency}$ through the intermediacy of W_μ's; but there is no interaction between $J^{leptons}$ and J^{color}. Due to spontaneous symmetry breaking valency and color gauge mesons mix. This generates (through diagonalisation of fields) the massless photon A_μ as a mixture of W_3 and the color-gluon $U° = 1/2\ (\sqrt{3}\ V_8 + V_8)$, but inevitably also the orthogonal color-gauge partner $\tilde{U}_\mu$ (with mass m_U), both of which contribute to lepton-color-interaction. The two contributions would exactly cancel each other, except for the difference between the photon and $\tilde{U}$-propagators. This has the consequence that while lepton-valency interaction has the matrix element $J^{lepton}\ (1/q^2)\ J^{valency}$, the corresponding expression[38] for lepton-color-interaction equals $J^{lepton}\ (1/q^2 - 1/(q^2-m_U^2))$ J^{color}, where the local symmetry is spontaneously broken. It is the crucial negative sign between the two propagators in the second case, resulting from a diagonalization of fields, which suppresses color-brightening effects. The resulting matrix element for lepto-production of color thus acquires a new kinematic factor within the gauge-theory approach (compared to the matrix-element for lepton-valency-interaction):

$$\Delta(q^2) \equiv q^2 \left(\frac{1}{q^2} - \frac{1}{q^2 - m_U^2}\right) = - \frac{m_U^2}{q^2 - m_U^2} \tag{15}$$

It is this kinematic Δ^2-factor (in the cross-section) that provides the sharp distinction between lepto-production of color and valency. Note that asymptotically $\Delta^2(q^2) \to 0$. In other words, leptons having started life as color-singlets, are not efficient to probe the color-octet part of quark-charges.

The familiar arguments outlined in subsection 2 were based upon the presumption that lepto-production of color (like that of valency) proceeds through one-photon-exchange only. They do not, therefore, take into account the cancellation effect between the photon and the $\tilde{U}$-gluon contributions. In the next section, we discuss in detail the consequences of this cancellation-effect on the color-contribution to structure functions.

(4) Lepto-Production of Color Through Quark and Gluon Partons:

A. Though the results stated in the previous subsection are general, they may explicitly be verified within the minimal symmetry model[2] $G = SU(2)_L \times SU(2)_R \times SU(4)'_{L+R}$, for which the mass matrix and the eigen-states are available in detail. We write below the coupling of the two low-mass eigenstates[39] -- the photon A_μ and the color-gluon $\tilde{U}_\mu$ -- which are coupled both to leptons and to the color-octet current (see Eq. (9); we have dropped $\mathcal{O}(\epsilon^2)$-corrections):

$$L_I = e\,A_\mu\,[J_\mu^{val} + J_\mu^{col}]$$

$$+ \tilde{U}_\mu\,[(\sqrt{3}/2)\,f\,J_\mu^{col} - (2/\sqrt{3}\,)\,(e^2/f)\,J_\mu^{val}] \qquad (16)$$

where the fermionic contents of the currents are:

$$J_\mu^{val} = \sum_{\alpha=r,y,b} \{2/3\,\bar{p}_\alpha p_\alpha - 1/3\,\bar{n}_\alpha n_\alpha - 1/3\,\bar{\lambda}_\alpha\lambda_\alpha + 2/3\,\bar{\chi}_\alpha\chi_\alpha\}_{L+R}$$

$$- (\bar{e}e + \bar{\mu}\mu)_{L+R}$$

$$J_\mu^{col} = \sum_{q=p,n,\lambda,\chi} \{-2/3\,\bar{q}_r q_r + 1/3\,\bar{q}_y q_y + 1/3\,\bar{q}_b q_b\}_{L+R} \qquad (17)$$

Note α runs over three colors (red, yellow, blue) = (r,y,b) and q runs over four valency-indices (p,n,λ,χ). The contributions to J_μ^{val} and J_μ^{col} arising from the Yang-Mills coupling of the gauge fields $\vec{W}$ and the color-gluons V(8) respectively are not exhibited, but should be understood.

Note that on account of the $W_{valency}$-component, the physical gauge particle $\tilde{U}_\mu$ becomes directly coupled to the electron. Such a coupling must be present within the gauge theory approach, if the photon must couple to J^{col}. The strength and relative sign of such a coupling are determined by the composition of the massless photon and the renormalized effective gauge coupling parameters.

Note the crucial feature of Eq. (16). The strength factor for the current correlation $(J_\mu^{lep}(x)\;J_\nu^{col}(x'))$ arising due to photon-interaction in second order is $-e^2$, while that arising due to $\tilde{U}$-interaction is $(2/\sqrt{3}\,)(e^2/f)(\sqrt{3}/2)f = +e^2$. The two contributions, as mentioned before, thus exactly cancel each other except for the difference between the photon and the $\tilde{U}$-propagators.

Treating the leptons (but not hadrons) perturbatively and including the contribution of only one photon-exchange[40] for color-singlet (valency)-production, but photon as well as $\tilde{U}$-exchanges[41] for color-production, we obtain

$$M(e + N \to e + X_{val}) = -e^2 \; (\bar{e}\gamma_\mu e) \; D_\gamma(q^2) \langle X_{val} | J^{val} | N \rangle$$

$$M(e + N \to e + X_{col}) = -e^2 \; (\bar{e}\gamma_\mu e) D_\gamma(q^2) \; \Delta(q^2) \; \langle X_{col} | J^{col} | N \rangle$$

where,

$$\Delta(q^2) = (D_\gamma(q^2) - D_U(q^2))/D_\gamma(q^2) = -\frac{m_U^2}{q^2 - m_U^2} \tag{19}$$

Here $D_\gamma(q^2)$ and $D_U(q^2)$ are the renormalized propagator functions for the photon and $\tilde{U}$ respectively. Thus there is the additional kinematic factor $\Delta(q^2)$ for the matrix element for lepto-production of color relative to that of valency production. (Quite clearly an identical factor would arise for time-like processes ($q^2 > 0$) such as e^-e^+-annihilation).

We thus deduce that for integer-charge quarks the structure-functions $F_{1,2}^{eN}(q^2,\nu)$ (representing cross sections) are sums of two pieces:

$$F_i^{eN} (q^2,\nu) = F_i^{val} (q^2,\nu) + \bar{F}_i^{col} (q^2,\nu) \tag{20}$$

where,

$$\bar{F}_i^{col} (q^2,\nu) = \Delta^2(q^2) \; F_i^{col} (q^2,\nu) \tag{21}$$

$F_i^{val} (q^2,\nu)$ and $F_i^{col} (q^2,\nu)$ are defined in the usual manner by the Fourier transforms of the current correlation matrix elements.

Above threshold for color-production (i.e. $M_x^2 \equiv M_N^2 + 2M_N\nu - |q^2| > M_{col}^2$) $\bar{F}_i^{col}$ should be non-vanishing; however we do not expect parton-model (light-cone or asymptotic-freedom) considerations to apply to F_i^{col} unless $|q^2|$ and $M_N\nu$ are sufficiently above characteristic color-octet masses. Noting that characteristic masses for valency and color transitions (in the q^2-variable) are of order m_ρ^2 and m_U^2 respectively, and using the empirical fact that $F_i^{val} (q^2,\nu)$ are damped[36] for small $|q^2|$ and that they acquire their "full weight" (scaling value) for $|q^2| > 2m_\rho^2$, we may thus expect parton-model considerations for $F_i^{col}\ (q^2,\nu)$ to apply only when $|q^2|$ and $M_N\nu \gtrsim 2m_U^2$. (For lower values of $|q^2|$, $F_i^{col}\ (q^2,\nu)$ should be damped, the damping becoming progressively more severe as $|q^2| \to 0$.).

Including the Δ^2-factor, the net asymptotic contributions of quark as well as charged spin-1 gluon-partons[42] to electro-production structure functions and to the R-parameter (for e^-e^+-annihilation) are given by:

$$F_1^{eN} = \left\{F_1^{val}\right\} + (1+\xi)^{-2}\left[F_1^{col}\right]$$

$$= \left\{\frac{1}{2}\sum_{q_i=p,n,\lambda,\chi} Q_{val}^2\ (q_i)\ [q_i(x) + \bar{q}_i(x)]\right\} + (1+\xi)^{-2}\left[\frac{1}{3}\sum_{q_i}\ \{q_i(x) + \bar{q}_i(x)\} + \frac{16}{3}\ (1+\frac{\xi}{4})\ v(x)\right] \quad (22)$$

$$F_2^{eN} = \left\{F_2^{val}\right\} + (1+\xi)^{-2}\ [F_2^{col}]$$

$$= \left\{\sum_{q_i} x\ Q_{val}^2\ (q_i)\ [q_i(x) + \bar{q}_i(x)]\right\} + (1+\xi)^{-2}\left[\frac{2x}{3}\sum_{q_i}\{q_i(x) + \bar{q}_i(x)\} + x\ v(x)\ (4 + \frac{4}{3}\xi + \frac{\xi^2}{3})\right] \quad (23)$$

$$R \equiv \sigma(e^-e^+ \to \text{hadrons})/\sigma(e^-e^+ \to \mu^-\mu^+)$$

$$= \{R_{val}\} + (1+\xi)^{-2}\ [R_{col}]$$

$$= \left\{\sum_{q_i} Q_{val}^2\ (q_i)\right\} + (1+\xi)^{-2}\left[\frac{2}{3}\cdot \text{(No. of quark valencies)} + (1/8)\ (1-\frac{4}{\xi})^{3/2}\ (12 + 20\xi + \xi^2)\right] \quad (24)$$

where, $\xi = |q^2|/m_U^2$; $x = 1/\omega = |q^2|/2M_N\nu$

$q_i(x)$ = momentum-distribution function for the ith type quark (within nucleon)

$v(x)$ = momentum-distribution function for any one of the octet of color-gluons (within nucleon).

$$Q_{val}^2(q_i) = \frac{4}{9}, \frac{1}{9}, \frac{1}{9}, \frac{4}{9} \quad \text{for } q_i = (p,n,\lambda,\chi)\ . \quad (25)$$

Note the curly bracket denotes color-singlet (valency production), while square-brackets denote color-octet production. The first and second terms inside the square bracket of Eq. (24) denote respectively the contributions to the R-parameter from the color-octet part of quark-charges and the gluon-charges (If there are valency-quarks in addition to (p,n,λ,χ), the sum over q_i should include these valencies as well.)

Asymptotically ($\xi \gg 1$, i.e. $|q^2| \gg m_U^2$), the color-structure functions $\bar{F}_i^{col}(q^2,\nu)$ and $\bar{R}_{col}$ reduce to:

$$\bar{F}_1^{col}\ (eN) \to 0$$

$$\bar{F}_2^{col}\ (eN) \to (1/3) \times v(x)$$

$$\bar{R}_{col} \to \bar{R}(\text{Gluons}) = 1/8 \qquad (26)$$

Thus asymptotically (i.e. for $|q^2| \gg m_U^2$), the contributions of the color-octet part of quark-charges to electro-production structure functions as well as to R-parameter dies out because of the Δ^2-factor. The contributions of the charged spin-1-color-gluons ($V_\rho^\pm$, $V_{K*}^\pm$) to F_1, F_2 and R (instead of growing like $|q^2|$, q^4 and q^4 respectively) scale (again because of the Δ^2-factor). Thus color-production survives in a scale-invariant manner only due to the gluon-parton-contributions.

It should be stressed that the damping due to the kinematic factor Δ^2 is not quite as rapid in the time-like-process ($q^2 > 0$, i.e. for e^-e^+-annihilation) as it is in the space-like-processes ($q^2 < 0$) especially for intermediate or semi-asymptotic $|q^2|$ (i.e. $2m_U^2 < |q^2| < 4m_U^2$); and in fact for $|q^2| < 2m_U^2$ the Δ^2-factor acts as an enhancement[43] for time-like processes; where as it provides a damping for space-like processes for all q^2. Thus, in the intermediate region ($m_U^2 < q^2 < 3m_U^2$), the color-contribution to R-parameter (just from the quark-charges) would still be significant. Exactly what is the contribution to R from color-part of quark-charges in this semi-asymptotic region and how it varies depends upon the thresholds of color-resonant states (which should be opening in this range) and their dynamics. Note, however, that the color-gluon $\tilde{U}$ is produced through its direct coupling to the leptons ($e^-e^+ \to \tilde{U}$) without the intermediacy of the photon, and thus without the Δ^2-factor.

It should also be noted that because of the relatively large coefficients associated with the non-leading terms in ξ compared to the leading term (see Eqs. (22), (23) and <u>especially</u> Eq. (24)), the asymptotic values of the color-gluon contributions to F_1, F_2 and R (Eq. (26)) are not attained until ξ is much much greater than unity. This is especially true of their contribution to the R-parameter, as it may be judged from the following numerical values of $\bar{R}_{col}$ for $\xi > 4$ (taken from Eq. (24)):

$$\begin{aligned}\bar{R}_{col} &= \bar{R}_{col}\,(\text{quarks}) + \bar{R}_{col}\,(\text{gluons}) \\ &\simeq 1/6 + (.8)(1/8) \simeq .27 \quad (\xi = |q^2|/m_U^2 = 5) \\ &\simeq .03 + (1.95)\,(1/8) \simeq .27 \quad (\xi = 10) \\ &\simeq .00 + (1.2)\,(1/8) \simeq .15 \quad (\xi = 100)\end{aligned} \tag{27}$$

Thus the net contribution[44] from color-gluon-pairs ($V_\rho^+ V_\rho^-$ and $V_{K^*}^+ V_{K^*}^-$) to R-parameter is at most about 1/4 (for $\xi \simeq 10$), which drops to 1/8 for sufficiently large ξ. We later point out distinct signatures for charged color gluons (or, in general, lowest lying charged color-octet states).

B. Rise in Structure Functions Due to Color-Production:

The contributions from quark and gluon-partons to the color-part of the structure functions, as listed in sec. A, should lead to a rise in these functions at energies (and $|q^2|$ and $M_N\nu$) sufficiently above color-threshold. Numerical values of such a rise depend upon the values of the kinematic variables $\xi = |q^2|/m_U^2$ and $x = |q^2|/(2M_N\nu)$ on the one hand, and the quark and gluon momentum-distribution functions $q_i(x)$ and $v(x)$ on the other. While the quark-distribution functions are somewhat known;[45] there is at present no information on the detailed properties of v(x). It is, however, known[46] (using momentum-conservation) that the gluons carry about 50% of the nucleon's momentum. Thus,

$$\int_0^1 8x\, v(x)\, dx \simeq 0.5 \tag{28}$$

Subject to this condition, we made two models for v(x).

Model I: v(x) has a shape similar to that of the neutron-quark distribution function within the proton. This yields:

$$\begin{aligned} x\, v(x) &\simeq 0.04 \quad (x \simeq .5) \\ &\simeq 0.1 \quad (x \simeq .2) \\ &\simeq 0.11 \quad (x \simeq 0) \end{aligned} \tag{29}$$

Model II: v(x) has a shape similar to that of the sea (being very small for x > .1 and rising steeply as $x \to 0$). This gives:

$$\begin{aligned} x\, v(x) &\simeq 0.01 \quad (x \simeq .5) \\ &\simeq 0.05 \quad (x \simeq .2) \\ &\simeq 0.65 \quad (x \simeq 0)\, . \end{aligned} \tag{30}$$

Below, we present numerical values of $(\bar{F}_2^{col}/F_2^{expt})$ for ep-scattering for models I and II for two values of x and two values of ξ. The quark-distribution functions are taken from Ref. 45.

TABLE I

		Model I		Model II	
		$\xi=3$	$\xi=5$	$\xi=3$	$\xi=5$
$F_2^{ep}(col)/(F_2^{ep})_{expt.}$	$x = .5$	.25	.18	.14	.10
	$x = .2$	.30	.20	.20	.14

We notice qualitatively the following features:

(i) In either model (I or II), the rise in F_2 due to color-production is expected to decrease as $|q^2|$ (or ξ) increases until fairly large values of $\xi >> 1$, which would manifest as a scale-violating contribution. These arise due to non-leading terms (in ξ) in the structure functions.

(ii) For large $x > .1$ (i.e. $\omega < 10$), the rise in F_2 due to color-production is expected to be more prominent for model I (for which the rise is of order 25 to 30% for $\xi = |q^2|/m_U^2 = 3$) than it is for model II (for which the rise lies between 10 to 20%).

(iii) For low values of $x < .1$ (i.e. $\omega > 10$) on the other hand, we would expect model II to exhibit a significant rise (much more so than model I) reflecting the relatively high concentration of gluons at low x for this model.

As we explained before, we, of course, do not expect color-contributions to structure-functions to acquire their scaling "weight" in any case until $|q^2|$ and $M_N\nu \geq 2$ to $3m_U^2$ (i.e. $\xi \geq 2$ to 3). Thus if $m_U^2 \simeq 9$ to 17 $(GeV)^2$ (corresponding to J/ψ-particle masses), the above considerations for color-contributions to structure-functions are expected to apply only for $q^2 \gtrsim 20$ to 30 $(GeV)^2$. Thus in order to verify the expected large rise ($\gtrsim 50\%$) of F_2 (for model II) at small $x \lesssim .05$ (say) (i.e. $\omega \gtrsim 20$), one would need energies such that $2M_N\nu \gtrsim 400$ to 600 $(GeV)^2$. For presently explored regions of q^2 and ω, the expected rise of F_2 for model II (and probably even for model I) is consistent with the data.

To summarize; the color-gluon contributions to structure functions, though not uncomfortably large for presently explored regions of q^2 and ω at MIT, SLAC and Fermilab, are observable at present energies, assuming that color-threshold and color-gluon-masses are $\lesssim 5$ GeV. It is conceivable that the observed[47] decrease in structure functions with increasing q^2 for $\omega < 6$ could be partly or entirely attributed to the color-gluon-contributions. More refined data would be needed to separate colour-gluon contributions from the logarithmic scale-violating terms, expected on the basis of asymptotically free theory.[48] However a large increase ($\sim 50\%$) in structure function at very large ω (>20) with $|q^2| \gtrsim 20$ $(GeV)^2$, if observed, is likely to signal colour-gluon contribution favouring their "sea-like"-momentum distribution within the nucleon (model II) over "quark-like"-momentum distribution (model I).

C. The Ratio σ_L/σ_T:

As is well known (σ_L/σ_T) is expected to receive finite contributions from charged spin-1 partons. With a Yang-Mills type coupling, the contributions of these spin-1 partons to (σ_L/σ_T) should in fact have grown with $|q^2|$. Including the Δ^2-factor, the ratio (σ_L/σ_T) becomes a function of the scaling variable x only. Using Eqs. (22) and (23), we obtain,

$$\frac{\sigma_L}{\sigma_T} = \frac{F_2^{ep} - 2x\, F_1^{ep}}{2x\, F_1^{ep}}$$

$$= \frac{(xv(x))}{(1+\xi)^2} \left[(4 + \frac{4}{3}\xi + \xi^2/3 - \frac{32}{3}(1 + \xi/4)\right] \Big/ \qquad (31)$$

$$\left[\sum_{q_i} x\ Q^2_{val}\ (q_i)\ \{q_i(x) + \bar{q}_i(x)\}\right.$$

$$\left. + \frac{1}{(1+\xi)^2}\ \frac{2}{3} \sum_{q_i} x(q_i(x) + \bar{q}_i(x)) + \frac{32}{3}(1+\xi/4)\ xv(x)\right]$$

$$\xrightarrow[(\xi>>1)]{} \frac{1/3\ xv(x)}{\sum_{q_i} xQ^2_{val}\ (q_i)\ \{q_i(x) + \bar{q}_i(x)\}} \qquad (32)$$

Substituting typical values of $x = .2,\ .5$ and $\xi = |q^2|/m_U^2 = 2,\ 3,\ 5$, we obtain (σ_L/σ_T) lying between 0.1 and 0.2 (for either model I and model II) consistent with the data. In other words, the cancellation-factor $\Delta^2 = (1/1+\xi)^2$ softens the colour-gluon contribution thereby allowing lepto-production data to be compatible with the twin notion that color-threshold is not very high ($\lesssim$ 5 GeV) and that photon carries color (quarks are integer-charged). There is, of course, the strong experimental prediction of this notion that asymptotically (σ_L/σ_T) can not vanish (especially at low x) and that it is a function of x only.

D. Neutrino-Production of Color:

Within the gauge theory approach, as stressed in sec. II, if the gauge-meson-mass matrix induces a mixing between the neutral members $(W_3)_{L,R}$ and $U°$ to make the photon, it must inevitably induce either $W_L^{\pm}$ or $W_R^{\pm}$ (or both) to mix with the charged members of the color-gluon-octet $V_\rho^{\pm}$ (or $V_{K*}^{\pm}$ or both).

For the case, where V_ρ^+ mixes with W_R^+ (rather than W_L^+), there will be no neutrino-production of color in charged current interactions (barring the tiny $W_L^{\pm}$-$W_R^{\pm}$-mixing), since ν_{eL} and $\nu_{\mu L}$ do not couple to W_R.

The case, where $V_\rho^{\pm}$ and $V_{K*}^{\pm}$ mix with $W_L^{\pm}$, is given in sec. II (see Eq. (11)). For this case, exchanging V_ρ^{+} and neglecting W_L^{+}-exchange insofar as color-production is concerned (since $m_{W_L}^2 >> m_{V_\rho}^2$), we obtain (using the mixing angles given in Eq. (12) and $G_F/\sqrt{2} = g^2/2m_W^2$):

$$M(\nu_\mu + N \rightarrow \mu^- + X_{col}) = \frac{G_F}{\sqrt{2}} \cos(\theta_L + \phi_L)\,(\bar{\mu}\gamma_\mu(1+\gamma_5)\nu_\mu) \left(\frac{m_{V_\rho}^2}{q^2 - m_{V_\rho}^2}\right) \langle X_{col}|J_{col}^{+}|N\rangle \quad (33)$$

Similarly V_{K*}^{+}-exchange contributes an amplitude proportional to $\sin(\theta_L+\phi_L)$. To compare, with the above, the amplitude for color-singlet (valency)-production is given by:[49]

$$M(\nu_\mu + N \rightarrow \mu^- + X_{val}) = \frac{G_F}{\sqrt{2}} (\bar{\mu}\gamma_\mu(1+\gamma_5)\nu_\mu) \langle X_{val}|J_{val}^{+}|N\rangle \quad (34)$$

Adding the two <u>separate</u> contributions (due to V_ρ^{+} <u>and</u> V_{K*}^{+} exchanges) to cross-sections, we note the emergence of the same $\Delta^2 = (m_V^2/q^2-m_V^2)^2$-factor for neutrino-production of color in charged-current interactions, as it appears for electro-production of color. Hence the discussion of sec. C (see Table) regarding the relative importance of color versus valency-production in the various kinematic regions ($|q^2|$ and x) would apply to neutrino-production of color, as it does to electro or muon-production of color (assuming, of course, that it is $W_L^{\pm}$ (rather than $W_R^{\pm}$), which mix with $V_\rho^{\pm}$ and $V_{K*}^{\pm}$). Thus, once again, the rise in structure functions due to color-production in neutrino-reactions are also limited to (10 to 30)%-level for the presently explored regions of q^2 and x.

Taking an average value $\langle\epsilon_{col}\rangle \lesssim (1/5)$ for an estimate of the suppression of color relative to valency-production, we may now evaluate the ratio r (Eq. (13)):

$$r = \frac{4G_F^2 M_N E_\nu}{2\pi(\sigma_\nu+\sigma_{\bar\nu})} \int F_2^{\gamma N} dx = \frac{\langle Q^2_{val}\rangle_{eN} [1 + \langle\varepsilon_{col}\rangle_{eN}]}{\langle Q^2_{val}\rangle_{\nu N} [1 + \langle\varepsilon_{col}\rangle_{\nu N}]} \tag{35}$$

where $\langle Q^2_{val}\rangle_{eN} = 5/18$; $\langle Q^2_{val}\rangle_{\nu N} = 1$. Even though $\langle\varepsilon_{col}\rangle_{eN}$ need not be exactly equal to $\langle\varepsilon_{col}\rangle_{\nu N}$ (since the data is folded for the eN and νN scatterings for different kinematic regions), since both are small ($\lesssim 1/5$), it is clear that r will not deviate from 5/18 by more than 10%. Thus, within the gauge theory approach to color, we expect the value of r (as extracted from the eN and νN-data) to be $\simeq .28 \pm .03$ for the case of integer charge-quarks and $5/18 \simeq .28$ for the case fractionally charged quarks. A value of r anywhere in the range .25 to .31 is certainly consistent with the data.[36] The moral is that, contrary to familiar expectations (based on phenomenological approach to color), this sum rule (Eq. (35)) involving ratio of eN to νN-data is potentially not sensitive to quark-charges within the gauge theory approach[1,2] to color.

The novel qualitative features, which emerge due to the cancellation effect between the photon and $\tilde{U}$-contributions may now be summarized below:

(a) For lepto-production experiments, integer-charge quarks behave asymptotically just as if they carried their fractional valency-charges.

(b) Color-gluon-contributions to structure functions, instead of growing like q^2 or q^4, scale. The ratio (σ_L/σ_T), though non-vanishing is small and becomes a function of x only.

(c) Color does not brighten on par with valency. The rise in structure functions due to color-production is limited to the (10-20)% level at least for presently explored regions of q^2 and ω. Correspondingly, the parton-model based sum rules, do not acquire modifications any larger than

(10-20)% to take account of color production.

These new results take away the only objection that existed on experimental and theoretical grounds to the hypotheses that color is physical (quarks are integer charged) and that color-threshold is "low" ($\lesssim 5$ GeV). There are two striking predictions of the integer-charge-hypothesis (within the gauge theory approach): (i) Asymptotically (σ_L/σ_T) should be nonvanishing and a function of x only; (ii) there should be scaling violations at 10-20% level due to color-production in the semi-asymptotic region $\xi \leq 8$, which should decrease with $|q^2|$ for fixed $\omega \lesssim 10$, above and beyond the logarithmic violations expected from asymptotic freedom-considerations. By contrast, for fractionally charged quarks $(\sigma_L/\sigma_T) \rightarrow 0$ (Callan-Gross sum rule) and scaling violations should be limited to the said logarithmic terms only. Here lies an experimental method of distinguishing between the hypotheses of unphysical versus physical color. I now turn to a discussion of the distinctive signatures for color, which should be helpful in its search.

E. Distinctive Signatures for Color: Di-muons:

One expects to produce charged members of the color-gluons (or similar color-octet states) carrying color-quantum numbers (I_3' and Y') in pairs either in hadronic collisions through strong interactions (the cross section in this case may lie in the range of 10^{-31} - 10^{-32} cm^2 at Fermilab energies for $m_V \simeq 3$ GeV), or in e^-e^+-annihilation:

$$p + p \rightarrow V_\rho^+ + V_\rho^- + \text{Hadrons} \tag{36}$$

$$e^-e^+ \rightarrow V_\rho^+ + V_\rho^- + (\text{Known Hadrons}) \tag{37}$$

In addition for the case of $W_L^{\pm}$ mixing with $V_\rho^{\pm}$ and $V_{K*}^{\pm}$, they may also be produced <u>singly</u> in charged-current neutrino interactions with an amplitude of order G_{Fermi} (see Eqs. (33), (22) and (23)), the net effect of the total color-production (using the Δ^2-factor) being limited to the 10~20% level relative to the <u>total</u> valency (color-singlet)-production in accordance with Table I. Thus, for example,

$$\nu_\mu + N \rightarrow \mu^- + V_\rho^+ + (\text{Hadrons}) \tag{38}$$

For e^-e^+-annihilation, the net contribution[50] to R from pair-production of charged color-gluon-partons ($V_\rho^+V_\rho^-$, $V_{K*}^+V_{K*}^-$) at $q^2 > 4m_{V_\rho}^2$ is at most 1/4 (for $q^2 \simeq 10m_{V_\rho}^2$), which drops to (1/8) for sufficiently large q^2 (see Eqs. (24) and (27)). Even though this is a small percentage of R_{total} varying from 2 to 5.5 for q^2 varying from 9 to 60 $(\text{GeV})^2$, the charged color-gluons possess <u>distinct decay modes</u>, which can help identify their production especially through a detailed study of non-collinear lepton-pair production in e^-e^+-annihilation (and, of course, they may also be identified by a study of their production in pairs in p+p-collisions). Note that in contrast to heavy leptons, the charged color gluons may be produced (in pairs) in e^-e^+-annihilation <u>together</u> with other known hadrons (such as pions) above threshold.

The charged color-gluons ($V_\rho^{\pm}$, $V_{K*}^{\pm}$), if they are the <u>lightest</u> color-octet states (transforming like (1,8) under $SU(3) \times SU(3)'_{color}$ in the absence of their small components), would decay into leptons and known color-singlet hadrons only through their small components $W_L^{\pm}$ (or $W_R^{\pm}$) as given by Eq. (11). Thus, their allowed decay modes[13,51] are:

$$(V_\rho^+, V_{K*}^+) \rightarrow \mu^+\nu_\mu,\ e^+\nu_e$$

$$\left.\begin{array}{l}\rightarrow \pi\pi e\nu \\ \rightarrow K\bar{K}e\nu \\ \rightarrow \eta\eta e\nu\end{array}\right\}$$ with hadrons in I=0 state and also in SU(3)-singlet state to the extent SU(3) is a good symmetry

$$\rightarrow \pi\pi,\ 3\pi,\ 4\pi,\ K\bar{K},\ \text{etc.} \qquad (39)$$

The leptonic and semi-leptonic decay modes are expected to be a significant fraction of all decay modes ($\simeq$ 20 to 40% for the electron-modes and similarly for the muon-modes). The life time of the charged gluons is roughly proportional to the inverse fifth power[52] of the mass of the gluon. We obtain[53] (for the strong gauge coupling parameter $f^2/4\pi \simeq 10$):

$$\begin{aligned}\tau(V_\rho^+) &\simeq 2 \times 10^{-14}\ \text{sec.} && (m_V \simeq 1.5\ \text{GeV})\\ &\simeq 5 \times 10^{-16}\ \text{sec.} && (m_V \simeq 3\ \text{GeV})\\ &\simeq 10^{-16}\ \text{sec.} && (m_V \simeq 4.1\ \text{GeV}) \qquad (40)\end{aligned}$$

Note the important selection-rule that the semi-leptonic decay modes of $V_\rho^\pm$ and $V_{K*}^\pm$ can not involve a single π, a single K or a single η (i.e. $V_\rho^+ \rightarrow \pi^\circ\mu^+\nu_\mu$, $K^\circ\mu^+\nu_\mu$ and $\eta\mu^+\nu_\mu$ are forbidden to $O(\alpha)$ in the matrix element relative to the allowed decay modes exhibited in Eq. (39)). By contrast such decay modes are allowed for color-singlet mesons carrying new valency quantum numbers such as charmed D and F-mesons. Thus, if one discovers new charged short lived objects, which on the one hand have large (leptonic + semileptonic) decay-branching ratios, and on the other hand exhibit semi-leptonic decays involving only two pions, or two kaons, or two η's etc. in the final state (but no single π, or single K or single η), then beyond doubt they must either represent color-gluons (or analogous color-octet states lighter than the color-gluons). Of course, other characteristic features for color-gluons (and analogous color-octet states) are that there must be eight[54] of them (nearly degenerate with each other within few to 50 MeV (say)) with charges (0,0,0,0,+,+,-,-). They may all be produced in pairs in hadronic collisions; the $\tilde{U}$-member may be produced singly in e^-e^+-annihilation through its direct coupling to leptons.

IV. THE NEW PHENOMENA AND PHYSICAL COLOR.

(1) Dimuons: The dimuons[19] of opposite charge ($\mu^-\mu^+$) recently produced in neutrino and anti-neutrino reactions can be attributed to production of charged color-gluons (or analogous color-octet particle) as in reaction (38), followed by their (leptonic + semileptonic)-decays as in (39). Given that the total color-production is of order 10 to 20% compared to non-color-production (see above) and that the net (leptonic + semileptonic)-decay-branching-ratio of the lightest charged color-octet members to be of order (20 to 40%), which is divided between electron and muon modes, we expect a dimuon event rate of order (1 to 4)% compatible with the observed 1% rate compared to the single muon-events.

(2) $\bar{\mu}e$-Events: Production of Integer-Charge-Quarks and Color-Gluons by e^-e^+-Annihilation

We expect integer-charge quarks and color-gluons to be produced in pairs in general in association with other known hadrons (sufficiently above threshold)

$$
\begin{aligned}
e^-e^+ &\to p^{o,+,+}_{r,y,b} + \bar{p}^{o,-,-}_{r,y,b} \\
&\to n^{-,o,o}_{r,y,b} + \bar{n}^{+,o,o}_{r,y,b} \quad \text{etc.} \\
&\to V^+_\rho + V^-_\rho \\
&\to V^+_{K^*} + V^-_{K^*} \qquad (41)
\end{aligned}
$$

A large or fair fraction of the quark-pairs treated as partons would recombine to form known hadrons, while the color-gluon-pairs (see. Sec. III.E) would recombine to form color-octet hadrons together with known hadrons. The total contribution to R from the various quark-parton-pair production (which gets divided between many available channels) may be obtained by noting that their charges receive contributions from two sources $Q_{val} = (2/3, -1/2, -1/3, 2/3)$ for (p, n, λ, χ) and $Q_{col} = (-2/3, 1/3, 1/3)$ for (red, yellow, blue) (See Eq. (8)). The charges from the second source do not asymptotically contribute to R (because of the Δ^2-factor, sec. III). Thus, interestingly enough, asymptotically all three proton-quark-pairs (including the neutral pair $p^\circ_r \bar{p}^\circ_r$) contribute the same amount to R:

$$R(p^\circ_r \bar{p}^\circ_r) = R(p^+_y p^-_y) = R(p^+_b p^-_b) = 4/9$$

Similarly,

$$R(n^-_r n^+_r) = R(n^\circ_y \bar{n}^\circ_y) = R(n^\circ_b \bar{n}^\circ_b) = 1/9$$

$$R(\lambda^-_r \lambda^+_r) = R(\lambda^\circ_y \bar{\lambda}^\circ_y) = R(\lambda^\circ_b \bar{\lambda}^\circ_b) = 1/9$$

$$R(\chi^\circ_r \bar{\chi}^\circ_r) = R(\chi^+_y \chi^-_y) = R(\chi^+_y \chi^-_b) = 4/9 \qquad (42)$$

Some of the allowed decay modes[55,56] for the yellow and blue quarks arising due to W-X mixing are listed below:

$$\left.\begin{aligned} p^+_{y,b} &\to \nu_e + \text{pions} \\ &\to \nu_\mu + K^\circ + \text{pions} \end{aligned}\right\} \quad \text{(dominant)} \qquad (43)$$

$$\to e^- + \pi^+ + \pi^+ \quad \text{(likely to be suppressed}^{56}\text{)}$$

$$\left.\begin{aligned} &\to \nu_e + (e^+ + \nu_e) \\ &\to \nu_e + (\mu^+ + \nu_\mu) \end{aligned}\right\} \quad \text{(suppressed}^{58}\text{)} \qquad (44)$$

$$\not\to \mu^- + \pi^+ + \pi^+ \quad \text{(forbidden}^{56}\text{)}$$

$$n^{\circ}_{y,b} \rightarrow \nu_e + \text{pions} \qquad \rightarrow \nu_\mu + K^{\circ} + \text{pions} \quad \text{(dominant)} \tag{45}$$

$$\rightarrow e^- + \pi^+ \quad \text{(likely to be suppressed}^{56}\text{)}$$

$$\lambda^{\circ}_{y,b} \rightarrow \nu_\mu + \eta \qquad \rightarrow \nu_e + \bar{K}^{\circ} + \text{pions} \quad \text{(dominant)}$$

$$\rightarrow p^{+}_{y,b} + \pi^- \quad \text{(ordinary weak decay)} \tag{46}$$

$$\chi^{+}_{y,b} \rightarrow \nu_e + (\text{charmed } D^+)$$

$$\rightarrow \nu_\mu + (\text{charmed } F^+)$$

$$\rightarrow p^{+}_{y,b} + \pi^{\circ} \qquad \rightarrow \lambda^{\circ}_{y,b} + \pi^{+} \quad \text{(ordinary weak decay)} \tag{47}$$

While for the red quarks (whose decays may need to utilise both V^{+}_{ρ}-W^{+} and W^{+}-X^{+}-mixings), the allowed decay modes are:

$$p^{\circ}_{r} \rightarrow n^{\circ}_{y} + \gamma \quad (\text{If } m_{p_r} > m_{n_y}) \qquad \rightarrow V^{-}_{\rho} + \pi^{+} + \nu_e \quad (\text{If } m_{p_r} > m_{V_\rho} + m_{\pi^+}) \quad \text{(dominant modes}$$

$$\rightarrow \nu_e + \pi^{\circ},\ e^- + \pi^+ \text{ etc.} \tag{48}$$

$$n^{-}_{r} \rightarrow V^{-}_{\rho} + \nu_e \quad (\text{If } m_{n_r} > m_{V_\rho}) \quad \hookrightarrow (\mu^-\bar{\nu}_\mu) \text{ or } (e^-\bar{\nu}_e) \quad \text{(dominant mode)} \tag{49}$$

$$\rightarrow \nu_e + \pi^-,\ e^- + \pi^{\circ} \tag{50}$$

$$\lambda^{-}_{r} \rightarrow V^{-}_{\rho} + \nu_\mu \quad (\text{If } m_{\lambda_a} > m_{V_\rho}) \quad \text{(dominant mode)}$$

$$\rightarrow \nu_e + K^-,\ \mu^- + \eta \tag{51}$$

$$\chi^{\circ}_{r} \rightarrow \lambda^{\circ}_{y} + \gamma \tag{52}$$

In above, we have exhibited few quark-number-conserving-decay modes arising via familiar weak interactions (e.g. $\lambda \to p + \pi$, $\chi \to \lambda + \pi^+$ and $\chi \to p + \pi^\circ$ etc.), which may have rates in the range of 10^{10} to 10^{11} sec^{-1} depending upon quark-mass differences. Such decays may be the dominant decay modes for the charm-quarks, but they are not important for the lighter quarks, especially the p and n-quarks.

In the basic model, quarks of <u>all</u> <u>three colors</u> (red, yellow and blue) with mass as low as 2 to 3 GeV can be relatively shortlived ($\tau \lesssim 10^{-11}$ sec.) without conflicting with the proton lifetime provided[57] that the quarks are heavier than some color-octet states like the color-gluons. For this model, the semileptonic decay modes of the yellow and blue quarks (e.g. $q \to \ell$ + pions, see (43) and (45)) strongly dominate[58] over their pure leptonic decay modes ($q \to \ell+\ell+\bar{\ell}$, see (44)). Thus at least within the basic model, the pair production of yellow and blue quarks with their predominant semileptonic decays cannot account for the ($\bar{\mu}e$)-events seen at SPEAR. On the other hand, if quarks are heavier than the color-gluons, the charged red-quarks (n_r^- and λ_r^-) would predominantly decay into (color gluon + lepton) (see (49)), which followed by rapid decay of the color-gluon into a lepton-pair (see Eqs. (39) and (40)) would appear like a three body leptonic-decay[59] of the quarks. <u>Thus, pair-production of ($n_r^- n_r^+$) and ($\lambda_r^- \lambda_r^+$)-quarks, followed by their sequential decays as above, can provide within our basic model a consistent explanation of the anomalous ($\bar{\mu}e$)-events seen at SPEAR</u>:

$$
\begin{array}{l}
e^- + e^+ \to n_r^- + n_r^+ \\
n_r^+ \to V_\rho^+ + \bar{\nu}_e, \quad V_\rho^+ \to (e^+ + \nu_e) \\
n_r^- \to V_\rho^- + \nu_e, \quad V_\rho^- \to \mu^- + \bar{\nu}_\mu
\end{array}
\tag{53}
$$

An important question arises: What are the sources of these anomalous events? Are they <u>hadronic constituents</u> such as[60] quarks, or are they heavy leptons?

To distinguish between these possibilities, among other means, one must search for pair-production of these objects in hadronic collisions. While neutral as well as charged quark-pairs (with quark-mass $\simeq$ 2 to 3 GeV) are expected to be produced via strong interactions with cross sections $\approx 10^{-32}$ cm^2 at Fermilab energies; the neutral and charged heavy lepton-pairs are expected to be produced via weak and electromagnetic interactions respectively (with cross sections $\lesssim 10^{-38}$ cm^2 for the neutral pair and $\approx 10^{-34}$ cm^2 for the charged pair at Fermilab energies for heavy-lepton-mass $\simeq$ 2 GeV). We, therefore, urge a search for anomalous dilepton-production (involving e^+e^- $\mu^\pm e^\mp$, $\mu^+\mu^-$-pairs) in hadronic collisions. Such a search may help decide whether the $(\bar{\mu}e)$-events seen at SPEAR have hadronic or leptonic parents.

Furthermore, above threshold, quark-pairs can be produced by e^-e^+-annihilation in association with other known hadrons, something not possible for heavy leptons. Secondly, the semileptonic decay modes of the red quarks[61] (i.e. $n_r^- \rightarrow V_\rho^- + \pi^+ + \pi^- + \nu_e$) or of the color-gluons (i.e. $V_\rho^- \rightarrow \pi\pi e\nu$) can give rise to anomalous semileptonic signals (e.g. $e^- + e^+ \rightarrow (\bar{\mu}e) + (\pi^+\pi^-)$ + Missing Momentum), which might be smaller within a factor of ten compared to the leptonic $(\bar{\mu}e)$-signal. Allowing for separate conservations of electron and muon numbers such semileptonic signals would not be produced via decays of heavy leptons. We urge a search for such semileptonic $(\bar{\mu}e)$-signal. Finally a third necessary consequence of the quark-hypothesis for the $(\bar{\mu}e)$ events is that a significant fraction of the total hadronic crosssection must involve real quark-anti-quark pair production.[62] Noting that the yellow and blue quarks (as well as χ_r° and p_r°) disappear into neutrinos plus mesons, this may on the one hand help understand the energy crisis and on the other the jet structure[63] observed at SPEAR.

3. Direct Leptons: We expect quarks as well as color-gluons to be produced in pairs in hadronic collisions

$$
\begin{aligned}
p + p &\rightarrow q + \bar{q} + \text{Hadrons} \\
&\rightarrow V_\rho^+ + V_\rho^- + \text{Hadrons} \\
&\rightarrow \tilde{U} + \tilde{U} + \text{Hadrons}
\end{aligned}
\tag{55}
$$

Pair production of quarks (mass ≈ 2 GeV) and/or color-gluons (which may even be lighter[53] than the quarks), followed by their predominant semi-leptonic or leptonic decays (Eqs. (44)-(52) and (39)) thus provide an obvious source of excess direct leptons (electrons as well as muons), which may account for the observed direct leptons[19] especially at higher incident energies ($\sqrt{s} \geq 10$ GeV). For reasons mentioned in the previous subsection, it is important to search for anomalous varying invariant mass dileptons ($\mu^\pm e^\mp$, $\mu^+\mu^-$ and e^+e^-) in (p+p)-collisions.

4. The J/ψ-Particles

At present the multitude of the J/ψ-particles ($J/\psi_1(3.1)$, $\psi_2(3.7)$, $\psi_3(4.1)$, $\psi_4(4.4)$ and possibly others) together with the C-even states (3.5, 3.4, 2.8) recently discovered at DESY and SPEAR allow alternative interpretations under the new quantum numbers: color, charm, and mirror (or heavy-quark-valencies). As we have stressed elsewhere,[13] it appears, however, that none of these quantum numbers by themselves (especially neither color nor charm alone) can account for the multitude of phenomena, which include not only the J/ψ-particles and their C-even analogs. but also dimuons (of both unlike and like charges) and the anomalous ($\bar{\mu}e$)-events .

Here we note one distinct possibility: $J/\psi_1(3.1)$ is the ground-state $3S_1$ mirror-antimirror quark composite with mirror-isospin[64] $I_m = 1$; $\psi_2(3.7)$ is its radial excitation; while $\psi_3(4.1)$ (which may represent a superposition of several resonances) and $\psi_4(4.4)$ represent color (such as the color-gluon $\tilde{U}$) and/or charm-anticharm-composites ϕ_c.

The C-even ^1S_o and $^3P_{o,1,2}$-composites of mirror (heavy) quarks $\bar{p}'p'$ and $\bar{n}'n'$ are expected to accompany the C-odd $^3S_1$-states analogous to the charmonium-picture, except that for the present case additional states are expected, their mirror-isospin I_m being either 0 or 1.

With the above assignment, the mirror analogs of the charmed D and F should lie around 2 GeV, while the D and F themselves should lie around 3 GeV or higher. The decays of mirror-[72] D and F-particles will not preferentially involve K-mesons (unlike D and F).

In this picture color and charm threshold start above 4 GeV. Alternatively an intriguing possibility is that color-threshold and color-gluon $\tilde{U}$ lies below[53] 3 GeV. Depending upon where lies $\tilde{U}$, there should of course exist its seven partners around the mass of $\tilde{U}$. If the ($\bar{\mu}e$)-events seen at SPEAR are to be attributed to quark-decays, it is preferable that color-gluon mass lies below 2 GeV (See Sec. IV). We urge an exhaustive search for narrow resonant states[65] in (e^-e^+)-system in the 1 to 3 GeV region.

We stress that a clear choice between the allowed interpretations of the new particles (including the one mentioned above) can be made only after a proper search for particles carrying the new quantum numbers (color, charm and/or mirror) is carried out. To this end, we have noted in sec. III that there are distinct decay modes of color-carrying particles (such as $V_\rho^\pm$, $V_{K*}^\pm$) which should enable one to distinguish experimentally between particles carrying color on the one hand and those carrying new valency-quantum numbers (such as charm and mirror) on the other.

V. SUMMARY AND CONCLUDING REMARKS

The main theoretical remark of this talk is that the commonly held objections to the concept of integer-charge quarks with relatively low-mass physical color do not hold within the gauge-theory approach to color.[1,2]

On the experimental side, we point out:

(i) The dimuon-events ($\mu^-\mu^+$) with their observed rate receive a simple explanation in terms of charged-color-gluon-production (or production of similar color-octet states).

(ii) Production of quark-pairs followed by their leptonic decay provides a viable explanation of the anomalous ($\bar{\mu}e$)-events observed at SPEAR. In order to distinguish between quark and heavy lepton-hypothesis for these events, one needs to study the rate of production of analogous events in hadronic collisions on the one hand and the production of hadron-associated ($\bar{\mu}e$)-events in e^-e^+-annihilation at $E_{CM} \geq 5$ GeV on the other.

(iii) Direct lepton production in hadronic collisions especially at higher energies finds an explanation in terms of quark and/or gluon-pair production. Such a hypothesis may be tested by studying production of anomalous like and unlike lepton-pairs in pp-collisions.

(iv) Asymptotically, for lepto-production experiments, integer charge charge quarks (treated as partons) would behave as though they carried their fractional valency-charges. A major prediction of the integer-charge-quark hypothesis (within the gauge theory framework) is that

(σ_L/σ_T) should be nonvanishing and asymptotically a function of x only; in the semi-asymptotic region ($|q^2| \lesssim 8\ m_U^2$), there should be scaling violations due to color-production at the 10 to 20% level, which should decrease with q^2 for fixed $\omega \lesssim 10$, above and beyond logarithmic scaling violations expected from asymptotic freedom considerations.

(v) Charged color-gluons ($V_\rho^\pm, V_{K*}^\pm$), which may be produced in pairs both by pp-collision and by e^-e^+-annihilation, possess distinct decay modes -- Their semileptonic decays (Eq. (39)) would involve two pions, $K\bar{K}$ and $\eta\eta$, but no single pion, or single kaon, or single η in sharp contrast to the decay modes of particles carrying new valency quantum numbers (such as charm).

(vi) Last but not the least, within the gauge-theory approach, one intriguing feature is that asymptotically the neutral red-quark-pair ($p_r^\circ\bar{p}_r^\circ$) will be produced by e^-e^+-annihilation on par with the charged quark pairs ($p_y^+p_y^-$ and $p_b^+p_b^-$), all three pairs possessing charged-particle decay modes (see sec. IV).

To conclude, there appear to be two major issues confronting particle physics at present: (1) What are the basic quark-flavors?, and (2) Is color confined or is it physical?. Correspondingly, are quark-charges fractional or integral? The existence of new flavors is crucial to the unification-concept and their discovery would be exciting. However the knowledge of quark-charges is of fundamental importance as depending upon their charges there are two alternative paths. The hypothesis of integer-charge quarks has the advantage, as mentioned before, that it provides a possible resolution[1,2] of the missing-quark-mystery without having to assume even partial confinement. Considering that the familiar objections to the concept of integer-charge-quarks and low-mass physical color have disappeared and that within our hypothesis dimuon events, anomalous $(\bar{\mu}e)$-events as well as direct leptons receive viable testable explanations, all of which utilize directly the concept of physical color, we urge an all-out search for color -- in particular unstable integer-charge quarks, and color gluons; all of these possess distinctive signatures.

I am grateful for several helpful discussions to O. W. Greenberg, G. A. Snow, J. Sucher and C. H. Woo.

REFERENCES AND NOTES

1. J. C. Pati and Abdus Salam, Phys. Rev. D8, 1240 (1973); Phys. Rev. Lett. 31, 661 (1973).

2. J. C. Pati and Abdus Salam, Phys. Rev. D10, 275 (1974); ibid D11 (E) 703 (1975).

3. By "natural" I mean a gauge-symmetry, in which the basic lagrangian conserves fermion number F, baryons number B and lepton number L, as well as left ↔ right symmetry, violations of these originating entirely from spontaneous symmetry breaking. Examples of such natural extensions are noted below.

4. Such as the approach of H. Georgi and S. L. Glashow, Phys. Rev. Lett. 32, 438 (1974) based on the gauge symmetry SU(5), which does not satisfy the criteria listed in Ref. 3.

5. J. C. Pati, Abdus Salam and J. Strathdee, Nuovo Cimento 26, 72 (1975). J. C. Pati, Proceedings of the Second Orbis Scientiae at the University of Miami, (Coral Gables), page 235, 1975. Here the possibilities of gauging fermion number and its spontaneous violation are noted.

6. Ref. 2; R. N. Mohapatra and J. C. Pati, Phys. Rev. D11, 2558 (1975); G. Senjanovic and R. N. Mohapatra, Phys. Rev. D12 (1975).

7. R. N. Mohapatra and and J. C. Pati, Phys. Rev. D11, 566 (1975); J. Frenkel and M. E. Ebel, Nuclear Physics (1975). (The second paper does not satisfy left ↔ right symmetry in the lepton sector).

8. The angles $\theta_{L,R}$ denote Cabibbo rotations for the left and right-handed fermions, while $(\delta_L - \delta_R)$ is the CP-violating phase angle.

9. See Ref. 2 and an analogous illuminating discussion in H. Fritzsch, M. Gell-Mann and P. Minkowski, CALT-6977 (1975).

10. This is assuming that we generate the desirable milliweak CP violation as discussed above (Ref. 7), from which, knowing $|\eta_{+-}| \approx 10^{-3}$, we deduce $(m_{W_R}/m_{W_L})^2 \geq 10^{-3}$. Experimentally V+A may be as big as 10% of V-A in amplitude.

11. This would not have been possible, if proton-stability constrained the masses of the X-like particles to be greater than 10^{15} GeV (Ref. 4).

12. J. C. Pati and Abdus Salam, Phys. Rev. Lett. 32, 1083 (1974); Phys. Rev. D11, 1137 (1975).

13. J. C. Pati and Abdus Salam; "Quarks, Leptons and Pre-Quarks" (IC/75/106), Invited Talk presented at Palermo International Conference, June 23-28, 1975 (To appear in the Proceedings).

14. J. C. Pati and Abdus Salam, Phys. Rev. D11, 1137 (1975); J. C. Pati, Abdus Salam and J. Strathdee, Nuovo Cim. 26, 72 (1975); Phys. Rev. D11, 2558 (1975) and H. Fritzsch and P. Minkowski, Annals of Physics 93, 222 (1975).

15. Vector-like Interactions and their phenomenological implications have recently been emphasized by H. Fritzsch, M. Gell-Mann and P. Minkowski, Caltech Preprint - 6977 (1975) and R. L. Kingsley, F. Wilczek and A. Zee (Princeton Univ. Preprint). Such interactions with new quarks and leptons were suggested as a necessary step towards higher unification schemes in Ref. 14.

16. See J. C. Pati and Abdus Salam (IC/75/73); Physics Letters (To appear) and G. Branco, T. Hagiwara and R. N. Mohapatra, CCNY-HEP 75-8 (1975) for the general possibilities arising within $SU(2)_L \times SU(2)_R \times SU(4)'_L \times SU(4)'_R$. It is good to note that the Fermi-mass matrix is fairly restrictive within a unifying symmetry such as $SU(4)_L \times SU(4)_R \times SU(4)'_L \times SU(4)'_R$ and it is not clear whether the degree of flexibility being advocated phenomenologically (see for example Ref. 15) is permissible within a unifying symmetry.

17. J. C. Pati and Abdus Salam, Trieste Preprint IC/75/95, Phys. Rev. Lett. (To appear); and Univ. of Maryland Tech. Rep. 76-061.

18. G. Rajasekharan and P. Roy (TIFR Preprints TH/75-38 and TH/75-42).

19. A. Benvenuti et al, Phys. Rev. Lett. 34, 419, 597 (1975).

20. M. L. Perl et al, SLAC-PUB 1626, Phys. Rev. Letters, 35, 1489 (1975).

21. See L. Lederman, Rapporteur's talk at SLAC-Conference, August (1975), To appear in the Proceedings.

22. These are renormalized effective coupling constants relevant for low energies $E \lesssim 1$ GeV. The two "effective" constants g and f appearing in weak and strong gauging of G reflect the passage from the higher unifying symmetry such as $SU(4)_L \times SU(4)_R \times SU(4)'_L \times SU(4)'_R$ with one coupling constant down to G through finite renormalization effects. Note that finite renormalization effects, though logarithmic (in lowest order), can be significant partly through summing over all orders and partly because of the large mass ratio (m_X/m_W)--see later.

23. R. N. Mohapatra, J. C. Pati and Abdus Salam; "Color as a Classification Symmetry and Quark-Charges," Md. Tech. Rep. No. 76-005 (1975) (Submitted to Phys. Rev.).

24. The hypothesis of quark-confinement has been advocated by H. Fritzsch, M. Gell-Mann and H. Leutwyler, Phys. Lett. B47, 365 (1973); S. Weinberg, Phys. Rev. Lett. 31, 494 (1973) and D. J. Gross and F. Wilczek, Phys. Rev. D8, 3633 (1973).

25. D. Gross and G. Wilczek, Phys. Rev. Lett. 30, 1343 (1973) and D. Politzer, ibid, 30, 1346 (1973).

26. In the sense of Politzer (Physics Reports 1974).

27. R. N. Mohapatra, J. C. Pati and P. Vinciarelli, Phys. Rev. D8, 3652 (1973) and S. Weinberg, Phys. Rev. Lett. 31, 494 (1973). There appears to be a frequent confusion in the literature, which conveys that electromagnetism must commute with SU(3)'-color in order to obtain "natural" violations of parity and strangeness. This is incorrect; the only requirement for "natural" violations of parity and strangeness is that $[G_{weak}, SU(3)'_{col}] = 0$; whether photon emerging after spontaneous symmetry breaking acquires a color-octet piece or not is immaterial in this regard.

28. M. Y. Han and Y. Nambu, Phys. Rev. 139, B 1006 (1965); O. W. Greenberg and D. Zwanziger, Phys. Rev. 150, 1177 (1966) and H. Lipkin.

29. One may still wish to include an additional multiplet A' = (2+2, 2+2, 15) to obtain appropriate Fermi-masses (see Ref. 2).

30. We present the case where $W_L^{\pm}$ mixes with $V_\rho^{\pm}$ and $V_{K^*}^{\pm}$. In general, ($V_\rho^{\pm}$ and $V_{K^*}^{\pm}$) can mix with $W_R^{\pm}$ as well.

31. W-X mixing effect is of course crucial for baryon number violation.

32. For example, the contribution to $(g-2)_\mu$ due to $\tilde{U}$-exchange is $(1/8\pi^2)(2e^2/\sqrt{3}\, f)^2\ (m_\mu/m_U)^2 \simeq 2 \times 10^{-9}$ for $m_U \simeq 2$ GeV and $f^2/4\pi \simeq 10$.

33. See for example H. Harari, Rapporteur's Talk at SLAC-Lepton-Photon-Symposium, August 1975 (To appear in the Proceedings). Harari's arguments against color-interpretation are confined to the case where the only new quantum number excited is color.

34. J. C. Pati and Abdus Salam (IC/75/73), Physics Letters (To appear) and Ref. 13.

35. See for example F. Gilman, Proceedings of the 17th International Conference, London (1974); C. H. Llewellynsmith, Rapporteur's talk at SLAC-Lepton-Photon Symposium, August 1975; A. deRujula, H. Georgi and S. L. Glashow, Phys. Rev. 12, 147 (1975); S. L. Glashow, after dinner talk at Northeastern University Conference on Gauge Theories and Modern Field Theory (Sept. 1975).

36. See for example, B. C. Barish, Invited Talk at the American Physical Society (Div. of Particles and Fields), Sept. 1974.

37. See for example, J. D. Bjorken, Proceedings of 1973 Bonn Conference, Page 25 (1974).

38. We neglect the logarithmic corrections to the propagators, which are not important at present energies (see Ref. 17).

39. We do not exhibit V° (see Eq. (9)), since it is not coupled to leptons. Even if $\tilde{U}$ and V° mix, our conclusions in this section are not affected.

40. $\tilde{U}$-exchange-contribution to color-singlet-production is smaller than the one-photon-contribution by a factor = $(2/\sqrt{3})(e^2/f^2) << 1$.

41. Note that multiple $\tilde{U}$-exchanges for color-production may be neglected to the same extent as multiple-photon-exchange.

42. The results for the present case may be obtained straightforwardly (by invoking the Δ^2-factor) from those of N. Cabibbo and R. Gatto, Phys. Rev. 124, 1577 (1961) and M. A. Furman and G. J. Komen, Nucl. Phys. B84, 323 (1975).

43. If one of the J/ψ-particles is identified with the $\tilde{U}$-gluon, it is possible to verify that no undue enhancement takes place even though $\Delta^2 >> 1$ as $q^2 \to m_U^2$, provided $\tilde{U}$ is the lowest mass color-octet state with $J^{PC} = 1^{--}$. (See Ref. 17 for details.)

44. This net contribution would exhibit as a sum over several possible color-octet states.

45. See for example J. D. Bjorken, Proceedings of the second International Conference on Elementary Particles, Aix-en-Provence, 1973. Aposteriori, since color-contributions (with the Δ^2-factor) turns out to be small $\lesssim$ 10 to 20% compared to valency-contribution (see later), the determination of the quark-distribution functions does not alter significantly with the inclusion of color-production.

46. C. H. Llewellyn-Smith, Phys. Rev. D4, 2392 (1971).

47. C. Chang et al, Phys. Rev. Lett. 35, 901 (1975), and R. Taylor, Report of MIT and SLAC data at the SLAC-Lepton-Photon-Symposium (August, 1975).

48. Asymptotic freedom applies if spontaneous symmetry breaking is dynamical, or else "temporarily" if quartic scalar couplings are small as mentioned before.

49. For simplicity of writing, we do not exhibit the Cabibbo-angle factors ($\sin \theta_c$ and $\cos \theta_c$) in Eqs. (33) and (34), which are immaterial for total cross sections.

50. Strictly speaking this is divided between different color-octet final states, not all of which need contain a pair of charged gluons.

51. These selection rules and decay modes in fact apply to the lightest color-octet states with quantum numbers of $V_\rho^\pm$ and $V_{K^*}^\pm$.

52. This is because mixing angle is proportional to m_V^2 (Eq. (11)) and phase space $\alpha\, m_V$. We take $\sin(\theta_L + \phi_L) \approx \cos(\theta_L + \phi_L) \approx 1/\sqrt{2}$ for simplicity.

53. With the new results on color-brightening (sec. 3), there is the intriguing possibility that color-gluons may in fact be relatively light ($m_V \approx$ 1 to 2 GeV) (J. C. Pati, J. Sucher and C. H. Woo (forthcoming preprint)).

54. For decays off ($\tilde{U}$, V°, $\tilde{V}^\circ_{K*}$) see Ref. 2, 13 and W. R. Franklin, Nucl. Phys. B91, 160 (1975).

55. See Ref. 13 and W. R. Franklin (Ref. 54).

56. In particular, see J. C. Pati, S. Sakakibara and Abdus Salam (Trieste Preprint IC/75/93, To appear). The semileptonic decay modes such as $p^+_{y,b} \to e^- + \pi^+ + \pi^+$ involving emission of a charged lepton require that one of the pions be emitted from a quark-line inside the loop, which would be suppressed by two large masses (m_X^2 and $m_{W_L}^2$), if pion-emission is associated with a form-factor. (Note pions are composites in the theory). In this case neutral lepton-emission (i.e. $p^+_{y,b} \to \nu_e$ + pions etc.) would be the dominant modes. A second point worth noting is the intricate selection rules, which arise for quark-decays. For example transitions such as $p^+_{y,b} \to \mu^- + \pi^+ + \pi^+$, $n^\circ_{y,b} \to \mu^- + \pi^+$, $\lambda^\circ_{y,b} \to e^- + \pi^+$ etc. are forbidden (neglecting corrections of order G_{Fermi}).

57. Otherwise, at least the red-neutron quark (n_r^-) would be longer lived ($\tau(n_r^-) \approx 10^{-6}$ to 10^{-7} sec. for $m(n_r^-) \approx 2$ GeV); even though the yellow and blue-quarks would still be shortlived ($\tau \lesssim 10^{-11}$ sec.)

58. The rates of semileptonic-decay modes ($q \to \ell$ + Mesons), when allowed (see Ref. 56), exceed those of leptonic decay modes ($q \to \ell + \ell + \bar{\ell}$) by a factor ~ 0 ($m_W^4/m_q^4$) > 10^5 within the basic model. This is because the former receive contributions from (convergent) loop-diagrams, while the latter receive contributions from tree-diagrams only.

59. Although with sufficient data, the sequential decay (53) might be distinguishable from the genuine three-body leptonic decays of the parent particles.

60. In addition to theproduction of $q\bar{q}$-pairs, by e^-e^+-annihilation, production of charged color-gluon pairs ($V^+_\rho V^-_\rho$ and V^+_{K*} V^-_{K*}), which is limited by the net contribution from color-gluons to R = 1/8 (see Sec. III), followed by their two-body leptonic decays would also contribute to the leptonic (μe)-events. The available SPEAR data is not inconsistent with three and two body-decays of parent particles (see M. L. Perl, SLAC-PUB-1664, Nov, 1975).

61. If the charged lepton-emission from the yellow and blue quarks (i.e. $n^\circ_{y,b} \to e^- + \pi^+$ etc.) were not suppressed by form-factors (see Ref. 56) one would have expected to see strong semileptonic signals (i.e. $e^- + e^+ \to e^- + e^+ + (\pi^+\pi^-\pi^+\pi^-)$ + Missing Momentum) with anomalous (e^-e^+)-pairs in the final state (but, due to selection rules no anomalous ($\bar{\mu}e$) or ($\mu^-\mu^+$)-pairs would arise from yellow and blue-quark-decays).

62. This requires in turn that the quark-electromagnetic form-factor must be unity or close to unity at SPEAR energies. Whether this may be realized theoretically remains to be seen.

63. G.Hanson, et. al., Phys. Rev. Lett. 35, 1609 (1975).

64. The heavy-quark-(mirror)-interpretation with $I_m=1$ for the lowest lying states has the advantage (R. M. Barnett, Phys. Rev. Lett. 34, 41 (1975) that their decays into hadrons would be suppressed by mirror-isospin selection rule as well as by the Zweig-rule-factor. This provides a natural explanation of their extreme narrowness without invoking an unusual Zweig-suppression-factor (as is needed for the charm-anticharm interpretation of 3.1).

65. The $\tilde{U}$-color-gluon may be searched for both in e^-e^+-annihilation and in photo-production experiments allowing for good resolution.

DISCUSSION

S. WEINBERG (Harvard)

If the weak, electromagnetic and strong interactions are unified in a gauge group with a single coupling constant, and if the masses of the vector bosons produced when this group is spontaneously broken are no larger than about 10^5 GeV, then how can the strong couplings be so different from the weak and electromagnetic couplings at a few GeV?

PATI

The main remark I will make is that I do not disbelieve in some ultimate scale provided we can unify gravity but that is a long way off. At the moment we haven't done it.

S. WEINBERG (Harvard)

I'm sorry. You discussed a unified group in which all the intermediate vector mesons had masses not bigger than about 10^5 GeV (low energy region) - and that included the strong interactions as well as the weak and electromagnetic interactions so I presume that at that point you were talking about a single coupling constant.

PATI

That is quite right, but I do not say that this scale is set by these methods necessarily.

S. WEINBERG (Harvard)

That's not the question. If there's a single coupling constant in that theory, why then is the strong interaction coupling constant that we observe at a few GeV so different from the weak and electromagnetic couplings?

PATI

I was thinking in a different way then you are. All the coupling constants, even if they start off in a similar manner, have finite renormalization effects which are different in different sectors because spontaneous symmetry breaking is giving different masses to the different sectors. In principle, there could be large differences which might be even calculable between different sectors due to finite renormalization effects. That is the only answer I can give you at the moment.

V. TEPLITZ (VPI)

With color nonconservation in quark decay, and strangeness changing decays, is there also violation of muon number conservation?

PATI

The muon number is not sacred again. Yes that's right. The muon number gets effectively violated. I showed you for example the λ is coupled to μ^- in this low energy effective gauge symmetry, and the n is coupled to e^-. For example, $\lambda + \bar{n}$ will go into $\mu^- + e^+$. So effectively muon number is being violated. But from the point of view that I am taking it is not being violated because λ-ness is being carried by the muon and the $\bar{n}$-ness is being carried by the electron.

MASS-SHELL INFRARED SINGULARITIES OF GAUGE THEORIES

John M. Cornwall*

*Department of Physics, University of California at Los Angeles,
Los Angeles, Cal. 90024.
Work supported in part by the National Science Foundation.

This work was done in collaboration with George Tiktopoulos, and has been briefly reported in Ref. 1. We are concerned with the structure of infrared singularities in S-matrix elements of gauge theories, with a view toward making progress on the confinement problem. Our work has its origin in perturbation theory, so it is necessary for us to add an infrared cutoff "by hand" in order that the theory exist. In a more sophisticated approach this cutoff would be supplied by the confinement process itself.

Since the cutoff we use is, for the most part, simply a mass term for the gauge mesons, it is necessary to emphasize that this mass has nothing to do with a genuine vector-meson mass which might be produced by breakdown of the gauge symmetry. In fact, we insist that this symmetry (to be identified with color SU(3), denoted $SU(3)_c$) remain exact; if it is broken, confinement as we envisage it will not take place. It is worth noting here that there is a simple class of theories in which $SU(3)_c$ is not broken by short-distance effects. Suppose the world is governed by an asymptotically-free Yang-Mills Lagrangian L symmetric under a simple group G (so that there is only one coupling constant g) and that the only fundamental fields are gauge mesons and fermions; in particular, there are no elementary scalar Higgs-Kibble fields. L is to undergo dynamical symmetry breakdown (with composite Goldstone bosons) in such a way as to keep $SU(3)_c$ unbroken. It has already been remarked[2] that this will be the case if only the fermion representations $3,\overline{3}$, and 1 appear in L , which is of course just the conventional structure. On the other hand, if G is SU(N) , then something more than $N,\overline{N}$, and 1 must be present. A mathematical example is Georgi and Glashow's[3] SU(5) model, which has 10-dimensional fermion representations which do lead to spontaneous symmetry breakdown leaving $SU(3)_c$ intact.

A. Graphical Experiments.

In what follows, μ stands for a vector-meson mass term and we work only in the Feynman gauge, so the propagator is $-ig^{\alpha\beta}(q^2-\mu^2)^{-1}$ There are several kinematical regimes where infrared singularities are important: (1) The fixed-angle regime where the kinematical invariants $(p_i \pm p_j)^2$ $(i \neq j)$ all obey $(p_i \pm p_j)^2 >> M^2, \mu^2$ where M is the fermion mass; (2) The infrared regime, where $(p_i \pm p_j)^2 \gtrsim M^2 >> \mu^2$; (3) the near-forward regime, where some invariant (say, t) is $O(\mu^2)$, while all the rest are at least as large as M^2 . In all three regimes, the p_i are momenta of external on-shell legs: $p_i^2 = m_i^2$ $(m_i = M, \mu)$. We extend the fixed-angle regime to the Sudakov regime in which the fermions are not exactly on shell, but

$p_i^2 - M^2 = O(\mu^2)$.

We have done a number of Feynman-graph experiments in these regimes, saving only leading-logarithm terms (the highest power of $\ln \mu$ for each graph). However, it can be shown in many instances (and possibly in all) that neglected terms are not just small by a power of $\ln \mu$, but are not even singular as $\mu \to 0$.[4] This is certainly true in the Abelian theory (QED). The contributions which can be neglected are: closed loops of massive particles, when summed to a gauge-invariant set; gauge terms ~ $q^\alpha q^\beta$ in the gauge-meson propagator; four-meson seagulls; ghost-loops. The absence of ghost-loops and propagator gauge terms suggests that the infrared-singular part of the theory obeys the naive Ward identities analogous to those of QED.

The results of these experiments will be given for the fixed-angle regime only, where they are very simple; the infrared regime will be handled by another method below. In the fixed-angle regime, a non-Abelian gauge theory is virtually Abelian, and all S-matrix elements are expressible as powers of certain form factors $F_i(t)$, which themselves exponentiate (at least to sixth order). $F_i(t)$ is defined as the asymptotic form factor ($t \gg M^2, \mu^2$) for the coupling of a particle of type i to a group-singlet (i.e., uncolored) external current carrying momentum transfer t , and

$$F_i(t) = \exp\left\{ \frac{-g^2}{16\pi^2} C_i \ln^2(-t/\mu^2) \right\} \quad \text{(fixed-angle regime)} \tag{Ia}$$

$$= \exp\left\{ \frac{-g^2}{8\pi^2} C_i \ln \frac{t}{p^2-M^2} \ln \frac{t}{p'^2-M^2} \right\} \quad \text{(Sudakov regime)} . \tag{Ib}$$

Here $t = (p-p')^2$, and C_i is the eigenvalue of the quadratic Casimir operator $\sum (t^a_{(i)})^2$, where $t^a_{(i)}$ is the group generator for particle i Eq. (I) has been verified to sixth order for fermions, to fourth order for gauge mesons. On-shell scattering amplitudes $T_{\{i\}}$ have the structure, in the fixed-angle regime:

$$T_{\{i\}} = T^B_{\{i\}} \Pi F_i^{1/2}(t) \tag{2}$$

where $T^B_{\{i\}}$ is the connected Born approximation, and the various invariants are epitomized by t. This has been calculated to sixth order for elastic fermion-fermion scattering, and the important property of factorization has been calculated for all one-loop processes.

To my knowledge, two other groups have carried out similar graphical experiments. Carrazone, Poggio, and Quinn[5] have done the sixth-order form factor in the Sudakov regime. After correcting some minor mistakes in their published work, these authors have informed me that they find exponentiation as in I(b). Schecter and coworkers (private communication) have also verified exponentiation of the form factor to sixth order. Poggio and Quinn (to be published) have verified the factorized form (2) for the fermion-fermion amplitude in the Sudakov regime through sixth order, and are working on the eighth-order amplitude. S. Brodsky has informed me that Lipatov and co-workers also find exponentiation, but I do not know their methods.

At this point it is amusing, but hardly conclusive, to note that in the limit $\mu \to 0$ form factors and fixed-angle amplitudes vanish for processes involving at least one colored particle. But totally colorless processes are independent of μ. It would appear that this has been shown only for elementary colorless particles, but in fact it is also true for composite colorless states. The physical reason is that the accumulation of powers of $\ell n\ \mu$ comes from soft gauge particles, and these long-wave probes can only see the total group charge of a state, not the constituent charges. This physical argument is reinforced by another graphical experiment. Define a "composite" state as one in which the constituents have essentially the same velocity, but are otherwise free, so that gluon exchanges between the constituents are not included in any fixed-angle graph. However, all other exchanges between constituents in different "composites" are allowed. The graphical experiments yield exactly the results (1-2), with C_i appropriate to the ith "composite." (Note that for form factors of composites, (1) should be multiplied by a connected Born term which contains the hard gluon exchanges.) This C_i is a square of a sum of group charges, which we term coherent addition of group charges.

One physical result about hadrons, as opposed to quarks, follows immediately from the graphical experiments. For hadron amplitudes, all the C_i are zero, and all that remains is the connected Born term.[6] This obeys the quark-counting rule[7] $T_B \sim t^{2-(N/2)}$, where N is the total number of quark lines. Then fixed-angle exclusive hadronic amplitudes (and form factors) should obey the quark-counting rule, modified only by short-distance corrections of the type which violate Bjorken scaling in electroproduction, that is, finite powers of $\ell n\, t$.

The results for the infrared regime are not quite as simple as those for the fixed-angle regime, and are best understood in terms of a differential equation in the cutoff μ, which can be used in all the infrared-singular regimes.

B. A Differential Equation in the Cutoff μ.

It is evident from (2) that fixed-angle amplitudes obey

$$\mu \frac{\partial}{\partial\mu} T_{\{i\}} = \sum_i \Gamma_i(t)\, T_{\{i\}} \tag{3}$$

$$\Gamma_i(t) = C_i \frac{g^2}{8\pi^2} \ell n(-t/\mu^2) \tag{4}$$

--an equation very similar to a renormalization group equation. Equations of the type (3) are the most powerful means of exploring infrared singularities, and we sketch their derivation here. The operation $\mu(\partial/\partial\mu)$ is, in effect, a mass-insertion term in the Callan-Symanzik equations and a little thought shows that

$$\mu \frac{\partial}{\partial\mu} T_{\{i\}} = \frac{-ig^2\mu^2}{(2\pi)^4} \int \frac{dk}{(k^2-\mu^2)^2} T_{\{i\}\alpha\alpha}^{\;aa}(k,-k) \tag{5}$$

where $T_{\{i\}\alpha\beta}^{\;ab}(k,q)$ is the amplitude for emission of gluon k_α, group index a, and gluon q_β, group index b. The sophisticated reader will observe that we have omitted a term involving mass derivatives of ghost loops; in the Feynman gauge, ghosts have mass μ. However, we have already remarked that ghost loops are non-leading. Partly because the ghost loops, which spoil normal Ward identities, are non-leading and partly because the important region of integration in (5) is for $k \simeq \mu$, $T_{\{i\}\alpha\beta}^{\;ab}$ is conserved on the indices α,β when k,q are $O(\mu)$.

Since μ is much smaller than any other momentum scale, we are motivated to look for a low-energy theorem for emission of gluons. Unfortunately, conventional low-energy theorems will not do, for they are useful when $k \ll \mu$, that is, when the gluon momentum is much less than any mass in the problem. In this case, emission from internal lines of the graph for $T_{\{i\}}$ can be neglected. When $k \simeq \mu$, internal emission <u>cannot</u> be neglected, but remarkably the answer is the same as the leading term in the Low theorem. When $k \simeq \mu$, the amplitude for emission of one soft gluon is given by

$$T_{\{i\}\,\alpha}^{\;a}(k) = \frac{\sum p_\alpha^i \eta_i t_{(i)}^a}{p\cdot k} T_{\{i\}}(k=0) \tag{6}$$

plus terms smaller by a power of μ. Observe that even though $k = 0$ in

$T_{\{i\}}$ on the right-hand side of (6). $T_{\{i\}}$ still depends on μ (as a given power of $\ell n\, \mu$, for a graph of given order in g). In (6), p^i is the momentum of external particle i (elementary or "composite"), $\eta_i = +1$ for ingoing particles, -1 for outgoing particles, and $t^a_{(i)}$ are the group generators for particle i. Note that $k_\alpha T^a_{\{i\}\alpha} = 0$, because $\sum \eta_i t_i^{(a)}$ is the ingoing total group charge minus the total outgoing group charge, and group charge is conserved.

The generalized Low theorem (6) can be proven in all generality for Abelian gauge theories by making use of the eikonal trick of dropping quadratic terms in k^2 in all propagators, and summing over the possible ways of inserting the soft photon onto charge-bearing lines before doing loop integrals. These ways include both internal and external brehmsstrahlung; saving only external-line emission gives a totally wrong answer. For non-Abelian gauge theories we have explicitly verified (6) only to fourth order in g, and a complete proof awaits a compact and coherent notation for the much more involved combinatorics. There is no reason to doubt that this will be forthcoming.

For the emission and reabsorption of two gluons, followed by summation over Lorentz and group indices, we need only to repeat (6). Then, performing the integration over k as indicated in (5),

$$\frac{-ig^2\mu^2}{(2\pi)^4}\int \frac{dk}{(k^2-\mu^2)^2}\, T^{aa}_{\{i\}\alpha\alpha}(k,-k) = -\frac{1}{2}\sum_{ij} \eta_i\eta_j\left(\mu\frac{\partial}{\partial\mu} B_{ij}\right) t^a_{(i)} t^a_{(j)}\, T_{\{i\}}(k=0) \tag{7}$$

$$B_{ij} = \frac{-ig^2}{(2\pi)^4}\int \frac{dk}{(k^2-\mu^2)}\, \frac{p_i\cdot p_j}{p_i\cdot k\, p_j\cdot k}\ . \tag{8}$$

Actually, (7) leaves out certain graphs in which the gluon is emitted and absorbed from the same point, but these either identically vanish or are non-leading.

The fixed-angle results may now be recovered. For $i \neq j$ in (7-8). $p_i \cdot p_j$ behaves like t, a large invariant, while the $i = j$ terms are non-leading. It follows that $B_{ij} = B(t)$ for $i \neq j$, and is zero for $i = j$. If B_{ij} were completely independent of i,j, current conservation would give zero for the right-hand side of (7); as it is, only the terms with $i = j$ contribute, and they are easily thrown into the form of the right-hand side of (3).

Eqs. (7-8) still hold in the <u>infrared regime</u>, but here all the B_{ij} are different. The resulting differential equations are, schematically,

$$\mu \frac{\partial}{\partial \mu} T_{\{i\}} = \sum \Gamma_{\{i\}\{j\}} T_{\{j\}} \tag{9}$$

where $\{i\},\{j\}$ are collective indices for the group labels of T. The matrices Γ are complicated, but have two simple and essential prcperties: (1) they are proportional to C_i, the Casimir operator, for elementary <u>and</u> "composite" states; (2) they have positive eigenvalues if $C_i \neq 0$. We observe below that these two properties furnish a signal for confinement. In view of the similarity of (9) to ultraviolet renormalization group equations, we call the Γ's <u>infrared anomalous dimensions</u>. These anomalous dimensions have a direct relationship to the large-N behavior of conventional (ultraviolet) anomalous dimensions γ_N in the Wilson expansion. The γ_N behave like $g^2 C_i \, \ell n \, N$, that is, of the form (4) when N is substituted by $-t/\mu^2$.

The <u>near-forward regime</u> is still largely unexplored. It is the most difficult, and is related to the dynamics of finding of true composite states. In the Abelian case, (7-8) yield the usual eikonal (impact-parameter) formula for scattering at fixed t, with the eikonal given by one-photon exchange. This formula also contains the information necessary to find the Coulomb bound states.

This differential-equation approach furnishes the most comprehensive statement of all infrared singularities in Abelian gauge theories, reproducing all the known results quickly and easily.

C. A Signal for Confinement.

We now show that (except for the unexplored near-forward regime) the S-matrix for any process involving at least one colored state vanishes in the limit $\mu \to 0$. A crucial step is the demonstration that soft real gluon emission cannot cancel the infrared singularities that we have already found stemming from soft virtual gluons; this is the only essential way in which non-Abelian gauge theories differ from Abelian ones.

This demonstration is simple, given the differential equation (9). Consider the amplitude $T_{(N)}$ for, say, fermion-fermion scattering with the emission of N real gluons. The solution to (9) will show that $T_{(N)}$ depends on μ as

$$T_{(N)} \sim T_{(o)} \exp(-N \, \ell n^2 \mu^2) \tag{10}$$

where some inessential positive factors have been omitted in the exponent. $T_{(o)}$ itself vanishes, for colored fermions (it vanishes like a power of μ in the infrared regime). The summed cross-section over all N looks schematically like

$$\sum_N \int |T_{(N)}|^2 \frac{(\ell n \, \mu)^N}{N!} \tag{11}$$

where the $(\ell n \, \mu)^N/N!$ comes from integrating over the phase space of the soft gluons. In Abelian theories, $T_{(N)} = T_{(o)}$ because the photons are uncharged (i.e., their $C_i = 0$) and real and virtual singularities exactly cancel. In non-Abelian theories, the larger N is, the more rapidly $T_{(N)}$ vanishes as $\mu \to 0$, and no such cancellation takes place.

Our argument makes use of an order of summation which defeats the Kinoshita-Lee-Nauenberg theorem. For a fixed number N of gluons, we sum to all orders in g, and then sum over N. If, in (11), only terms of a fixed order in g are kept, the KLN theorem says that there will be no infrared singularities. If that were so in the theory summed to all orders, it would be difficult to see how confinement could ever take place.

The second part of the signal for confinement is that if there are no colored particles in $T_{\{i\}}$ the right-hand side of (9) vanishes, and $T_{\{i\}}$ is independent of μ. Of course, this is only a signal, not a proof of confinement, for many reasons: the near-forward colored S-matrix has not been shown to vanish, non-leading terms have not been shown to be unimportant, the dynamics of composite states have been ignored, etc.

D. Extrapolation to the Future.

The infrared cutoff μ is not really a mass, and is not really a part of Yang-Mills theory at the Lagrangian level. The hope is that someday it will be possible to bring the theory to a point where the cutoff μ can be dropped, possibly to be replaced by a "natural" cutoff, such as a self-consistently determined confinement region (bag). In that case, μ might be reinterpreted as R^{-1}, where R is a characteristic size of the confinement region, and the limit $\mu \to 0$ will really mean $R \to \infty$. The vanishing of the colored S-matrix as $R \to \infty$ then means that any machinery devised to stretch a bag to infinite dimensions by pulling on the quarks in the bag will have to fail.

The most natural way of reinterpreting the differential equation (9) in space-time appears to be to replace $\mu \frac{\partial}{\partial \mu}$ by $x^{\alpha} \frac{\partial}{\partial x^{\alpha}}$. The result will be a scaling equation which describes what happens when a bag is "stretched" by rescaling. It remains to be seen whether it will be possible to describe, with such a scaling equation, the one-dimensional string-like deformations which are intuitively expected when a quark and an anti-quark are pulled apart. We also do not know how to describe exactly how the right-hand side of (9) will be changed in the future; our hope is that the basic properties of positivity and of coherent additivity of the Γ's remain, even though there may be many new effects. It is urgent to provide a usable space-time interpretation of our momentum-space results, and work is in progress in several directions. One or two things are clear: our formulas are close in spirit to the usual eikonal forms of QED, and we know that the eikonal describes essentially classical (point-like) fermions in a sea of spacelike gluons. (That the gluons are spacelike follows from direct computation of imaginary parts, which only receive contributions from fermion intermediate states in fermionic processes, before μ goes to zero.) Soft real gluons cannot be emitted, and the spacelike sea somehow becomes a bag.

At the moment we see no direct connection between the expected linear (in r) potential between quarks and the singularities in μ of Feynman graphs. Nor is there any obvious connection to Wilson's lattice-gauge theory,[8] which seems to be best suited for handling just the problems of bound-state dynamics that we have not yet explored. Some other lines of investigation, beginning from our Feynman-graph-oriented approach, which might be fruitful sooner include possible explanations of a number of features insoluble in a free-quark model: Zweig's rule, valence-quark purity (i.e., relative unimportance of the $q\bar{q}$ sea), and absence of exotics (these three topics are probably manifestations of one underlying phenomenon); the axial-vector baryon current; the $\Delta I = 1/2$ rule. It does not appear to us that Zweig's rule has any simple interpretation in terms of $q\bar{q}$ annihilation by two or three gluons, since all processes inevitably involve infinitely many gluons.

REFERENCES

1. J. M. Cornwall and G. Tiktopoulos, Phys. Rev. Lett. 35, 338 (1975).

2. J. M. Cornwall, Phys. Rev. D10, 500 (1974).

3. H. Georgi and S. Glashow, Phys. Rev. Lett. 32, 438 (1974).

4. This is not true graph by graph. When all graphs are summed, the leading logarithms can be factored out; neglected terms are then non-singular as $\mu \to 0$ in the remaining factor.

5. J. J. Carrazone, E. R. Poggio, and H. R. Quinn, Phys. Rev. D11, 2286 (1975); Poggio and Quinn (to be published).

6. There are also disconnected graphs, involving spectator quarks. The remaining scattering amplitude is colored, and hence vanishes in the limit $\mu \to 0$. A similar argument can be constructed to show the irrelevance of the Halliday-Landshoff pinch singularity.

7. S. J. Brodsky and G. Farrar, Phys. Rev. Lett. 31, 1153 (1973), Phys. Rev. D11, 1309 (1975); V. A. Matveev, R. M. Muradyan, and A. N. Tavkhelidze, Nuovo Cimento Lett. 5, 907 (1972), 7, 719 (1973).

8. K. Wilson, Phys. Rev. D10, 2445 (1974).

DISCUSSION

J. PATI (Maryland)

You said if fermions belong to $N,\bar{N}$ or 1 of the SU(N) subgroup, and if there is dynamical symmetry breaking, then the SU(N) subgroup is not necessarily broken. This then cannot apply to any gauge group containing the SU(2) subgroup for weak interactions, for which fermions are doublets or singlets of SU(2) and note that for most approaches such as $SU(n)_L \times SU(n)_R$ or SU(5), the above is the case.

CORNWALL

Yes, if you don't have scalars.

J. PATI (Maryland)

But then I would have wondered. Because let's consider SU(5) since that's the example you took. With respect to the weak so-called SU(2) group, they are doublets and singlets so that should not have been broken either.

CORNWALL

No, this is only a mathematical example. You shouldn't interpret it physically.

J. PATI (Maryland)

I didn't understand then. Are you saying that it isn't necessarily broken if you have dynamical symmetry breaking and if the representations are $N,\bar{N}$ or 1? If that's the case you can't consistently put it into this SU(5) picture.

CORNWALL

No, but I'm saying that's just mathematical. Don't try to draw physical conclusions from that, O.K.? I can say the same thing about SU(3) that if I try to break down SU(3) with only octets of "quarks" that it cannot break down the 27 direction. I'm simply saying that there are restrictions on the directions the symmetry can break down when dynamical symmetry breakdown is involved. The restrictions can always be beat if you put in scalars. Since I see a restriction I want, I'm going to pitch out the scalars for now.

I. BARS (Yale)

If I understand correctly, your treatment did not make a distinction between color singlet $q\bar{q}$ pairs which are bound in a small region of space, and color singlet $q\bar{q}$ pairs which are well separated in space. Should I understand that the S-matrix elements with unbound color singlet quark pairs is non-zero? Then, what does this imply about confinement?

CORNWALL

For us a composite simply means 2 particles going along with just about the same momentum - at just about the same velocity that they can be localized. But now, somebody else coming along with quite a different momentum obviously that's not localized with respect to the first state I wrote down. Am I getting at your question?

I. BARS (Yale)

I'm not sure. What I want to know is if I have a quark and an anti-quark well separated from each other but in a color singlet state.

CORNWALL

No. I want them to be close. I want you to think of this being a wave packet, and in a wave packet there is a momentum spread but it's not very big. It's of the order of the radius of the wave packet.

I. BARS (Yale)

How do you take that into account?

CORNWALL

I haven't yet.

I. BARS (Yale)

All right. So this does not mean still confinement?

CORNWALL

Well, I think it does.

B. WARD (Purdue)

Would you please comment on why you believe this analysis transcends the leading log approximation?

CORNWALL

First of all, in QED where these techniques of using this differential equation furnished what I think is the quickest way to all infrared effects in QED, it's known that those infrared logarithms that I add up contained all the mass singularities of QED, period. There are no others. And the reason for this is because in QED you can see that what you drop when you drop closed loops of fermions and so on, are things which are non-leading by powers of μ not by powers of logarithm but by powers of μ. And I have a strong suspicion that the same thing is true in the Yang-Mills theory, but I can't prove that.

H. FRITZSCH (Caltech)

It is possible to exchange two massless gluons in a color singlet state between two color singlet hadrons. In lowest order those exchanges lead to a long range force of the van der Waals type. How do you avoid these?

CORNWALL

You never get by with just putting in two. You've got to put in the whole works, millions and millions of them and by that time we don't know what the mass spectrum of the mesons is. It could be a long range van der Waals force, except we don't see that so we conclude that somehow the millions and millions of these things (which are really all near forward regime processes) prevent this from happening, but I have no details. I haven't done any of those calculations.

T. T. WU (Harvard)

You discussed the very interesting case where all the Casimir operators are zero. In this special case, is there no logarithm at all or does the double log series become a single log series?

CORNWALL

First of all, some of the double logarithms already turn into single logarithms when we went to the infrared regime (which I admitted as a technical complication) but in any event if you multiply C_i times $\log^2$ or you multiply C_i times log and you put C_i equal to zero you get zero.

T. T. WU (Harvard)

Suppose you start out by putting all C_i's to be zero to begin with. Then do you get any logarithm at all?

CORNWALL

No, I don't think you get any logarithms at all.

J. SCHECHTER (Syracuse)

Bob Cahelen and Dan Knight of Syracuse have independently verified the exponentiation of the form factor to sixth order in Yang-Mills theory.

CORNWALL

Oh, good.

P. MINKOWSKI (Caltech)

You have assumed that there is a quark mass M and went on to demonstrate what happens if one approaches this mass shell in the form of what can be called "confinement". But at the same time there cannot be a quark mass M either. Would you like to comment on that?

CORNWALL

Not really. I don't know what a quark mass is. If I did I would comment.

QUARKS AND STRINGS ON A LATTICE

Kenneth Wilson*

*Department of Physics, Cornell University, Ithaca, N.Y. 14850.
This transcript of K. Wilson's talk has been prepared by Y. Srivastava, Department of Physics, Northeastern University.

References

Some references for background are listed below:

1. K. Wilson, Phys. Rev. D10, 2445

2. J. Kogut and L. Susskind, Phys. Rev. D11, 395

3. R. Balian, J. Drouffe and C. Itzykson, Phys. Rev. D10, D11

4. F. J. Wegner, J. Math. Phys. 12, 2259 (1971).

5. C. Korthals-Alles, Marseille Conference 1974.

6. K. Wilson, Phys. Reports (to be published).

7. J. Willemsen and V. Baluni, Preprint.

8. G. Canning and D. Förster, Preprint.

9. J. Kogut, L. Susskind and Banks, Preprint.

10. 2 recent preprints by Migdal.

11. K. Wilson, Erice 1975 lectures.

This will not be a complete talk on the subject because it is not possible to do so in 40 minutes, and so what I want to give you is a status report.

STATUS OF LATTICE GAUGE THEORY

I am reporting, in particular, from the point of view that the lattice gauge theory is simply a way of defining to a continuum limit of a continuum color gauge theory, which, as many people have said, is the most probable theory of strong interactions at the present time. Unfortunately, the color gauge theory will remain in limbo until we learn how to solve it and in particular get the spectrum out of it. So, in particular, I wish to emphasize how one might solve the color gauge theory to get a spectrum, and I am going to go through three methods.

A. Weak Coupling Expansion

This, as you know, cannot give a spectrum. But I have to discuss it anyway - from the point of view of how it looks in the lattice gauge theory.

B. Strong Coupling Expansion

Lattice gauge theory allows other ways of solving it aside from perturbation theory, which is why (at least to me) it makes it an interesting approach. I will discuss in particular the strong coupling expansion, which is what I have been working on for the past year. Unfortunately, the situation there is not very encouraging. The output spectrum is lousy.

C. Block Spin Method

This is another, as yet untested method, which I shall comment upon later (see discussion).

To those of you who have heard me before this will sound like a pessimistic report. But I should say, before entering into details, that the nice thing about the lattice theory is that it is a completely well-defined theory (on the lattice). There is no infinite renormalization as long as it is on a lattice, there is no perturbation expansion and it is written down in closed form exactly. So, one can imagine going in many different ways about solving it. The problem is not that we have failed to solve it so far, but that there are so many things one can do with it that one has hardly scratched the surface. If I give a pessimistic talk, it's partly because I want to reserve this problem for myself!

COLOR QUARK GAUGE THEORY ON A LATTICE

I am going to talk about the regular color quark gauge theory but on a discrete space-time lattice instead of on a continuous space. The most important thing is that we will insist that there be an exact gauge invariance on the lattice. Of course, I will have to define what is gauge invariance on the lattice. However, the lattice is only an intermediate step in setting up the theory and to solve it - in the same way as one takes a partial differential equation and puts it on a discrete mesh. The theory we are interested in involves the limit as the lattice spacing goes to zero and I have no interest in what the theory looks like apart from that limit.

Now there is no problem with what the quark fields look like on a lattice - there is just a quark field at every lattice site. But it is a little more complicated when one considers the form of the gauge field. The form of the gauge field to be used here is entirely governed by the need to preserve gauge invariance on the lattice. What happens is that the basic gauge field on the lattice is not the gauge field A^a_μ itself, but is an exponentiated form which corresponds to the line integral of the gauge field between one lattice site and the next. That line integral is then approximated on the lattice by

$$U_{n\mu} = e^{iag_0 A_{n\mu}^b T^b} ,$$

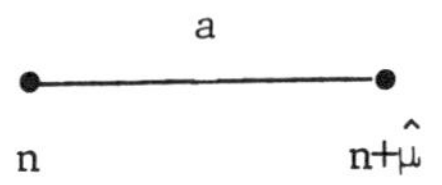

where a is the lattice spacing, $A_{n\mu}^b$ is the gauge field at the lattice site n, g_0 is the gauge coupling constant and T^b is the (hermitean) color matrices.

The <u>gauge transformation</u> laws on the lattice have the following form: The quark fields at site n, ψ_n, $\bar{\psi}_n$ transform as

$$\psi_n \to V_n \psi_n , \qquad \bar{\psi}_n \to \bar{\psi}_n V_n^\dagger ,$$

where V_n is the color (3×3 unitary, unimodular) matrix at site n (local gauge transformation). From the form of the line integral, one can deduce that

$$U_{n\mu} \to V_n U_{n\mu} V_{n+\hat{\mu}}^\dagger ,$$

where $n+\hat{\mu}$ denotes the nearest neighbor to n.

This operator U has a nice, simple physical interpretation. One can think that $U_{n\mu}$ creates (or destroys) a string bit connecting two adjacent lattice sites:

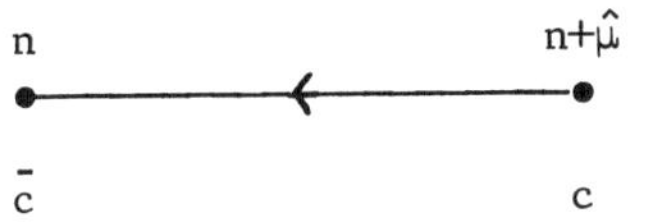

The peculiar feature of this string bit is that it has a color index at each end, which is reflected by the above transformation law. The string bit plays an important role in preserving the gauge invariance and they are also crucial in defining the spectrum in the strong coupling limit. We shall come back to it later.

ACTION

The lattice action (in Euclidean metric) has the following form:

$$A = - \sum_n \bar{\psi}_n \psi_n$$

$$+ K \sum_n \sum_\mu \bar{\psi}_n (1 - \gamma_\mu) U_{n\mu} \psi_{n+\hat{\mu}}$$

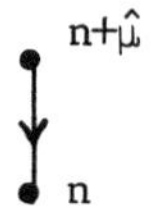

$$+ K \sum_n \sum_\mu \bar{\psi}_{n+\hat{\mu}} (1 + \gamma_\mu) U^\dagger_{n\mu} \psi_n$$

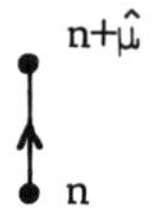

$$+ \frac{1}{2g_o^2} \sum_n \sum_\mu \sum_\nu \mathrm{Tr}[U_{n\mu} U_{n+\hat{\mu},\nu} U^\dagger_{n+\hat{\nu},\mu} U^\dagger_{n\nu}] \quad ,$$

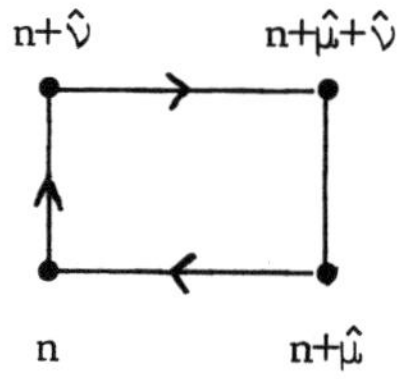

$$K = \frac{1}{8 + 2m_q a} \quad .$$

The first term in A denotes the free quark action. The second and third terms are the coupling of the quark to the gauge field. The 1 that appears in the parenthesis here is a special feature of the lattice theory and is present to ensure that, while the quark field has 4 components, the quark states must have only 2. When one works out the details that rule is satisfied with the above choice, because the $(1 \pm \gamma_\mu)$ are the projection operators -

especially the term jumping in the time direction which has a $(1 \pm \gamma_0)$. So there are only 2 quark states even though there are 4 quark operators, and I'm not going to go into details.

The last term represents the free gauge field action in the U-language and is the analog of the $F_{\mu\nu}F^{\mu\nu}$ term in the continuum case.

This action is designed to maintain gauge invariance. I will not give a complete argument that this represents the continuum theory but I will give an example later.

The above is written in the Euclidean metric so that, for example, the anticommutator between γ_μ matrices is $\delta_{\mu\nu}$ rather than $g_{\mu\nu}$, but it can be rigorously proven, under certain circumstances I will mention later, that the analytic continuation from the Euclidean metric to the Lorentz metric exists and gives an honest quantum theory. It is a little complicated to demonstrate and I can't give you any details here.

What one calculates here is propagators just as in any other quantum field theory. For example, the π-meson propagator looks like

$$D_n = \frac{\langle \bar{\psi}_n \gamma_5 \psi_n \bar{\psi}_o \gamma_5 \psi_o e^A \rangle}{\langle e^A \rangle} ,$$

where

$$\langle \ . \ . \ . \ \rangle = \underbrace{\prod_n \int d\psi_n \int d\bar{\psi}_n}_{\text{Fermion Integration}} \ \underbrace{\prod_\mu \int dU_{n\mu}}_{\text{Invariant Group Integration}} \quad .$$

This is done in a Feynman path integral framework so that to calculate the propagators just involves integrations - of a somewhat peculiar kind, either Fermion integrations as defined sometime ago by Berezin, or they are invariant group integrations over the U's. But it is still a genuine Feynman path integral and it can be shown to have all the properties of quantum mechanics.

CONTINUUM LIMIT AND ASYMPTOTIC FREEDOM

What about the continuum limit of the above theory? This limit has to be discussed in two parts, because it depends upon whether or not one worries about renormalization.

Now the classical continuum limit would just involve

$$a \to 0 \quad ,$$

and

$$K \to \frac{1}{8} \quad .$$

We will see the origin and significance of this number $\frac{1}{8}$ later. As soon as one does renormalization, one expects from asymptotic freedom that

$$g_o \to 0 \quad \left(\text{as } \frac{1}{\ell n a}\right) \quad \text{in the continuum limit,}$$

and I will come back to that.

Now, I've done some renormalizations of the fields to make things look elegant in the lattice formulation. In particular, the relation between the lattice ψ_n and the continuum ψ involves a normalization constant which depends upon the lattice spacing:

$$\psi_n = \left(\frac{a^3}{2K}\right)^{\frac{1}{2}} \psi(na) \quad .$$

The factor a^3 is essentially there to ensure that sums over lattice sites turn into integrals in the continuum limit which involve powers of a.

Now the important thing is that the $U_{n\mu}$ in the continuum limit - at least classically - is expected to be close to 1,

$$U_{n\mu} \sim 1 + ig_o a \sum_{b=1}^{8} A_\mu{}^b(na)T^b \quad ,$$

because of the path length _a_ between nearest neighbors. I will show you later that this path length _a_ has no relevance when you start discussing quantum mechanics. The coupling constant $\underline{g_o}$ on the other hand is going to be crucial. Also, in the continuum limit

$$A \to \int \mathcal{L} \, d^4x$$

with $x = na$ and $\mathcal{L}(x)$ the continuum color quark Lagrangian density.

<u>An Example</u>:

Let me discuss one example of the way the continuum limit develops. Consider the last term in the action, which involves a trace of four U's. They can be thought of as going round in a square if one thinks of them as going around a path. And the expression is gauge-invariant due to the presence of both a V and a $V^\dagger$ at each point under gauge transformations. One has

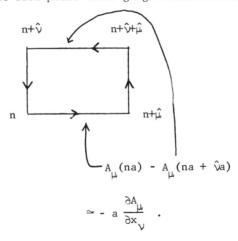

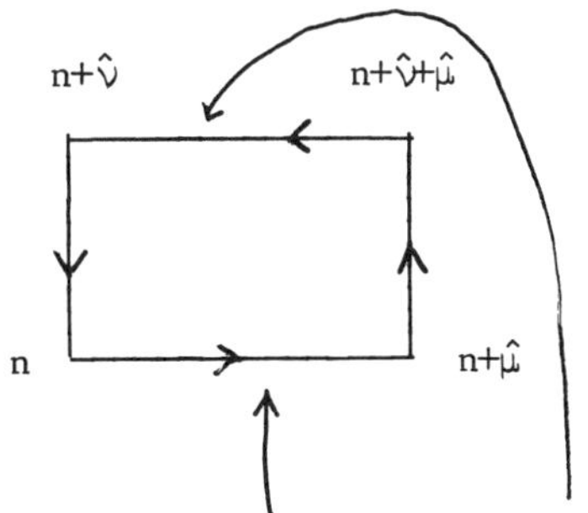

$$A_\mu(na) - A_\mu(na + \hat{\nu}a)$$

$$\simeq - a \frac{\partial A_\mu}{\partial x_\nu} .$$

It is a set of calculations of this sort which build up the complete $F_{\mu\nu}$ structure (including the $A \times A$ term in the non-Abelian case) in the continuum limit. The same holds for the quark terms as well which go to the standard continuum quark Lagrangian.

I am sorry to be moving rather rapidly in this section but I do wish to discuss the interesting features of the strong coupling limit of the theory.

RENORMALIZATION AND EXPANSION IN g_o

A careful, detailed renormalization for this form of lattice theory has not been performed. But we can reasonably imagine what the results are going to be.

About renormalization and power-series expansions, the first question to ask is: Can the $U_{n\mu}$ given by

$$U_{n\mu} = e^{ig_o a \sum_b A^b_{n\mu} T^b}$$

be expanded near 1? The crucial parameter then is the size of $A_{n\mu}$ that enters here. If one can get back to the continuum limit, we can certainly calculate the size of $A_{n\mu}$, namely, we can get at the expected size by calculating the expectation value of $\langle (A^b_{n\mu})^2 \rangle$ and then taking the square root. It is not difficult to imagine that the size of this expectation value will be of the same order of magnitude as the continuum propagator (for $x \sim a$), which we know behaves like

$$\langle A^b_\mu(x)\, A^b_\nu(0) \rangle \sim \frac{1}{x^2} \;.$$

On the lattice we are talking about the same point but that is roughly two things separated by a distance a in the continuum limit. Therefore, this expectation value is expected to be

$$\langle (A^b_{n\mu})^2 \rangle \sim \frac{1}{a^2} \;,$$

and this can be confirmed. Thus,

$$g_o a\, A^b_{n\mu} \sim g_o \;.$$

Also, we find that

$$\langle (A^b_{n\mu} - A^b_{n+\hat{\nu},\mu})^2 \rangle \sim \frac{1}{a^2} \;,$$

and we see that a has no relevance in the quantum limit since A has roughly the size of $\frac{1}{a}$.

Therefore, to get back to the polynomial form of the continuum theory, we need that g_o be small. But from asymptotic freedom we know that g_o is small and so that part is O.K. Now, it is not enough that g_o be small because a gauge transformation can take a U near 1 to any unitary matrix whatever! So, we need a gauge fixing term in the action. I did not put a gauge-fixing term in the original action because none was needed due to the compact nature of the U's as opposed to the non-compact A-integrations of the continuum theory. But, if one wishes to do perturbation theory in g_o, to hold U near 1, one needs a gauge fixing term. That can be done with no change in the continuum theory as was demonstrated by Fadeev and Popov, and investigations on the lattice indicate there is no difficulty here either.

CONCLUSIONS

i. $A^b_{n\mu}$ not of order 1.

ii. $(A^b_{n\mu} - A^b_{n+\hat{\nu},\mu})$ not of order a, hence the classical prescription of replacing it by a derivative is not valid; instead quantum fluctuations are very important. So, we have to keep the lattice in to discuss renormalization.

iii. $U_{n\mu}$ near 1 if g_o small, but a gauge fixing term is needed, e.g.,

$$-\frac{\alpha}{2}\sum_n \mathrm{Tr}[\{\sum_\mu (U^\dagger_{n\mu} - U^\dagger_{n-\hat{\mu},\mu})\}\{\sum_\mu U_{n\mu} - U_{n-\hat{\mu},\mu}\}].$$

iv. We cannot take the continuum limit $A^b_{n\mu} \to A_\mu$ (x = na) in any ordinary sense.

So, to do perturbation theory we add the gauge fixing term and add the necessary Fadeev-Popov ghost term to preserve equivalence to the gauge invariant action. We have then the usual Feynman expansion, but with a rather peculiar and non-covariant cutoff. E.g., the quark propagator is the usual $\frac{1}{\not p}$ for "ordinary momenta", i.e., $p \ll \frac{1}{a}$, but it is modified for $p_\mu \sim \frac{1}{a}$ to

$$p_\mu \to \frac{\sin p_\mu a}{a} .$$

All integrals stop at $p_\mu = \frac{\pi}{a}$. But one expects to deal with that sort of cutoff in the same way as one deals with any other cutoff and that such dependences will disappear into renormalization effects. This theory has the nice feature that even in the presence of a cutoff it is fully gauge-invariant. So, one does not have to go to 3.99 dimension to make the theory gauge invariant and one does not need to put in cutoffs which still leave some divergences. There are no divergences as long as the cutoff is present. Of course, it is still a pain to calculate anything which depends upon the cutoff.

EXPECTED RESULTS

So, in renormalized perturbation theory, the expected, though yet unconfirmed, results are:

(1) Usual asymptotic freedom. That is to say that renormalization requires that the bare coupling constant

$$g_o \to 0 \quad \text{proportional to} \quad \frac{1}{\ell n a} .$$

This is in order that the theory has the continuum limit in terms of the vacuum expectation values of operators.

For QED it is a completely different question since asymptotic freedom is crucial to this argument. One can obtain the continuum QED but the arguments would be quite different.

(2) m_{quark} is renormalized but

$$K = \frac{1}{8 + am_{quark}} \rightarrow \frac{1}{8} \quad \text{as} \quad a \rightarrow 0 \, .$$

(3) Infrared problem. This is the crucial question which has defeated all efforts to solve the gauge theory, namely that the renormalized coupling constant g_{eff}, relevant for large distances $\sim 10^{-13}$ cm, gets large and so there are serious infrared divergences which no one knows how to handle.

SECOND EXPANSION METHOD

We need to find another procedure to find the spectrum. In this respect lattice gauge theories are nice because procedures other than ordinary perturbation method can be formulated.

We can expand in powers of K and $\frac{1}{2g_o^2}$, i.e., treat all but the first term in the action as perturbations. The various resulting terms are represented diagrammatically, which are decomposed into 3 classes. All diagrams shown below are for the π-meson propagator.

Class I

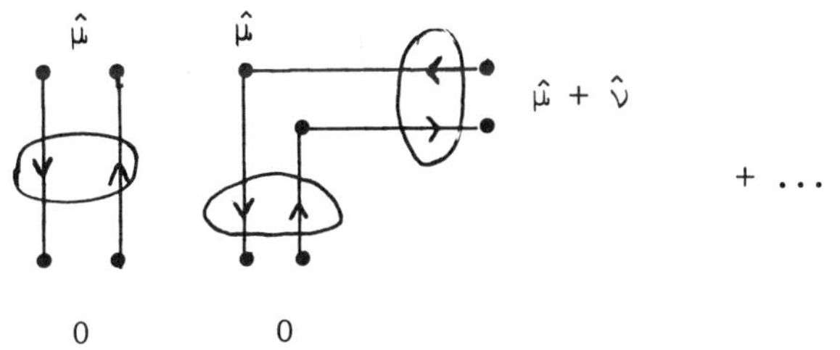

This is the simplest set of diagrams. This set can be summed analytically and gives poor results.

Class II

We combine links and squares. An example is

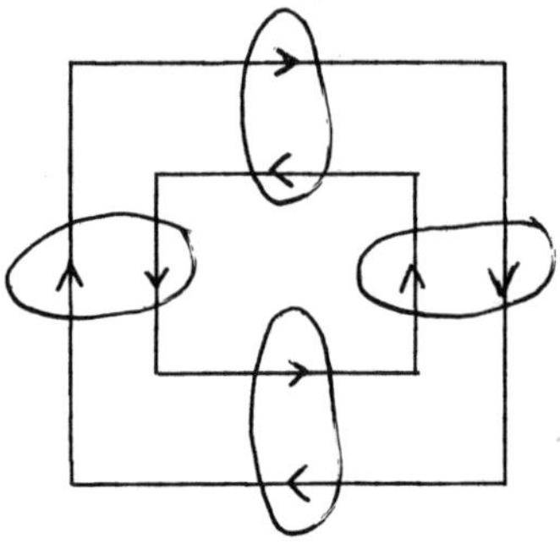

which can be done by hand. We can calculate more complicated cases (up to 8 links and 5 squares) by a computer program. This set also does nothing at all.

Class III

Everything else - which we cannot calculate. An example

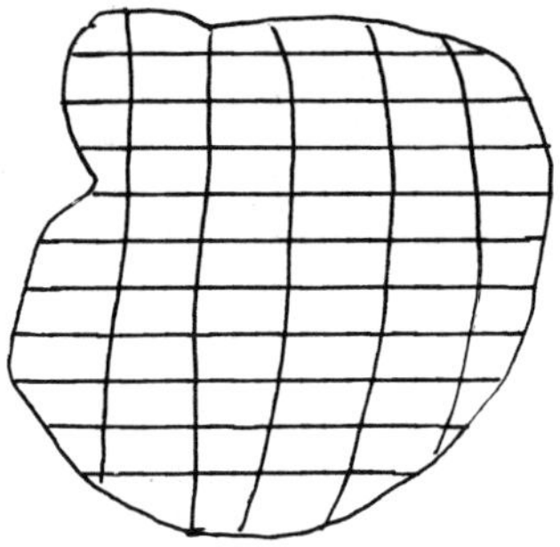

This is where the action is, and I don't know how to calculate it.

Rules for diagrams

These are rules resulting from doing various Feynman path integrals over all the variables in the problem.

(i) One has a factor $K(1+\gamma_\mu)$ or $K(1-\gamma_\mu)$

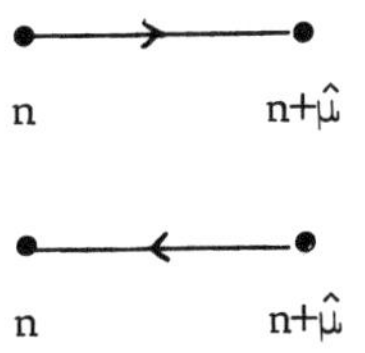

$K(1+\gamma_\mu)$ quark link

or

$K(1-\gamma_\mu)$ quark link

for nearest neighbor quark link, depending upon the direction.

(ii) The trace of four U's we denote by a square. One has

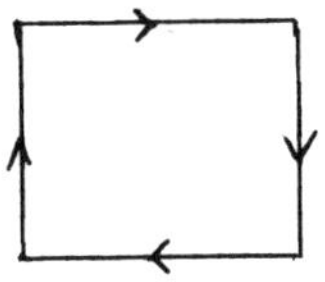

$\frac{1}{2g_o^2}$ for the gauge loop square.

The expansion in powers just involves writing down the number of links and squares to get any given term. [We might mention the origin of the $\frac{1}{2g_o^2}$ factor in the term in the action with a trace of four U's. This term limits back to the normal F^2 part of the Lagrangian. Now when the U's are expanded we get a g_oA and when these combine to form a curl we get $g_o^2F^2$ and so it's necessary to divide by g_o^2 to correctly limit to the continuum F^2. Inevitably, an expansion on the lattice involves $\frac{1}{g_o}$.]

(iii)

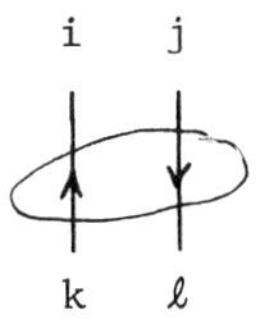

The integration over the gauge fields, i.e., integration over the U's are invariant group integrations over the color group for each pair of lattice sites, and produces these circles which stand for contraction. At end of each line is a color index. A <u>color contraction</u> arises from the integration over the U's. A simple example is:

$$\frac{1}{3}\delta_{ij}\delta_{k\ell} = \int_U U_{\ell j}\,(U^\dagger)_{ik} \quad .$$

As an example of the above Feynman rules we have a simple term from the π meson propagator:

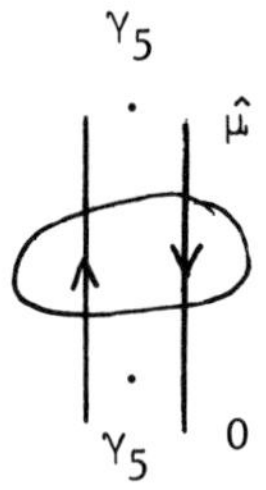

$$K^2 \cdot \mathrm{Tr}[\gamma_5(1 - \gamma_\mu)\gamma_5(1 + \gamma_\mu)] \times 3$$

↑ color trace

The important thing about the color contraction is that one needs at least two links to get a non-zero result because the invariant group integration over a single U, which would be the contraction of one line only, gives zero. These contractions forbid the propagation of isolated single quarks. I have discussed at length elsewhere the whole question of quark confinement, so this will not be repeated here.

In this strong coupling expansion then, we have local color invariance: only local color singlets propagate. Some examples are drawn below:

 $= 0$ since $\int_U U = 0.$

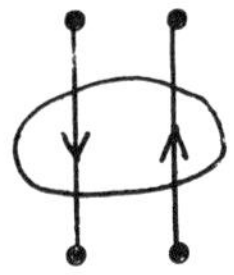

a $q\bar{q}$ pair can propagage.

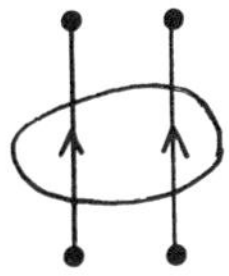

= 0 two q's do not propagate.

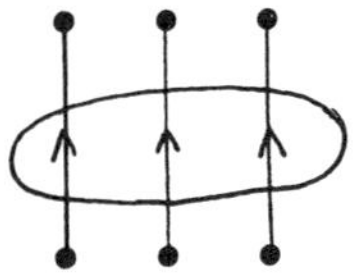

qqq baryon propagation (a special property of color SU(3) allows the contraction of three lines in same direction to be non-zero).

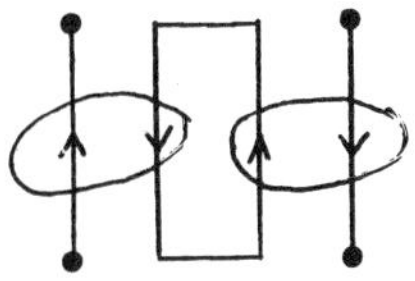

propagation of a string bit:

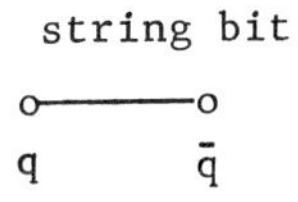

So this is a theory where the strong coupling expansion gives a spectrum with quarks and strings and thus one expects a complicated spectrum like any string model.

What are the results of performing this strong coupling expansion?

1. Class I Graphs

This set can be solved because they look like the simple self-energy series.

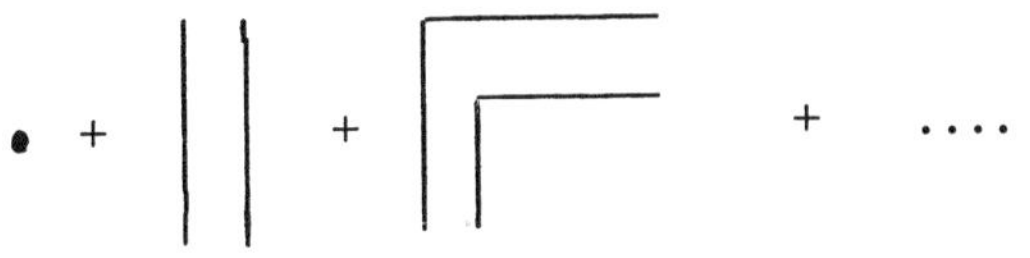

$$1 + \Sigma + \Sigma^2 + \ldots\ldots = \frac{1}{1 - \Sigma}$$

So this term gives rise to poles. If we fit the π and the ρ-mass, K and a get determined:

$$K \approx \frac{1}{4}$$

and

$$a \approx 1/\text{GeV}.$$

In this simple picture the η is identical to π and ω identical to ρ (in mass). This is not terribly satisfactory. But more disappointing is the nucleon and Δ mass relation.

For nucleons, we have a similar series with 3 quark propagation:

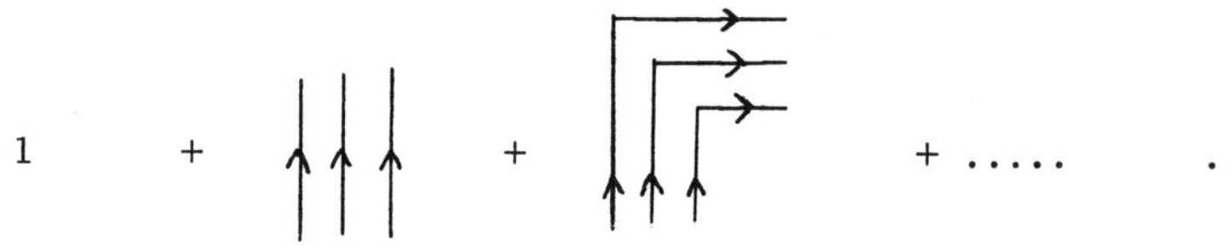

Again, it is a simple self-energy sum and we can calculate their masses since there are no free parameters (K and a being determined above already):

$$m_{\text{Nucleon}} = 1720 \text{ MeV} ,$$
$$m_{\Delta} = 1750 \text{ MeV} ,$$
$$m_{\Delta} - m_{N} = 30 \text{ MeV} .$$

Of course, there is no reason to expect this to agree with experiment because I have taken a theory where $g_o \to 0$ and have put $g_o \to \infty$.

Class II: My hope had been that by using the ability to do more complicated diagrams one could work away from the strong coupling limit and know what was going on for some g_o.

These are the remaining diagrams we can calculate and the trouble is they don't do anything. These diagrams give no qualitative changes, because the graphs are too small (at least for mesons). The reason is that K is too small (it comes out ¼). E.g.,

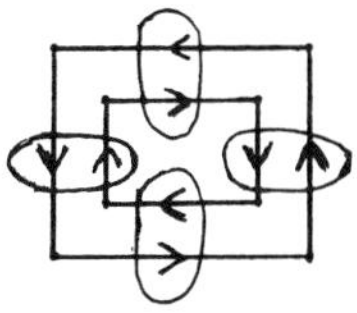

$\sim K^4 \sim \frac{1}{256}$ and therefore negligible.

The reason K is small is the large dimension of space-time. Namely, Σ contains the elementary graph:

$\rightarrow 2K^2 \times 8 = 16K^2$.

↑ # of directions

There is a factor 8 for the eight possible orientations of the graph in space-time. The result is that the meson pole occurs at zero mass for $16K^2 = 1$, i.e., $K = \frac{1}{4}$. For $K > \frac{1}{4}$ the pion mass is imaginary.

All calculations of class II diagrams yield at most 25% changes in the class I results.

Class III. Example:

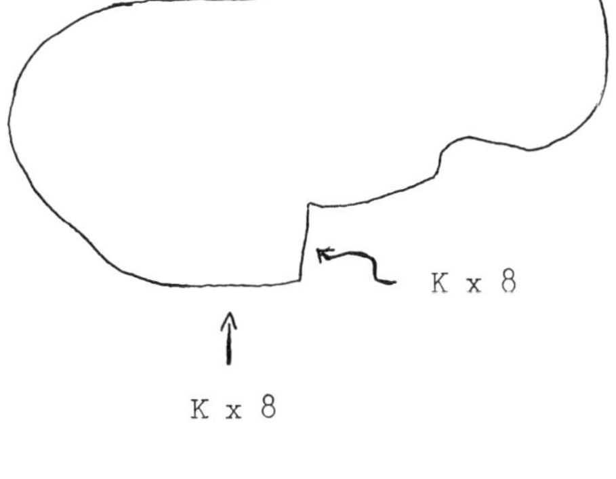

So, very complex diagrams are important for $K > \frac{1}{8}$, especially in K^{16}-K^{30} and higher order. Basically, for a very large quark line, each quark link can point in 8 directions and so there is a factor of 8 for each K. (One needs a long line to be able to orient each link independently.) Thus, as long as g_o is small, so one doesn't get a small factor for the garbage in the middle, the graph will be important. However, there are so many diagrams at K^{16} one doesn't know how to calculate.

To summarize the results so far:

(1) Expansion in g_o (weak coupling) fails to give spectrum because spectrum involves large g_{eff}.

(2) Expansion in $1/g_o$ fails because g_o is small.

However, the discussion shows that with the lattice one can try ways to solve the problem that were totally unimaginable from the continuum theory viewpoint. We've tried one way that hasn't worked very well, but one can think of more.

RESOLUTION

One should do a non-perturbative expansion of the theory with g_{eff} not g_o. A possibility is the "Block spin" approach of Kadanoff, Niemeyer and Van Leeuwen. See T. Bell and K. Wilson, Phys. Rev. B11, 3431.

DISCUSSION

B. ZUMINO (CERN)

If I understood you correctly, you get a value for the lattice spacing which is comparable to the inverse mass of an elementary particle. Now this looks very large. One would expect this to be much smaller if the theory should be insensitive to the size of the lattice spacing. Would you comment upon this?

WILSON

With these simple sets of diagrams, when I fit the π and ρ masses, I get $a \sim 1/\text{GeV}$, and the momentum goes up to $\pi/a \sim 3$ GeV. Now this is basically a strong coupling result. When the coupling is weak, I have to go Class III diagrams and then one will get a different lattice spacing. One's hope is that as g_o gets smaller so would a and eventually the two will get related by the asymptotic freedom formula $g_o \sim \frac{1}{\ell na}$. The trouble is that I can do only one calculation which is for $g_o \to \infty$ so I cannot see any g_o dependence on a, and thus cannot follow out to see how g_o depends on a.

H. S. TSAO (IAS)

Suppose you don't feed in the mass of the π and ρ. Can you derive a Gell-Mann Okubo like mass formula since you have only two parameters and lot of masses?

WILSON

There is a third parameter, K, for the strange channels. So in all we have three parameters plus the lattice spacing. If you are close to SU(3), then the breaking of SU(3) is of standard type - you get the standard type formula for SU(3) breaking. When the masses are sufficiently far from SU(3) there are non-linear effects and I don't know what they are.

B. LEE (FNAL)

Would you tell me what the "Block spin" is?

WILSON

The block spin idea is the following. We clearly would very much like to distinguish the large distance regime where the physics is the physics of bound quarks, from the short distance regime where the physics is that of the free quarks. Since the two physics are rather different we would like to have our mathematics distinguish the two cases. A logical idea which a lot of people are thinking of in one way or another is that we would like to interpose an intermediate stage where there is some kind of effective action or effective Lagrangian which describes the large distance behavior only. The Block spin procedure is one very specific way of doing that - of defining an effective action which depends upon a g_{eff} rather than g_o, where g_{eff} would describe just the large distance behavior of the theory. The idea would be to choose the lattice spacing a for the effective action just small enough so that we can calculate the action by perturbation theory and then we solve it by some other method. Presumably a lattice spacing around $a \sim \frac{1}{\text{GeV}}$ would be an ideal break point where we solve it by some non-perturbative method.

Now let me show you briefly how this may be done.

Block spin method:

It is called that because like most other methods I use, it was developed in statistical mechanics. There one talks about spin, now we are talking about fields.

We define a Block field, whose block size is $\sim 10^{-13}$ cm ($\sim$ 1/GeV). But we start with a very much smaller lattice spacing so that a block contains many lattice sites. We define one field variable which is a sum of all the fields inside this block. So that it is a field which is referring to wave-lengths of order block size rather than individual lattice spacings. Then we introduce a new field called φ which is related though not necessarily identical to the above, and the effective action A_{eff} is defined in terms of φ, which is an integration over the original action with a kernel which involves the new variables and the block variables.

Example of a scalar field:

$$e^{A_{eff}[\varphi]} = \prod_n \int d\varphi'_n \exp\{-\tfrac{1}{2} \sum_m (\varphi_m - \varphi'_{block,m})^2\} e^{A[\varphi']} \quad .$$

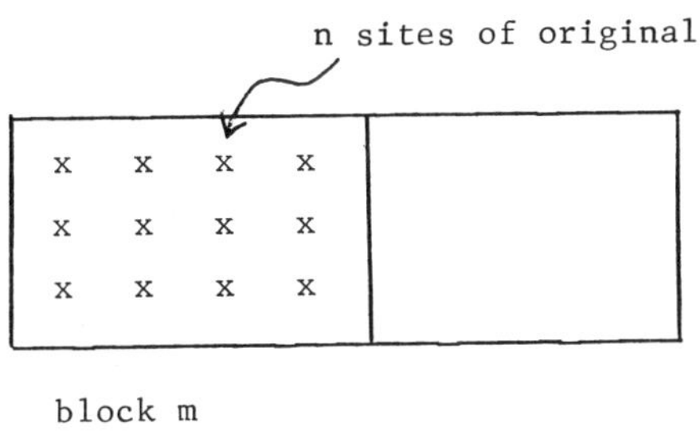

$$\varphi'_{block\ m} = \sum_{n \in block\ m} \varphi'_n \quad .$$

The kernel of the integration is defined such that the Feynman path integrals over A_{eff} and the original action A coincide.

$$\prod_m \int d\varphi_m \, e^{A_{eff}[\varphi]} = \prod_n \int d\varphi'_n \, e^{A[\varphi']} \quad .$$

Thus, under certain restrictions one can show that the two actions lead to identical S-matrices. The kernel adds an artificial mass term to the original action and that mass term is of the order of 1/(lattice spacing). It eliminates the infrared divergence. So calculations with A_{eff} do not suffer from infrared troubles while those of the complete propagators do. Thus perturbation theory can be used for A_{eff}. That is the idea.

S. WEINBERG (Harvard)

I have the feeling that strong interactions ought not to change dramatically as you turn off the quark mass. The success of soft pion theorems and all that seem to indicate it. Is it essential to what you are doing that there be a finite quark mass?

WILSON

No. In strong coupling I come out with $K = \frac{1}{4}$ which implies a non-trivial m_q - that is all a renormalization effect. The minimal renormalization theorem which states that the renormalization of the quark mass is proportional to the quark mass depends upon chiral symmetry. When I set up the theory on the lattice there is no chiral symmetry present. It may be there when you take the continuum limit, but is not present at any level on the lattice theory and therefore this renormalization theorem fails. So when I get a zero mass pion with $K \sim \frac{1}{4}$ instead of $K \sim \frac{1}{8}$ it is just because of quantum renormalization effects. Thus, you cannot talk about a zero and a non-zero quark mass except in the weak coupling case which I haven't been able to reach with my calculations.

V. BALUNI (MIT)

Could you elaborate on your comment about the gauge fixing term on the lattice?

WILSON

Faddeev and Popov have a standard technique whereby, if you start from a theory which is fully gauge invariant in the form of a functional integral, they have a procedure for introducing a gauge fixing term plus a ghost counter term such that the resulting functional integral is identical to the original functional integral. That is what they wrote down. Of course, they had to apply it to a case where the original invariant integral did not exist, and they had to weasel their way around that difficulty. But now we are in a situation where the original gauge invariant integration exists with no gauge fixing term and so the Faddeev-Popov technique works fine. That is what it is designed for - the lattice theory, even though they didn't know that.

So, you have to put a gauge fixing term for the lattice and the integrations look a little different. This has been done by my student and so far I can determine from his results it just goes through in the general fashion.

V. BALUNI (MIT)

But if you introduce a gauge fixing term, don't you lose confinement?

WILSON

If the original gauge-invariant theory has confinement then of course confinement will always be there, but it will appear in a very different fashion. We already know from electrodynamics what we expect to happen if we solve the theory with a gauge fixing term. With a gauge fixing term there will certainly be quarks and gluons present with some mass, a part of the spectrum of states. But they will decouple from the gauge invariant sector. So even if a π-meson is above the threshold to decay into a quark - antiquark pair it will not do so because there is a vanishing matrix element for it. That is the way it has to work. Just as the longitudinal modes of the EM field are present in the spectrum but are never produced from physical states.

C. THORN (MIT)

What happens to your numbers if you fit to the baryon sector instead of the meson sector?

WILSON

I haven't tried to do that because then $K > \frac{1}{4}$ and horrible things would happen in the π-meson propagator. The best it could be is a broken symmetry of some kind - because there would be a tachyon pole. There is also a result which I must mention. The theory I am writing down has a positive metric only for $K < \frac{1}{6}$ so I am already going beyond the legitimate range by putting $K = \frac{1}{4}$, and I am certainly not very happy about going any larger than $\frac{1}{4}$ which is what I have to do to fit the baryon spectrum.

D. WEINGARTEN (Rochester)

It looks like your theory should have non-vanishing propagators for any triality zero state, in other words 6 quarks, 9 quarks, 2 quarks and 2 anti-quarks, etc. It looks like, at least the Class I graphs would be calculable for that. Do you expect such things to survive in the continuum limit?

WILSON

There is an interesting calculation here which hasn't been done yet. If you take a 6 quark state for example, then when you do this expansion, this 6 quark state mass is twice the mass of two 3 quark states. That means those things become unstable as soon as one turns on the parameters K and $1/g_o$. If they are not identically zero, they are unstable. Of course, this does not imply that they disappear - one has to do a calculation to see whether they are absorbed in the continuum or still exist. But it is possible that such a calculation will show that all the exotic states get wiped out as soon as K is turned on and then there is no reason for them to come back in the continuum limit. But such calculations have not been done even though they are perfectly doable because they belong to the trivial Class I graphs.

GAUGE THEORIES AND NEUTRINO INTERACTIONS

Stephen Adler*

*Institute for Advanced Study, Princeton University, Princeton, N.J. 08540.
This transcript of S. Adler's talk has been prepared by Y. Srivastava, Department of Physics, Northeastern University.

I will talk about the interaction between neutrino physics and gauge theories in two contexts.

(A) First, in connection with gauge theories of weak and electromagnetic interactions, I will talk about the implications of neutrino physics for

1. Heavy leptons
2. Neutral currents
3. Intermediate Bosons, and
4. Phenomena associated with "charm" production.

(B) Then, in connection with gauge theories of strong interactions, I will discuss the implications for neutrino physics in relation to scaling violation in theories possessing "asymptotic freedom" or other possible patterns.

Let me begin with the first topic.

A. Gauge Theories of Weak and Electromagnetic Interactions.

As we know these theories are unitary in the tree approximation, which means for example, that the bad high energy behavior of a graph shown in Fig. (1) occurring in $\nu_\mu \bar{\nu}_\mu \to W^+ W^-$ has to be

(Fig. 1)

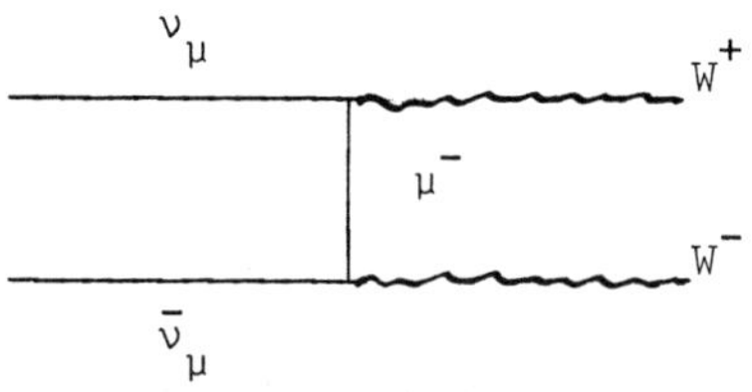

Normal μ^- exchange

cancelled by another tree graph - one (or both) of the following, shown in Fig. (2),

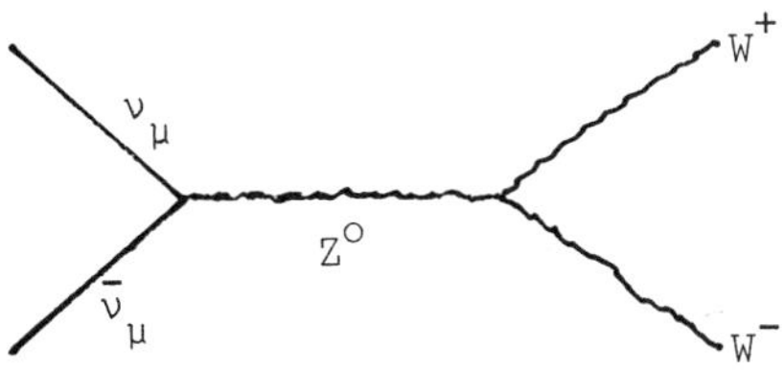

(a) Neutral current mediated by Z^0 exchange

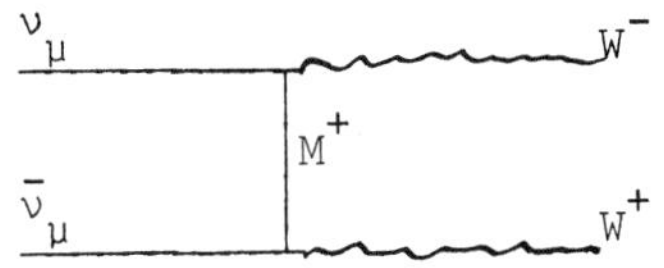

(b) Heavy lepton M^+ with lepton number of μ^-.

(Fig. 2)

Either a neutral heavy vector boson Z^0 is exchanged in the s-channel (Fig. 2a) or a heavy lepton M^+ with the same quantum numbers as the μ^- is exchanged in the t-channel (Fig. 2b) - or a combination of these mechanisms occur. Therefore, in gauge models of the weak and EM interactions as now known, one expects neutral currents or heavy leptons.

1. Heavy Leptons

Let me begin first with a discussion of heavy lepton phenomenology.

(i) Charged Leptons

Charged heavy leptons can be searched for, through their production and decay, in neutrino reactions

$$\nu_\mu + N \to M^+ + \text{hadrons}.$$

Typical signatures for the charged heavy lepton decay are as follows:

(a) Purely leptonic decays

$$M^+ \to \nu_\mu + \nu_\mu + \mu^+ \quad ,$$

Here the final lepton has a wrong sign which is a good signature, because in a normal interaction initiated by a ν_μ you expect to see a μ^- coming out.

(b) Another signature would be

$$M^{+} \rightarrow \nu_{\mu} + \text{hadrons}.$$

This looks like a neutral current, but on closer analysis when you calculate distributions you find that for this reaction about 75% of the initial ν_{μ} energy goes into hadrons whereas, for a true neutral current reaction - if you assume scaling and the usual things - the hadronic energy is $\lesssim$ 50%. Therefore, if you look at the high hadronic energy tail of the neutral current events, you can start to study processes like this.

I will now describe the experimental results from the Cal Tech - FNAL experiment. These are taken from lectures by B. Barish given at the Hawaii Summer School, which will be written up shortly.

μ^{+} - Decay Mode

For the charged mode (called (a) above) the results are presented in Table 1. For the ν_{μ} beam, they have 1522 events with the right sign and 8 events with the wrong sign muons. But, from an analysis of what the $\bar{\nu}_{\mu}$ contamination in their ν_{μ} beam is, and also from an estimate of what the probability of a wrong-sign determination in the muon-spectrometer is, they expect 11.4 ± 6 wrong sign events - so the measured 8 events are perfectly compatible with what is expected. As a check they ran also with the $\bar{\nu}_{\mu}$ beam where the wrong sign events are now produced by ν_{μ} contamination and the calculations in this case also agree with what they observe. So the number of wrong sign events are consistent with expected backgrounds.

Table 1

Cal Tech-FNAL results for $\left.\frac{\nu_\mu}{\bar{\nu}_\mu}\right\} + N \to \mu + X$

Beam	Right Sign Events	Calculated Wrong Sign Events	Observed Wrong Sign Events
ν_μ	1522	11.4 ± 6	8
$\bar{\nu}_\mu$	60	6.1 ± 3	6

They made a further check by looking at the y-distribution, where $y_{obs} = \frac{E_{hadron}}{E_{observed}}$. If the 8 wrong sign events were due to $\bar{\nu}_\mu$ contamination, they should show a $(1 - y)^2$ variation with y. On the other hand if the wrong sign events came from a heavy lepton, they should show a peaking towards large y. The experimental results are completely consistent with the $\bar{\nu}_\mu$ background.

So, the 8 events are not considered by the Cal Tech-FNAL group to be heavy lepton candidates, since the events are consistent in every respect with background. From this experiment then one gets

$$\boxed{\text{Mass } (M^+) \gtrsim 8.4 \text{ GeV}} \quad , \quad (\text{at } 90\% \text{ CL})$$

for usual coupling $G_M^2 = G_F^2$ and assuming a leptonic branching ratio $B_L \gtrsim 0.3$. Of course, if $B_L \sim 0$ this test is not of any use.

Hadron Decay Mode

Now we turn to the alternative mode (b) above. In a neutral current run one finds

796 charged current events (μ^-)

and 6 events with no identified muon and $E_{had} > 80$ GeV.

Now, 3 events would be expected from missing muons of the normal charged current events. On the other hand, one expects 10 such events for Mass(M^+) = 8 GeV and hadronic branching ratio $B_H = 1$.

Combining the two modes (a) and (b), we have independent of the branching ratios,

$$\text{Mass}(M^+) > 7.5 \text{ GeV} \quad .$$

Remark: You can still have lighter, charged heavy lepton(s) with non-muonic quantum number. If they have some new quantum number with their own neutrinos, then the above experiments provide no information, and maybe such a heavy lepton has now been seen around 2 GeV at SPEAR.

(ii) Neutral Leptons

Now let me turn briefly to neutral heavy leptons. Consider

$$\nu_\mu + N \to M^0 + \text{hadrons}$$

$$\hookrightarrow \nu_\mu + \mu^+ + \mu^-$$

$$\hookrightarrow \nu_e + e^+ + \mu^-$$

(the above two: Two leptons-good signature)

$$\hookrightarrow \mu^- + \text{hadrons.}$$

The characteristic feature of dimuons coming from an M^0 have been studied in model calculations by Chang, Derman, Ng and Albright. Also, some nice exact arguments were given by Pais and Treiman.

The key point I want to mention about dimuon events is the following. Call

$$< \;\; > \equiv \text{mean lab. energy},$$

and let us consider the ratio

$$R = \frac{< E_{\mu^-} >}{< E_{\mu^+} >} \quad .$$

Then Pais and Treiman show that this ratio must lie between

$$0.48 \leq R \leq 2.10 \quad ,$$

in the general weak coupling case.

From FNAL Expt. 1A or its extensions, we have

$$R_{EXP} = 3.7 \pm 0.65 \quad .$$

Thus FNAL dimuons are unlikely to be from M^o. They are something else.

2. Neutral Currents

One can develop general phenomenologies for neutral currents and the Hawaii lectures I gave, in fact, discuss semi-leptonic neutral currents with the general V, A, S, P, T phenomenological Lagrangian. We set up an apparatus and computer programs to correlate all the experiments that are going to be done with good statistics in the next year or two. But, to keep things simple and because this conference is on simple gauge theories, let me discuss neutral currents within the standard Weinberg-Salam SU(2) x U(1) model, and I will assume the Glashow-Illiopoulos-Maiani (GIM) construction for the hadrons.

Then, we have the following effective Lagrangians:

$$\mathcal{L}^{\text{charged}}_{\text{effective}} \quad \text{arising from } W^{\pm} \text{ exchange}$$

and

$$\mathcal{L}^{\text{neutral}}_{\text{effective}} \quad \text{arising from } Z^{0} \text{ exchange.}$$

For the neutral current part of the Lagrangian, we have explicitly

$$\mathcal{L}^{n}_{\text{eff}} = \frac{G_F}{\sqrt{2}} \left[j^{\lambda}_{n} + J^{\lambda}_{n} \right] \left[j_{n\lambda} + J_{n\lambda} \right] \quad ,$$

with the leptonic neutral current given by

$$\begin{aligned} j^{\lambda}_{n} = & - \tfrac{1}{2}\, \bar{e}\gamma^{\lambda}(1 - \gamma_5)e - \tfrac{1}{2}\, \bar{\mu}\gamma^{\lambda}(1 - \gamma_5)\mu \\ & + \tfrac{1}{2}\, \bar{\nu}_e\gamma^{\lambda}(1 - \gamma_5)\nu_e + \tfrac{1}{2}\, \bar{\nu}_{\mu}\gamma^{\lambda}(1 - \gamma_5)\nu_{\mu} \\ & + 2x_w(\bar{e}\gamma^{\lambda}e + \bar{\mu}\gamma^{\lambda}\mu) \quad , \end{aligned}$$

and its hadronic counterpart by

$$J^{\lambda}_{n} = \mathcal{F}^{\lambda}_{3} - \mathcal{F}^{5\lambda}_{3} - 2x_w J^{\lambda}_{EM} + \Delta\mathcal{J}^{\lambda} \quad ,$$

where

$$J^{\lambda}_{EM} = \mathcal{F}^{\lambda}_{3} + \frac{1}{\sqrt{3}} \mathcal{F}^{\lambda}_{8} \qquad \text{and}$$

$\mathcal{F}^{\lambda}_{3,8} = 3^{rd}$ (8th) component of the vector octet

$\mathcal{F}^{5\lambda}_{3} = 3^{rd}$ component of the axial-vector octet

$$x_w = \text{Sin}^2\theta_w$$

where θ_w is the Weinberg angle. $\Delta\mathcal{J}^{\lambda}$ is the GIM piece, which is the isoscalar V, A charm and strangeness current. The usual assumption made in the phenomenologies is that they should have small matrix elements in processes involving low-mass, non-strange hadrons. In processes involving high-mass particles or where strange particles are explicitly produced one may not be able to throw them away.

Let me now give a channel by channel survey of what the status of the experiments was at the time of the SLAC Conference 1975.

Leptonic Reactions

Let me write the cross-sections for the three reactions for which data exists in the following way:

$$\begin{bmatrix} \sigma(\bar{\nu}_e e^- \to e^- \bar{\nu}_e) \\ \sigma(\nu_\mu e^- \to e^- \nu_\mu) \\ \sigma(\bar{\nu}_\mu e^- \to e^- \bar{\nu}_\mu) \end{bmatrix} = \begin{bmatrix} C_{\bar{\nu}_e e^-} \\ C_{\nu_\mu e^-} \\ C_{\bar{\nu}_\mu e^-} \end{bmatrix} \times (10^{-41} cm^2) \times \left[\frac{E}{GeV}\right] ,$$

where E is the initial neutrino energy and we characterize the strengths of these reactions in terms of the dimensionless constants C's. The theoretical predictions are shown in Table 2. The coefficients $C_{\nu_\mu e^-}$ and $C_{\bar{\nu}_\mu e^-}$ come

Table 2 : Theoretical Values for C

Coefficient	Weinberg-Salam	Old V-A
$C_{\bar{\nu}_e e^-}$	$0.14 - 2.9$ $(0 \le x_W \le 1)$	0.57
$C_{\nu_\mu e^-}$	0.11 at $x_W \approx 0.35$	0
$C_{\bar{\nu}_\mu e^-}$	0.22 at $x_W \approx 0.35$	0

purely from neutral currents and hence are zero in the old V-A model. Their characteristic values are given in Table 2 for the "best fit" value of $x_W \approx 0.35$. The experimental situation is as follows.

For the first coefficient we have from Gurr, Reines and Sobel - Savannah River,

$$\frac{C_{\bar{\nu}_e e^-}}{0.57} = 1.5 \pm 0.7 \quad .$$

The quotation of an error here should not be taken as if the effect has actually been seen but rather should be interpreted as a two standard deviation upper limit. A new version of the experiment is being run now which should be able to see the effect.

For the seond coefficient, with <u>no</u> candidates, CERN Gargamelle gives

$$C_{\nu_\mu e^-} < 0.26 \quad , \quad \text{(at 90\% CL)}$$

which is perfectly OK for the model.

There are three events from CERN Gargamelle for the last process which gives then

$$C_{\bar{\nu}_\mu e^-} = (0.13 \pm .08)^*$$

which is again within one standard deviation of the model prediction. The star denotes that W-S model angular distribution has been utilized to make efficiency corrections. Thus, this value will change somewhat if other models are invoked to make efficiency corrections though $C_{\bar{\nu}_\mu e^-}$ will stay in this range.

<u>Semi-leptonic Phenomena</u>

Let me begin with the inclusive reactions where neutral currents were first discovered by the Gargamelle group. As usual, we define the neutral to charged current inclusive ratios

$$R_\nu = \frac{\sigma(\nu_\mu + N \to \nu_\mu + X)}{\sigma(\nu_\mu + N \to \mu^- + X)} \quad \text{and} \quad R_{\bar\nu} = \frac{\sigma(\bar\nu_\mu + N \to \bar\nu_\mu + X)}{\sigma(\bar\nu_\mu + N \to \mu^+ + X)} \quad ,$$

where N = average nucleon.

Neglecting isoscalar contributions and using the empirical fact (the evidence for which shall be reviewed later in connection with scaling) that

$$\frac{\sigma(\bar\nu_\mu + N \to \mu^+ + X)}{\sigma(\nu_\mu + N \to \mu^- + X)} \approx \frac{1}{3} \quad ,$$

we get the simple inequalities

$$R_\nu \geq \frac{1}{6} [1 + (1 - 2x_W) + (1 - 2x_W)^2] \geq 0.17 \quad \text{for } x_W \leq 0.5$$

$$R_{\bar\nu} \geq \tfrac{1}{2} [1 - (1 - 2x_W) + (1 - 2x_W)^2] \geq 0.38 \quad \text{for } x_W \leq 0.5 \quad .$$

If one goes one step further and assumes the quark-parton model along with the neglect of the anti-parton content and strange and charmed parton content in the nucleon, then these inequalities get replaced by equalities because the isoscalar part has been taken into account.

The experimental situation as of Fall 1975 is shown in Table 3.

Table 3

Data on R_ν and $R_{\bar{\nu}}$

R_ν	$R_{\bar{\nu}}$	Group	Comment
0.22 ± 0.02	0.43 ± 0.12	CERN Gargamelle	Fits $x_W \sim 0.3$ to 0.4
0.11 ± 0.05	0.32 ± 0.09	FNAL EXPT. 1A	$R_{\bar{\nu}}$ O.K. ; R_ν low, outside bounds
0.22 0.21	0.33 (Run1) 0.43 (Run2)	Cal Tech-FNAL	"Raw data" - no errors quoted yet. y-distributions in Run 2 compatible with SU(2) x U(1).

Now let us turn to Exclusive channels.

(a) νp Scattering

Define,

$$R_{el} = \frac{\sigma(\nu_\mu + p \to \nu_\mu + p)}{\sigma(\nu_\mu + p \to \mu^- + p)}$$

To get a simple formula, let us approximate Rel by the ratio of $\frac{d\sigma}{dk^2}$'s at zero momentum transfer, $k^2 = 0$. Then

$$R_{el} \simeq \tfrac{1}{4} \; \frac{(1 - 4x_W)^2 + g_A^2}{1 + g_A^2}$$

where g_A is the usual axial vector constant ($\simeq 1.24$) and the ¼ comes from the isospin Clebsch-Gordon coefficient. One gets

$$0.15 \leq R_{el} \leq 0.25 \qquad \text{for } x_W \leq 0.5$$

Here at present (Fall 1975) there is only one published experiment by Cundy et al. at CERN, which is very old (done in 1970) and should not be taken as a measurement but only as a two standard deviation upper limit. They give

$$R_{el} = 0.12 \pm 0.06 \quad .$$

A good new measurement is underway now by a Harvard-Penn collaboration at Brookhaven.

(b) <u>Weak π Production</u>

This is the simplest inelastic channel. π^0 production turns out to be particularly interesting mainly because it decays into photons and photons give a good signature, whereas the charged pions, in the absence of magnetic fields, get readily confused with protons. Hence, the focus in these experiments so far has been on the π^0 channel.

If we define,

$$R_0 \equiv \frac{\sigma(\nu_\mu + p \to \nu_\mu + p + \pi^0) + \sigma(\nu_\mu + n \to \nu_\mu + n + \pi^0)}{2\sigma(\nu_\mu + n \to \mu^- + p + \pi^0)}$$

then we find the predictions shown in Table 4. In the W-S model there is a big isovector component and so one expects $\Delta(1236)$ production to be prominent.

Table 4 : Theoretical Values of R_o

	Δ(1236) only	Δ(1236) + non-resonant background	x_W
R_o	0.56	0.40	0.3
	0.46	0.33	0.4

Here the relevant experiments are done on light-to-medium nuclear targets. For example, Brookhaven spark chamber experiment uses three-quarters ${}_{13}A^{27}$ and one-quarter ${}_{6}C^{12}$, and CERN Gargamelle uses freon (CF_3Br) as targets. So, we define

$$R'_o = \frac{\sigma(\nu_\mu + T^A \rightarrow \nu_\mu + T' + \pi^o)}{2\sigma(\nu_\mu + T^A \rightarrow \mu^- + T'' + \pi^o)} ,$$

which is the comparable ratio but with nuclear targets. One observes an event with a π^o and no μ's , T^A is the nuclear target and T' and T" denote any nuclear final state (including breakup).

A simple semi-classical model for charge-exchange effects, which occur, when the produced pion rescatters inside the nucleus, gives a substantial reduction in the predicted ratio, as shown in Table 5. It is principally due to the fact that Δ^{++} production via the charged current is very big and even a 20% probability of a π^+ (from the Δ^{++} decay) converting into a π^o doubles this denominator. This was first pointed out by Perkins and we have done quantitative calculations which verify it.

Table 5 : Predictions for R'_o

	$\Delta(1236)$ + non-resonant background + charge Exchange effects	x_W
R'_o	0.23	0.3
	0.18	0.4

The experiments as of Fall 1975 are given in Table 6.

Table 6 : Experimental Values for R'_o

R'_o	Group
0.17 ± 0.04	Columbia-Illinois-Rockefeller (BNL)
0.15 ± 0.05	CERN (Gargamelle)

Thus, the neutral current model is O.K. and gives the right-order of magnitude value of R'_o . Now, the next important qualitative question which the experiments are focussing on is this: Is the $\Delta(1236)$ excited by the neutral current, as expected in the $SU(2) \otimes U(1)$ model? Here the data is very preliminary at present and I have not yet done a computer analysis. By eye, it seems that the BNL and CERN data show ~ 1 standard deviation excess in the Δ region for the incident neutrino's, and a possible larger excess (but with poorer statistics) in the CERN $\bar{\nu}$ data. Statistics are not good enough yet. The neutrino distributions are equally consistent with phase space at this point. Very much more statistics is needed to settle this question.

Conclusion:

$SU(2) \otimes U_1$ model is O.K. so far for $x_W \sim (0.3 \text{ to } 0.4)$.

Much hard experimental work is needed to nail down the neutral current structure and even to settle some simple qualitative questions like what the level of the Δ-excitation is.

3. Intermediate Bosons.

The most important diagram for the production of intermediate bosons (from neutrinos) is shown in Fig. 3.

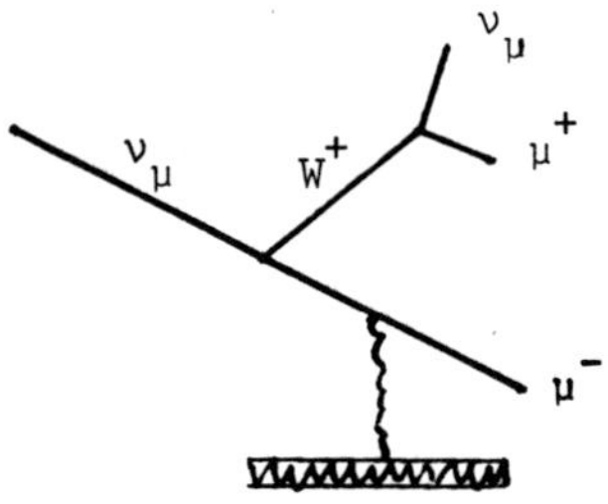

Fig. 3. Intermediate vector boson production

The main characteristic of this diagram is that the Weizsacker-Williams approximation is valid for it: the photon wants to have low invariant $(\text{mass})^2$ and it also tends to transfer very little energy to the hadrons. So, the primary signature of these events is:

(i) Little or no hadronic energy, i.e., "quiet" events.

(ii) Have μ^+ and μ^- in the final state, i.e., dimuons.

(iii) For ν_μ incident, μ^+ has most of the visible energy.

Based on the above we can immediately conclude that the NAL dimuons are not from intermediate bosons. The NAL dimuons are "noisy" events. In intermediate boson models we find typically $\sim$ 1 GeV going into hadrons whereas in the NAL dimuon events typically 20-40 GeV goes into hadrons.

Also, for the NAL events the μ^- has most of the visible energy. Now one can construct funny scalar intermediate bosons where the μ^- has most of the energy but again even those get killed just on the hadronic energy criterion.

What are the bounds on the intermediate boson mass? There are two ways to estimate this for the Cal Tech-FNAL experiment. The first is to use the fact that of 1500 ν events, there are no W^+ events: that is to say there are no events which are noisy and for which the μ^+ has most of the visible energy. This gives us

$$M_{W^+} \gtrsim 8 \text{ GeV} , \quad \text{at } 90\% \text{ CL}$$

assuming a leptonic branding ratio

$$\text{B.R.} = \frac{\Gamma(W^+ \to \mu^+\nu)}{\Gamma(W^+ \to \text{all})} = 0.25 \quad .$$

Another way is to assume scaling (and that of course may not be valid) and fit the charged current ν_μ inclusive data to

$$\frac{d\sigma}{d\,x\,dy} \simeq F_2(x)\,\frac{1}{(1 - k^2/\Lambda^2)^2}$$

This is not a good way to study scaling breakdown because in asymptotically free theories this is not the way deviations from scaling occur. But this is just the way the intermediate boson would act, so such fits are still relevant. With the 1027 ν events we find

$$\text{i.e.} \quad \begin{aligned} \Lambda &> 10.3 \text{ GeV} \\ M_W &> 10.3 \text{ GeV} \end{aligned} \qquad (95\% \text{ CL})$$

Of course, in SU(2) ⊗ U(1) model, we expect

$$\left.\begin{aligned} M_{W^\pm} &= \frac{37 \text{ GeV}}{\text{Sin } \theta_w} \approx 63 \text{ GeV} \\ M_{Z^o} &= \frac{37 \text{ GeV}}{\text{Sin } \theta_w \text{ Cos } \theta_w} \approx 78 \text{ GeV} \end{aligned}\right] \quad \text{for } x_w \approx 0.35$$

So, these bounds are no problem and direct observation is a long way off - at least not until Isabelle is built.

4. Phenomena associated with "charm" production

In the Glashow-Illiopoulos-Maiani (GIM) mechanism, we have 4 quarks:

		p	n	λ	p'
		u	d	s	c
charge	Q	$\frac{2}{3}$	$-\frac{1}{3}$	$-\frac{1}{3}$	$\frac{2}{3}$
strangeness	S	0	0	- 1	0
charm	C	0	0	0	1

The usual Cabibbo weak current

$$J^\lambda_{ch} = \text{Cos}\theta_c \; \bar{u}\gamma^\lambda(1 - \gamma_5)d + \text{Sin}\theta_c \; \bar{u}\gamma^\lambda(1 - \gamma_5)S \; ,$$

picks up an extra piece

$$\Delta J^\lambda_{ch} = - \text{Sin}\theta_c \; \bar{c}\gamma^\lambda(1 - \gamma_5)d + \text{Cos}\theta_c \; \bar{c}\gamma^\lambda(1 - \gamma_5)S \; .$$

Hence, charged current neutrino physics is affected.

(a) Charm production will produce apparent violations of the $\Delta S/\Delta Q$ selection rule. For, schematically we have

Production: $\nu_\mu + d \rightarrow \mu^- + c$. This would be proportional to $\sin^2\theta_c$ and hence a small (a few percent) effect.

Decay: $c \rightarrow s + u + \bar{d}$. This is a non-leptonic decay and is proportional to $\cos^2\theta_c$ and hence not suppressed at all. Overall therefore, we have

$$\nu_\mu + d \rightarrow \mu^- + \underbrace{s + u + \bar{d}}$$

$$\begin{array}{ccc} \uparrow & \uparrow & \\ Q = -\frac{1}{3} & Q = \frac{2}{3} & \Delta Q = 1 \\ S = 0 & S = -1 & \Delta S = -1 \end{array}$$

Hence

$$\left[\frac{\Delta S}{\Delta Q}\right]_{\text{apparent}} = -1 \quad .$$

There is one (incredible) candidate of this type seen by the Samios group at BNL:

$$\nu_\mu + p \rightarrow \mu^- + \Lambda^0 + \pi^+ + \pi^+ + \pi^+ + \pi^- \quad .$$

They had good luck and all the tracks can be fit here. There is an ambiguity as to what is a μ^- and what is a π^- . If you assume that the μ^- track has been correctly identified, they find for the mass of the hadronic part

$$M(\Lambda\pi^+\pi^+\pi^+\pi^-) = 2426 \pm 12 \text{ MeV} \quad ,$$

which could be the mass of a possible charmed baryonic state.

There are non-charmed explanations - principally due to a missing forward K_L^o. These have a probability of about 3×10^{-5}.

It is feared that there is not a commensurate number of candidates of this type at FNAL. This suggests that the Samios event, if correctly interpreted, represents a very improbable event itself - both the energy and the momentum transfer are in the tail of the BNL neutrino distributions. So, we cannot reliably extrapolate rates from it because it is a statistical fluctuation.

(b) Charm production with subsequent leptonic decay will produce dimuons.

Production

This will go two ways

(i) $$\nu_\mu + d \to \mu^- + c \quad .$$

This is from the "valence" quarks, present for large x, and is proportional to $\mathrm{Sin}^2\theta_c$.

(ii) $$\nu_\mu + s \to \mu^- + c$$

$$\bar{\nu}_\mu + \bar{s} \to \mu^+ + \bar{c} \quad .$$

This is from the s or $\bar{s}$ quarks which are "sea" quarks (with $x \leq 0.1$) and is proportional to $\mathrm{Cos}^2\theta_c$.

Decay (leptonic)

Will proceed as

$$c \to s + \mu^+ + \nu_\mu$$

$$\bar{c} \to \bar{s} + \mu^- + \bar{\nu}_\mu$$

Hence, we expect

$$\nu_\mu \to \mu^- + \mu^+ + --- \quad ,$$

Here ν_μ produces dimuons both via "valence" as well as "sea" quarks, whereas

$$\bar{\nu}_\mu \to \mu^+ + \mu^- + --- \quad ,$$

would proceed only via "sea" quarks. This is because we do not have $\bar{d}$ as a valence quark. (The antiquark content in the nucleon is very small at large x).

Now, we have already seen that the dimuon characteristics are inconsistent with either the weak boson or the heavy lepton interpretations. In the charm model, events are "noisy" - the action occurs at the hadronic vertex - and we can get $\frac{< E_{\mu^-} >}{< E_{\mu^+} >}$ big for incident neutrinos. The models are very non-specific, since there are no definite Feynman diagrams to compute here, but there is no inconsistency.

In the Cal Tech-FNAL experiment for $\nu_\mu \to \mu^+ + \mu^- + ---$, since they have a monoenergetic ν beam, they can actually measure x (determined by the ν_μ , μ^- kinematics) and it looks like the usual deep inelastic x-distribution:

$$< x >_{obs} = 0.21 \quad \text{for the dimuon events}$$

$$< x >_{obs} = 0.24 \quad \text{for all charged current events,}$$

which is consistent with "valence" quark production being predominant for the ν dimuons. Ben Lee will have more to say about dimuons.

(c) GIM charm will lead to large charge symmetry violations.

At $\theta_c = 0$, the weak charged current

$$J^\lambda_{ch} = \bar{u}\gamma^\lambda(1 - \gamma_5)d$$

and its adjoint

$$(J^{\lambda}_{ch})^{+} = \bar{d}\gamma^{\lambda}(1 - \gamma_5)u \quad ,$$

are in the same I = 1 multiplet and are related by a rotation about the 2-axis in the isospin space

$$e^{-i\pi I_2} J^{\lambda}_{ch} e^{+i\pi I_2} = -(J^{\lambda}_{ch})^{+} \quad .$$

But also at $\theta_c = 0$, the extra charm piece

$$\Delta J^{\lambda}_{ch} = \bar{c}\gamma^{\lambda}(1 - \gamma_5)s \quad ,$$

is an isoscalar and hence goes into itself under rotation

$$e^{-i\pi I_2} \Delta J^{\lambda}_{ch} e^{+i\pi I_2} = \Delta J^{\lambda}_{ch} \neq -(\Delta J^{\lambda}_{ch})^{+}$$

Thus, above charm threshold, charge symmetry relations, such as

$$W_i^{\nu N} = W_i^{\bar{\nu} N} \quad , \quad N = \tfrac{1}{2}(n + p) \quad ,$$

may be strongly violated. One gets the same result even when $\theta_c \neq 0$, but I have set $\theta_c = 0$ to emphasize that this effect is there even without the Cabibbo rotation and hence is a big violation.

(d) Something about which I have nothing specific to say is that charm thresholds may lead to temporary scaling violations in deep inelastic neutrino reactions. Presumably, scaling will reappear at energies sufficiently far beyond thresholds. A bit more on this later, when I talk about scaling and asymptotic freedom.

(e) The presence of charm changes neutrino reaction sum rules.

The sum rules are typically derived by considering commutators of the type

$$[J^{o}_{ch}(\vec{x},o)\ ,\ J^{o+}_{ch}(\vec{y},o)] \quad ,$$

which is no longer equal to

$$[J^{o}_{ch}(\vec{x},o) + \Delta J^{o}_{ch}(\vec{x},o)\ ,\ J^{o+}_{ch}(\vec{y},o) + \Delta J^{o+}_{ch}(\vec{y},o)] \quad ,$$

and hence the relevant sum rules are changed. E.g., the Adler sum rule

$$\frac{1}{M_N^2}\int_o^\infty d\nu\ [W_2^{\bar{\nu}} - W_2^{\nu}] = <4\mathrm{Cos}^2\theta_c I_3 + (3Y + 2I_3)\mathrm{Sin}^2\theta_c> \qquad \underline{\text{3 Quark-Model}}$$

gets modified to simply

$$= <4I_3> \qquad \underline{\text{GIM}}$$

Similarly, in the Gross-Llewellyn-Smith sum rule

$$\lim_{k^2\to-\infty}\int_{-k^2/2}^{\infty} d\nu \left[\frac{k^2}{2M_N^2}\right][W_3^{\bar{\nu}}+W_3^{\nu}] = <4B+Y(2-3\mathrm{Sin}^2\theta_c)+2I_3\mathrm{Sin}^2\theta_c> \qquad \underline{\text{3 Quark-Model}}$$

changes to

$$= <4B + 2Y> \qquad \underline{\text{GIM}}$$

These differences have been numerically tabulated by Beg and Zee. For an Iron target:

	Adler Sum Rule	GL-S
3 Quark Model	- 1.12	330
GIM	- 8	336

Present sum rule tests are all extrapolations up from the low energy (Gargamelle) data. They indicate consistency with GL-S and provide no test of the Adler sum rule. Since they are extrapolations up, they obviously say nothing about the possibility of a jump at some threshold - any more than extrapolating R in SPEAR up from $\sqrt{s}$ = 2 GeV, would have told that one would see something interesting at 3.1 GeV. So, there is no test here.

B. Gauge Theories of Strong Interactions.

I will now discuss "asymptotically free" strong interaction theories, which were invented to help understand scaling and which predict specific forms for scaling violations.

1. Scaling

Let me begin with a quick review of scaling and evidence for scaling in neutrino physics

$$\left.\begin{matrix} \nu(k_1) \\ \bar{\nu}(k_1) \end{matrix}\right] + N(p) \rightarrow \mu(k_2) + X$$

The basic kinematical variables are $k = k_1 - k_2$, $\nu = k \cdot p$ and the momentum transfer $-k^2$. The usual scaling variables are

$$x = -\frac{k^2}{2\nu} \qquad\qquad o \le x,y \le 1 \quad .$$
$$y = \frac{\nu}{M_N E}$$

Before scaling is assumed the most general form of the differential cross-section is given by

$$\frac{d^2\sigma^{\nu,\bar{\nu}}}{dxdy} = \frac{G_F^2 M_N E}{\pi}\Big[(1 - y - \tfrac{1}{2}xy\frac{M_N}{E})G_2^{\nu,\bar{\nu}}(x,k^2/M_N^2)$$
$$+ xy^2 G_1^{\nu,\bar{\nu}}(x,k^2/M_N^2) \mp xy(1 - \tfrac{1}{2}y)G_3^{\nu,\bar{\nu}}(x,k^2/M_N^2)\Big] \quad .$$

In the scaling limit, $G_i(x,k^2/M_N^2) \to F_i(x)$, so that we have

$$\frac{d^2\sigma^{\nu,\bar{\nu}}}{dxdy} = \frac{G_F^2 M_N E}{\pi}\Big[(1-y)\, F_2^{\nu,\bar{\nu}}\ (x) + xy^2\ F_1^{\nu,\bar{\nu}}\ (x) \mp xy$$
$$(1-\tfrac{1}{2}y)\, F_3^{\nu,\bar{\nu}}\ (x)\Big]$$

Obviously, if we integrate over x and y, we predict

$$\sigma^{\nu,\bar{\nu}} = {}_{\nu,\bar{\nu}}\, E,$$

i.e., a linearly rising cross-section.

Cal Tech - FNAL now have first normalized cross-sections from 0 to about 100 GeV, which show no obvious deviations from scaling at the 20% level. Data at about the 10% level are now being taken from 30 to 250 GeV and should be available soon.

What they have on linearity, is the following:

From Cal Tech - FNAL, with data taken at two points, one at about 40 GeV and one at about 100 GeV, the sum of the two proportionality constants is:

$$\alpha_\nu + \alpha_{\bar{\nu}} = (1.11 \pm 0.12) \times 10^{-38}\ \mathrm{cm}^2 / \mathrm{GeV},$$

to be compared with the Gargamelle value

$$\alpha_\nu + \alpha_{\bar{\nu}} = (1.02 \pm 0.05) \times 10^{-38}\ \mathrm{cm}^2 / \mathrm{GeV},$$

which is lower energy data taken between 1 to 10 GeV. The same straight line which goes through Gargamelle essentially goes through the Cal Tech - FNAL data. The ratio $\sigma^{\bar{\nu}} / \sigma^{\nu}$ is given to be

Cal Tech - FNAL : $\alpha_{\bar{\nu}} / \alpha_\nu = 0.33 \pm .08$

Gargamelle : $\alpha_{\bar{\nu}} / \alpha_\nu = 0.38 \pm 0.02$,

so the 1/3 ratio, within larger errors, still seems to be holding up.

Also, Cal Tech - FNAL find <u>no evidence</u> for the anomaly in y - distribution reported in FNAL Expt. 1A. In the simple quark picture, if we put in the anti-quark density, albeit small, in the nucleon, we expect:

$$\frac{d^2\sigma^\nu}{dxdy} = \frac{G_F^2 M_N E}{\pi}\left[q(x) + (1-y)^2 \bar{q}(x) \right]$$

$$\frac{d^2\sigma^{\bar{\nu}}}{dxdy} = \frac{G_F^2 M_N E}{\pi}\left[\bar{q}(x) + (1-y)^2 q(x) \right],$$

where $q(x)$, $\bar{q}(x)$ denote the quark and anti-quark parton densities from the Gargamelle experiment. We expect the shapes of $q(x)$ and $\bar{q}(x)$ to look as follows:

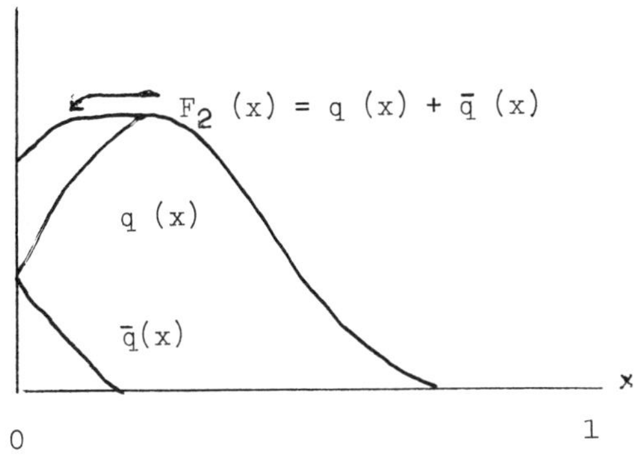

$\bar{q}(x)$ is small beyond x of 0.1 to 0.2 and at $x = 0$ is equal to $q(o)$ by the Pomeranchuk Theorem. So, for small x, where the anomaly was asserted to occur, the $\bar{q}$ component is important in discussing the phenomenology of the y- distributions.

The Cal Tech -FNAL fit for $x < 0.1$ is

$$\frac{d\sigma^\nu}{dy} = \frac{G_F^2 M_N E}{\pi}\left[1 + 0.3(1-y)^2 \right]$$

$$\frac{d\sigma^{\bar{\nu}}}{dy} = \frac{G_F^2 M_N E}{\pi}\left[0.3 + (1-y)^2 \right]$$

The anti-quark to quark ratio (for small x) is given by

$$\frac{\bar{Q}}{Q} = \frac{\int_0^{0.1} dx\ \bar{q}(x)}{\int_0^{0.1} dx\ q\ (x)} = 0.3 \text{ in Cal Tech - FNAL fit.}$$

The ratio of the differential cross-sections at y = 1 just measures this ratio.

$d\sigma^{\bar{\nu}} \Big/ d\sigma^{\nu}$ at y = 1 :

Value	Group
~0.29	Gargamelle
~ 0.3	Cal Tech - FNAL
~ 0.36	FNAL Expt. 1A

Thus all three experiments agree in small x, large y region with a simple parton picture. So, if there is an anomaly, it is a small y anomaly and it is here that Cal Tech - FNAL and FNAL Expt. 1A disagree. Expt. 1A finds that they do not fit with what is expected - even though I think that they did not put $\bar{q}$ (x) distributions in in such detail. Cal Tech - FNAL fit all their data (small y too) - within rather large errors - with theoretically reasonable q (x), $\bar{q}(x)$ which add up to $F_2(x)$ and similar to those inferred from Gargamelle.

Moral: No clear evidence exists yet for threshold effects. Perhaps models for dimuon production should not try to simultaneously produce a y-anomaly.

2. Asymptotic Freedom and Scaling Breakdown

We have seen so far that, within 20%, in neutrino physics scaling is in good shape. All interacting field theories predict scaling breakdown. The way to discuss scaling breakdown is to form the moments:

$$\int_0^1 dx\, x^{n-2}\, G_2\,(x,\, k^2/M_n^{\;2}) = \mathcal{M}_n\,(k^2/M_n^{\;2})$$

We have three interesting alternatives:

i) Exact scaling: $$\lim_{-k^2\to\infty} \mathcal{M}_n(k^2/M_n^{\;2}) = \int_0^1 dx\, x^{n-2}\, F_2\,(x)$$

ii) Anomalous dimension breakdown: $$\lim_{-k^2\to\infty} \mathcal{M}\,(k^2/M_n^{\;2}) \sim \text{Const.}(-k^2)^{-\frac{1}{2}\gamma_n}$$

iii) Asymptotic Freedom breakdown: $$\lim_{-k^2\to\infty} \mathcal{M}_n\,(k^2/M_n^{\;2}) \sim \text{Const.}\left[\ln(-k^2)\right]^{-\frac{1}{2}\gamma_n}$$

Form (2) is given in the general garden variety field theory, and γ_n is the "anomalous dimension". In general, one needs all orders of perturbation theory here to determine γ_n and so it is not known. Form (3) is obtained in non-abelian gauge field theories with certain restrictions on them. Here γ_n is determined by the underlying field theory in low order, so one knows it explicitly.

In general, for both case (2) and (3) γ_n is monotonically increasing with n and γ_2 is zero or small from the conservation of the energy-momentum tensor. We get the following qualitative picture, which is similar for both (2) and (3), with basically a difference of k^2 - scale:

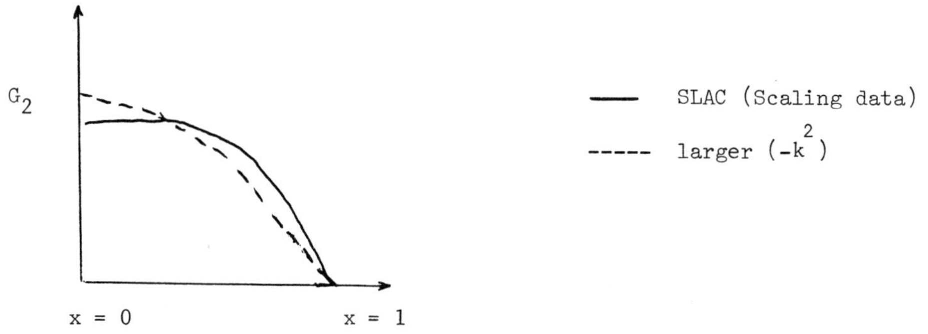

The curve (from either (2) or (3)) decreases near x = 1 (with respect to the scaling curve) to make the higher moments decrease with increasing $-k^2$, because obviously the moments are weighted most by x near 1. But then it has to increase near x = 0 to keep the area $\approx$ constant, so we have the above picture.

An alternative way of describing this is to draw G_2 vs. k^2 and plot a family of curves vs. x. (This has been discussed in detail by Wu-Ki Tung in PRL):

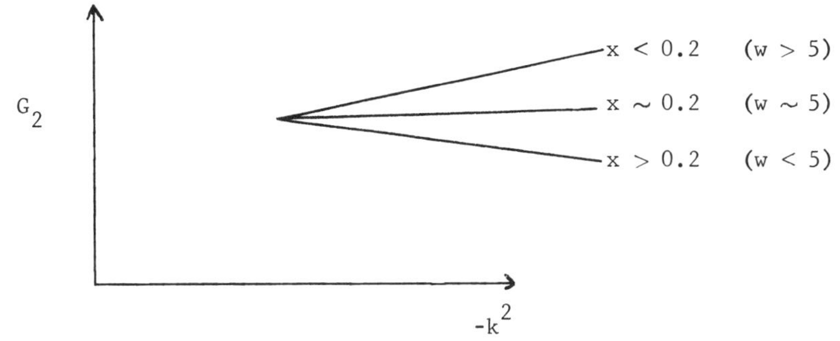

For $x \sim 0.2$, the curve is more or less flat - that is the cross-over point indicated in the earlier diagram. Then, for $x < 0.2$ as $(-k^2)$ increases, the curve has to rise to keep the area constant; for $x > 0.2$ $(w < 5)$ the curve has to fall to make the moments go down.

The new data from FNAL apparently do indicate such a trend.

There are problems however in establishing which is the correct pattern of scaling breakdown:

(a) Distinguishing $[\ln(-k^2)]^{\text{power}}$ from $[-k^2]^{\text{smaller power}}$ is very hard and will require a lot of statistics.

(b) We will have to worry about complications from charm-associated threshold effects.

This is an obvious problem in neutrino experiments, (where we have seen charm produces changes in sum rules) but may also be a problem in e and μ induced reactions. The EM current picks up an extra piece from charm

$$\Delta J^{\lambda}_{EM} = \bar{c}\, \gamma^{\lambda}\, c \quad ,$$

which we know from SPEAR has electromagnetic effects! From SPEAR, the value of $R = \sigma^{had}/\sigma(\mu^+\mu^-)$ jumps at $k^2 \sim 16$ GeV on the time-like side. There are also a lot of fine-structure effects, but the value of R changes from about 2 to about 5 in a rather narrow interval. So, if we use dispersion relations then the hadronic vacuum polarization must also correspondingly change in its asymptotic form on the space-like side, somewhere between $k^2 = -16\ \mathrm{GeV}^2$ and $k^2 \sim -50\ \mathrm{GeV}^2$. It is spread out, because one is doing an integral, but there has to be a change because a constant R eventually forces the space-like asymptotic form of the hadronic vacuum polarization to have the same constant R. Now, if we have space-like effects in one process, it is not at all obvious why we should not have them in others - e.g., in deep inelastic scattering. At NAL people are looking just in the range $k^2 \sim -16$ to $-50\ \mathrm{GeV}^2$ for scaling violations. I think that this is a problem in addition to just the statistical one, which may be apparent in scaling violations and may even appear in purely hadronic effects like rising total cross-sections. The issue can be stated without reference to charm. <u>If there are two hadronic mass scales (and we know now that there are) - a small scale and a large scale - how do we disentangle asymptotic phenomena associated with the first scale from onset phenomena associated with the second?</u>

DISCUSSION

S. WEINBERG (Harvard)

You have discussed the SU(2) x U(1) model within the context of only having a left-handed doublet. Do you have any comments about the status or viability of the pure V neutral current presently?

ADLER

I think that even pure V would be allowed now because the error bars in the data are big. But I would defer an answer till Barish puts error bars on his data. Barish says that if he analyzes with pure V, his y-distribution is changed but he won't yet say if it is ruled out. Again in pion-production, the error bars are too big to say anything definite. I gave you R_o for ν and for ν (Table 4) Brookhaven has a lot of events but their analysis has just begun. Let me make one other comment. Pure V theories make less ambiguous predictions about the N^*_{33} production than the V-A theories do and the reason is that there are certain ambiguities which I haven't discussed. These have to do with the first order corrections to the soft-pion theorems for the axial-π production. If one simply adds them as polynomials, they can have a rather large effect (even though not as large as Wolfenstein showed, because I found that I had not taken out effects of unitarizing the 3,3 multipoles in the original graph of my paper) and when you subtract out and don't double count, there is a sort of 50% ambiguity if you add them as polynomials. As is known to a number of people, one should add things in the hard pion form with proper propagators. That is still model dependent, but that is the sort of analysis which has to be done. There isn't much of an ambiguity associated with the vector part. There are not any corrections until you get order q·k, where q is the pion 4-momentum and k is the virtual 4-momentum. The predictions you get from vector-like theory are that there should be a clear, unambiguous 3-3 resonance. You could start doing things when an axial current is present to wash it out a bit.

P. MINKOWSKI (Caltech)

An updated value for inclusive $\left(\frac{\sigma_{\bar{\nu}h \to \bar{\nu}h}}{\sigma_{\bar{\nu}h \to \bar{\mu}h}} = 0.55 \pm 0.09\right)$ was reported as a preliminary result by Morphin from the CERN Gargamelle experiment at the Stanford Conference.

ADLER

One thing is clear, it will be a very long story before these numbers settle down. We have got our phenomenological programs on the shelf for a year.

B. WARD (Purdue)

You consider the S, T, P cases to be remote?

ADLER

Yes, there is no evidence requiring S, T, P now that the Argonne events have disappeared.

DIMUON EVENTS

Benjamin W. Lee*

*Fermi National Accelerator Laboratory, Batavia, Ill. 60510.

1. Introduction

I planned to talk about the role of anticommuting symmetry transformation in gauge theories.[1] However, I consider the recent discovery of dimuon events[2] in neutrino interactions so momentous that I should report on my understanding of these events, and discuss a preliminary interpretation. No doubt the understanding of this new phenomenon will have a profound impact on the future development of gauge theory of particle interactions and model making.

I shall first describe the reasons why I believe these events represent a new phenomenon, and I shall indulge in a theoretical interpretation on them based on the minimal gauge theory. Experimental data I shall present to you were provided to me by Professor David Cline of the Harvard-Pennsylvania-Wisconsin-Fermilab collaboration.

Table 1 shows the number of dimuon events observed by HPWF. There are altogether 84 dimuon events observed by this collaboration.

Table 1

Time	Beam	Number of Protons (10^{19})	Number of Dimuon Events
April 1973	400 GeV Bare Target	~0.1	1
Jan. 1974	400 GeV Horn (ν)	0.5	3
July 1974	300 GeV Quadrupole Triplet	0.3	1
Nov. 1974	300 GeV Horn ($\bar{\nu}$)	4.8	11
Feb. 1975	380 GeV Quad. Triplet	7.7	61
April 1975	300 GeV Double Horn ($\bar{\nu}$)	3	7

In addition, the Cal Tech-Fermilab collaboration[3] has observed 4 dimuon events with both muons going through the magnet which is used as a muon spectrometer.

The antineutrino horn beam is an antineutrino-enriched beam. It contains approximately the same number of $\bar{\nu}$'s as ν's. The double-horn antineutrino beam (with a plug) is about 90% pure. The other beams are mostly neutrino beams with about 90% purity.

2. What Is A Dimuon Event?

A typical dimuon event is schematically shown in Fig. 1. This particular event originates in the hadron calorimeter which contains a scintillating material in mineral oil. In this case two muons of opposite signs go through the steel hadron filter and are momentum analyzed in the muon spectrometer.

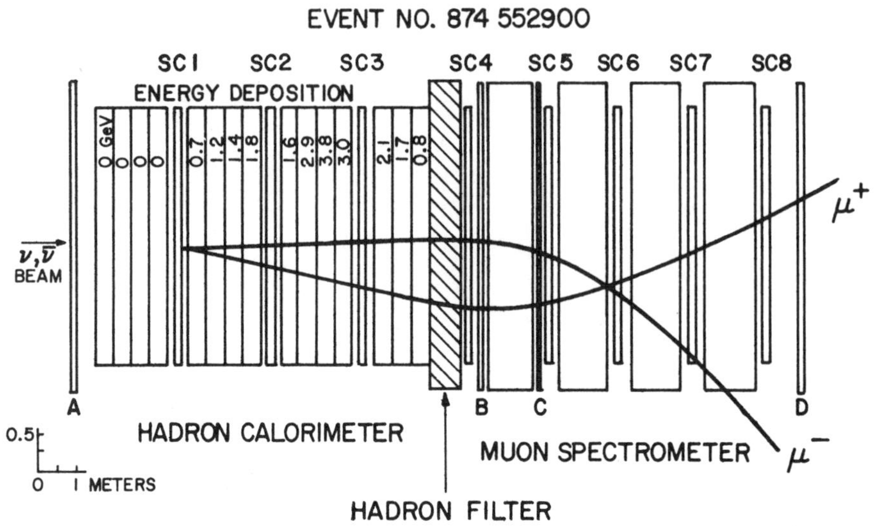

Fig. 1

That the two muons emanate from the same neutrino interaction can be verified by the spatial coincidence of the two muon tracks at the event vertex and the temporal coincidence of the two muon detections.

Figure 2 shows the distribution of diumuon events in the visible energy, E_{vis},

$$E_{vis} = E_H + E_{\mu_1} + E_{\mu_2},$$

where E_H is the hadronic energy deposition in the calorimeter, and $E_{\mu_1} + E_{\mu_2}$ is the sum of muon energies. The rate is proportional to the neutrino event rate, defined as the neutrino flux times the neutrino energy.

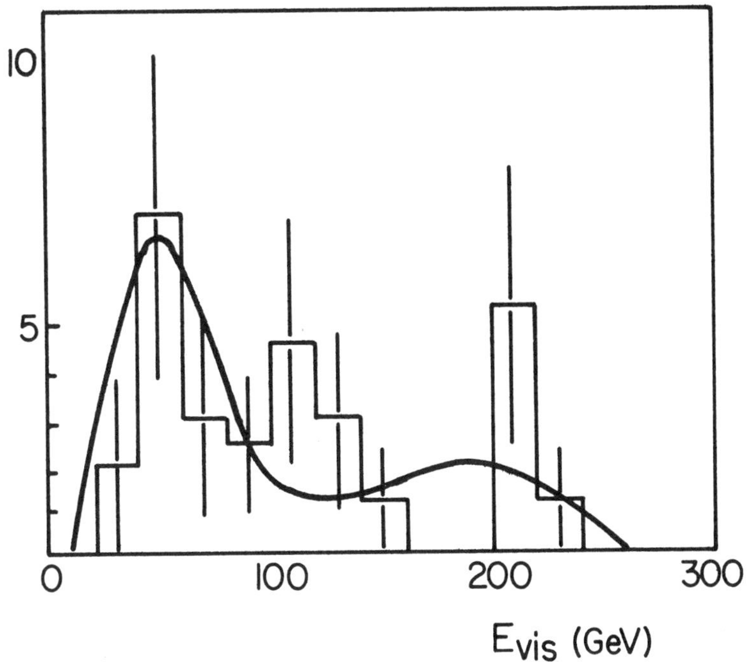

Fig. 2

The basic parameters of the dimuon events are summarized in Table 2. The strongest evidence that the second muon (μ^+ in ν-induced

Table 2

$\dfrac{\sigma(\nu \to \mu^- \mu^+)}{\sigma(\nu \to \mu^-)}$	$\simeq 10^{-2}$
$\dfrac{\sigma(\bar{\nu} \to \mu^+ \mu^-)}{\sigma(\nu \to \mu^- \mu^+)}$	$\simeq 0.8 \pm 0.6$
$\dfrac{\sigma(\nu \to \mu^- \mu^-)}{\sigma(\nu \to \mu^+ \mu^-)}$	$\simeq 0.1$

reactions, for example) does not come from a mundane source, such as from π or K decays, comes from the relative rates of dimuon events per density in the hadron calorimeter and in the hadron filter. The former is mostly carbon, the latter iron. Since the absorption length of pions (and kaons) differs vastly in the two materials, the relative dimuon event rates would differ vastly if the second muon came from pion/kaon decays. Figure 3 shows a more-or-less constancy of the relative event rates at two different absorption lengths.

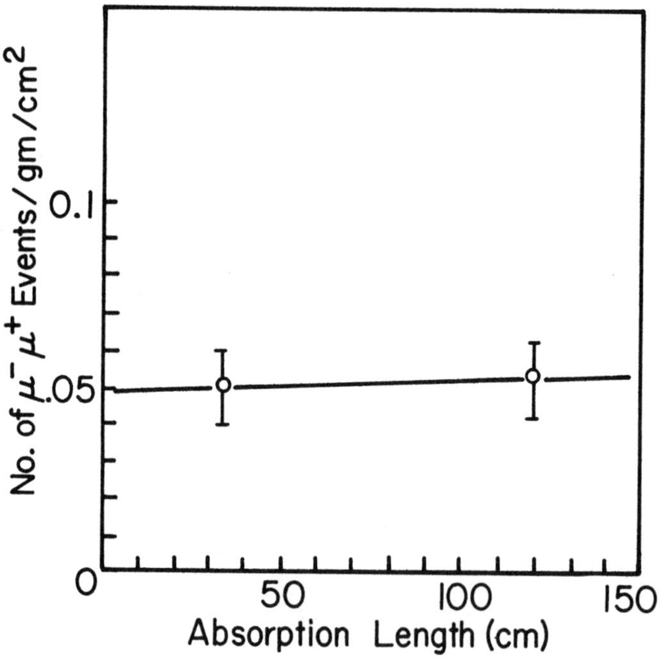

Fig. 3

3. Source of Dimuon Events

The rather large observed rates for dimuon events rule out the possibility that these are four-lepton interactions in nuclear Coulomb field. Figure 4 shows the distribution of ν-induced dimuon events as a function of the μ^- momentum P_- and the μ^+ momentum P_+. The straight line corresponds to $P_+ = P_-$. The numbers attached to events marked by triangles are the hadronic energy depositions for events originating in the hadron calorimeter.

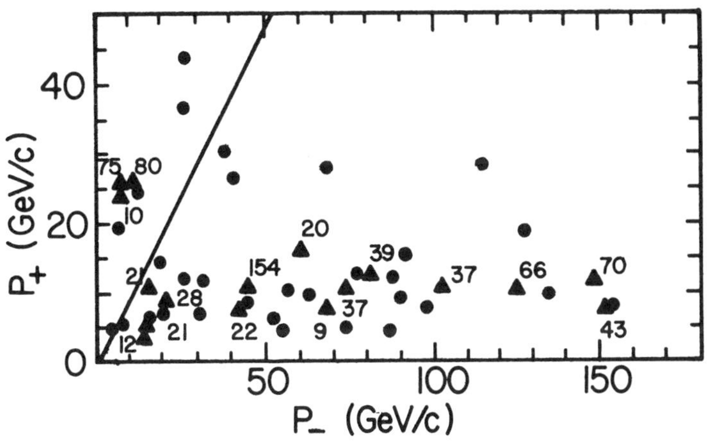

Fig. 4

One notes immediately the preponderance of events in which $P_- > P_+$. This is in opposition to the expectation[4] for the decay of the intermediate boson W^+

$$\nu + N \rightarrow \mu^- + W^+ + N$$
$$W^+ \rightarrow \mu^+ + \nu$$

which is produced in nuclear Coulomb field, in which case $P_+ > P_-$, and the hadronic shower is expected to be "quiet".

These events are not likely to come from decays of neutral heavy leptons that might be produced in ν-induced reactions:[5, 6]

$$L^0 \to \mu^+ + \mu^- + \nu.$$

In fact, Pais and Treiman[7] considered the ratio $\langle P_- \rangle / \langle P_+ \rangle$ assuming that the opposite sign muons have the same parent and the above decay is described by a local interaction (S, P, T, V, and A). Extremizing the ratio with respect to the velocity and polarization of the parent heavy lepton, they obtained the bounds

$$0.48 \leq \langle P_- \rangle / \langle P_+ \rangle \leq 2.1.$$

This ratio for the events shown in Fig. 4 is

$$\langle P_- \rangle / \langle P_+ \rangle = 3.7 \pm 0.7$$

which is well beyond the upper bound. The HPWF group further notes that it is statistically consistent to assume that events with $P_+ > P_-$ are caused by the $\bar{\nu}$ contamination. Excluding these events, they obtain

$$\langle P_- \rangle / \langle P_+ \rangle = 8.5 \pm 1.7.$$

It is therefore very unlikely that all of the dimuon events arise from the decay of a neutral heavy lepton.

Another piece of evidence, perhaps intuitively more appealing, is the dimuon mass distribution as the incident neutrino energy changes.

If the dimuons came from a common parent of well-defined mass, the mass distribution should be independent of the ν-energy. Figure 5 shows however that the dimuon mass distribution tends to be broadened as E_{vis} increases.

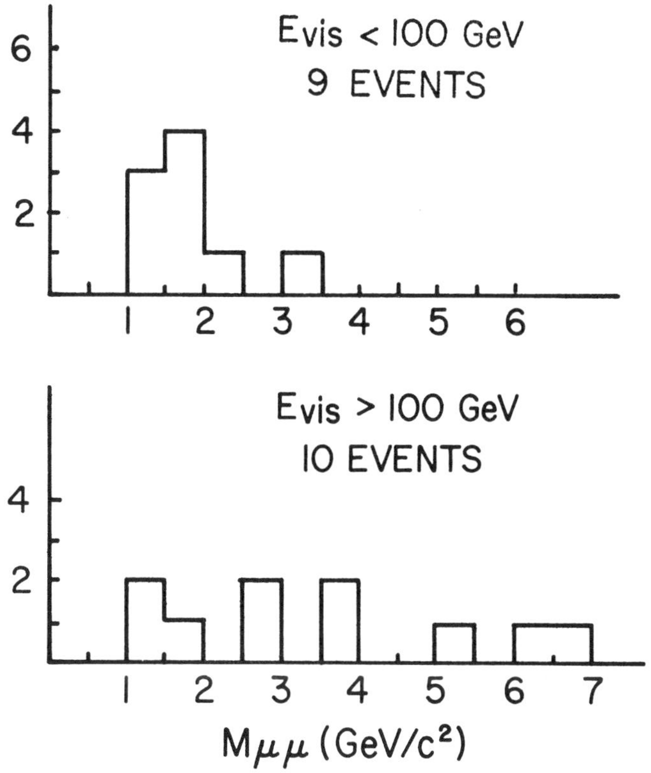

Fig. 5

It is therefore extremely plausible that the extra muon comes from decays of a new particle (or particles) produced at the hadron vertex. To explain the preponderance of opposite-sign dimuons, it is necessary to assume that the new particles carry a new quantum number, which we shall denote by C, and their semileptonic interactions interactions obey the rule $\Delta C = \Delta Q$ in the hadronic sector.

In the minimal model of gauge theory of weak and electromagnetic interactions which is based on the group $SU(2)\times U(1)$[8,9] and incorporates the GIM mechanism[10] with four flavors of quarks, the new particles may be charmed ones. In the deep inelastic region there are various mechanisms for exciting the charm degree of freedom above charm threshold, as we depict in Fig. 6.

We note that the first process, i.e., charm production off valence quarks, is not available for antineutrinos.

4. Issues

In the minimal model interpretation of dimuon events there are three issues we must pay attention to, to understand the gross features discussed in Table 2.

(1) What is the $s\bar{s}$ content of a nucleon?

(2) What is the $c\bar{c}$ content of a nucleon?

(3) What is the inclusive branching ratio of muon-yielding decays of the (generic) charmed particle?

Our knowledge on these matters is not sharp enough to answer these questions definitively. However, it is possible to set reasonable qualitative bounds on these quantities. First, the precise amount of $q\bar{q}$ pairs present in a nucleon is a matter of considerable debate. The

Elementary process	Quark diagram for $\nu + N \rightarrow \mu^- + C + X$	Cabibbo Suppression	Valence or sea quark
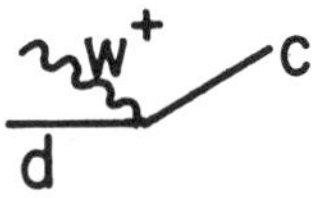		$\sin\theta_C$	valence
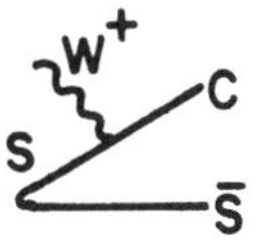	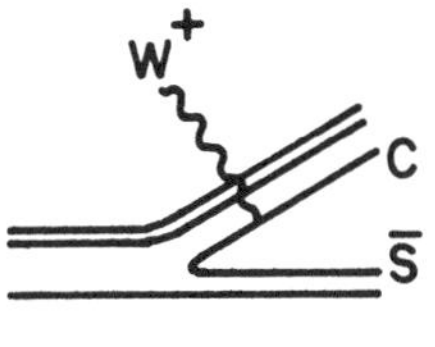	$\cos\theta_C$	sea ($s\bar{s}$)
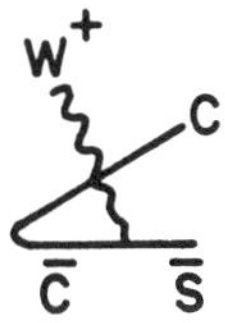	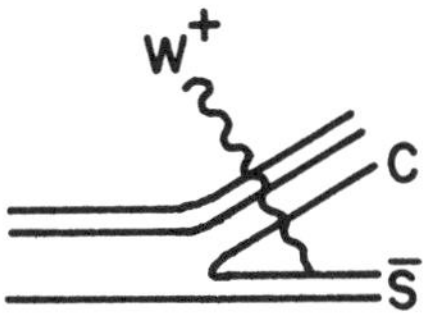	$\cos\theta_C$	sea ($c\bar{c}$)
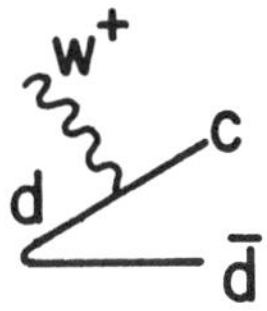	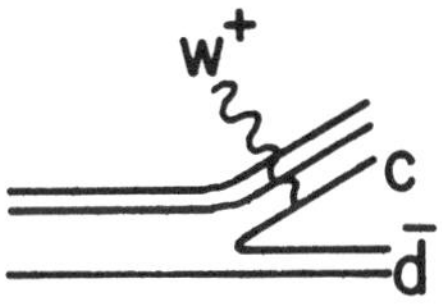	$\sin\theta_C$	sea ($d\bar{d}$)
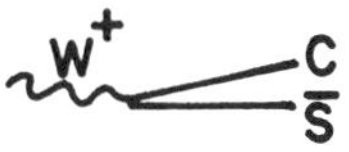	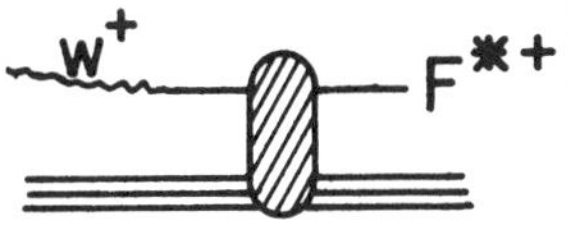	$\cos\theta_C$	diffractive
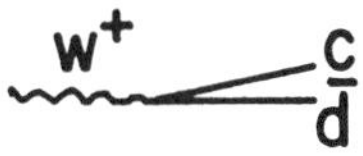	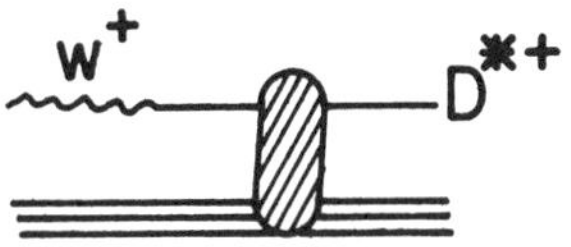	$\sin\theta_C$	diffractive

Fig. 6

$s\bar{s}$ content of a nucleon is limited by our preconception that it should not be larger than the $u\bar{u}$ or $d\bar{d}$ contents. However, say 5-10% contamination of $s\bar{s}$ pairs (in terms of the contribution to the F_2 function) in a nucleon seems reasonable and not contradicted by any known facts. We shall use 5% in the following discussion. As for $c\bar{c}$ pairs, we know much less. In the following discussion we will ignore them.

As for the branching ratio into muon channels, Gaillard, Lee, and Rosner[11] gave an estimate of a few percent based on a naive quark model and the notion that the $\underline{20}$ piece [in SU(4)] of nonleptonic Hamiltonian is enhanced uniformly by the same amount as the octet piece in nonleptonic decays of hyperons and K-mesons. On the other hand if selective enhancement of a particular nonleptonic channel is not operative, then the branching ratio into muon channels may be considerably bigger. In fact, if there is no selective enhancement, and if all ordinary quark masses can be neglected compared to the charmed quark mass, then the above ratio may be estimated by a simple quark counting:

$$c_\alpha \to s_\alpha + (u_\beta + \bar{d}_\beta) \qquad \beta = \text{red, blue, white}$$

$$c_\alpha \to s_\alpha + \mu + \bar{\nu}_\mu$$

$$\to s_\alpha + e + \bar{\nu}_e$$

where α and β are color indices. Thus,

$$\frac{\Gamma(C \to \mu + X)}{\Gamma(C \to \text{all})} = \frac{1}{3+1+1} = 20\%.$$

We consider this as a loose upper bound. We shall use the figure 10% in the following discussion.

5. Consequences - Predictions

Some of the implications of the assumptions made in the last section on dimuon productions have been discussed by Pais and Treiman,[12] Wolfenstein,[13] and Llewellyn-Smith.[14]

One of the most remarkable features of the dimuon events predicted from these assumptions is that there are two components in these events. The small x component, which reflects the sea $s\bar{s}$ content, yields predominantly $S = \pm1$, $C = \pm1$ final states. This component is present both in ν- and $\bar{\nu}$-induced events. The valence component, which arises from the elementary process $W^+ + d \to c$, reflects the valence d-quark distribution, and yields predominantly $S = 0$, $C = +1$ final states. This latter component is present only in ν-induced events.

The ratio of the charm production cross section to the "background" deep inelastic cross section is summarized in Table 3:

Table 3

	ν	$\bar{\nu}$
Small x component $\Delta C = \pm1$, $\Delta S = \pm1$	$\sim 5\%$	$\sim 5\% \times 3$
Valence component $\Delta C = \pm1$, $\Delta S = 0$	$\sim \sin^2\theta_c \sim 5\%$	-
Total	$\dfrac{\sigma(\nu+N\to\mu^-+C+---)}{\sigma(\nu+N\to\mu^-+--)} \simeq 10\%$	$\dfrac{\sigma(\bar{\nu}+N\to\mu^++C+---)}{\sigma(\bar{\nu}+N\to\mu^++--)} \simeq 15\%$

In this table we have used the empirical fact that

$$\sigma(\bar{\nu} + N \to \mu^+ + ---) \simeq \frac{1}{3}\, \sigma(\nu + N \to \mu^- + ---).$$

In this picture the relative rates of dimuon events are given by

$$\frac{\sigma(\nu + N \to \mu^- + C + ---)}{\sigma(\nu + N \to \mu^- + --)} \times \text{B. R. } (C \to \mu^+ + \nu + ---) \simeq 1\%$$

for ν-induced events, and

$$\frac{\sigma(\bar{\nu} + N \to \mu^+ + \bar{C} + ---)}{\sigma(\bar{\nu} + N \to \mu^+ + ---)} \times \text{B. R. } (\bar{C} \to \mu^- + \nu + --) \simeq 1.5\%.$$

Further,

$$\frac{\sigma(\bar{\nu} \to \mu^+ \mu^-)}{\sigma(\nu \to \mu^- \mu^+)} \simeq 0.5\%.$$

<u>The two-component nature of dimuon events is most important in verifying the present interpretation.</u> In Fig. 7 I have sketched the expected x- and y-distributions of dimuon events.

The experimental data bearing on the x, y distributions are shown in Fig. 8. Because incident neutrino energy is not known, x_{vis} and y_{vis} are defined as

$$y_{vis} \equiv \frac{E_H + E_{\mu_2}}{E_H + E_{\mu_1} + E_{\mu_2}} \leq y,$$

$$x_{vis} \equiv \frac{\nu}{y_{vis}} \geq x$$

where $\nu = xy$ can be measured in a flux independent way. Because the data are still statistically poor, I will not draw any conclusions.

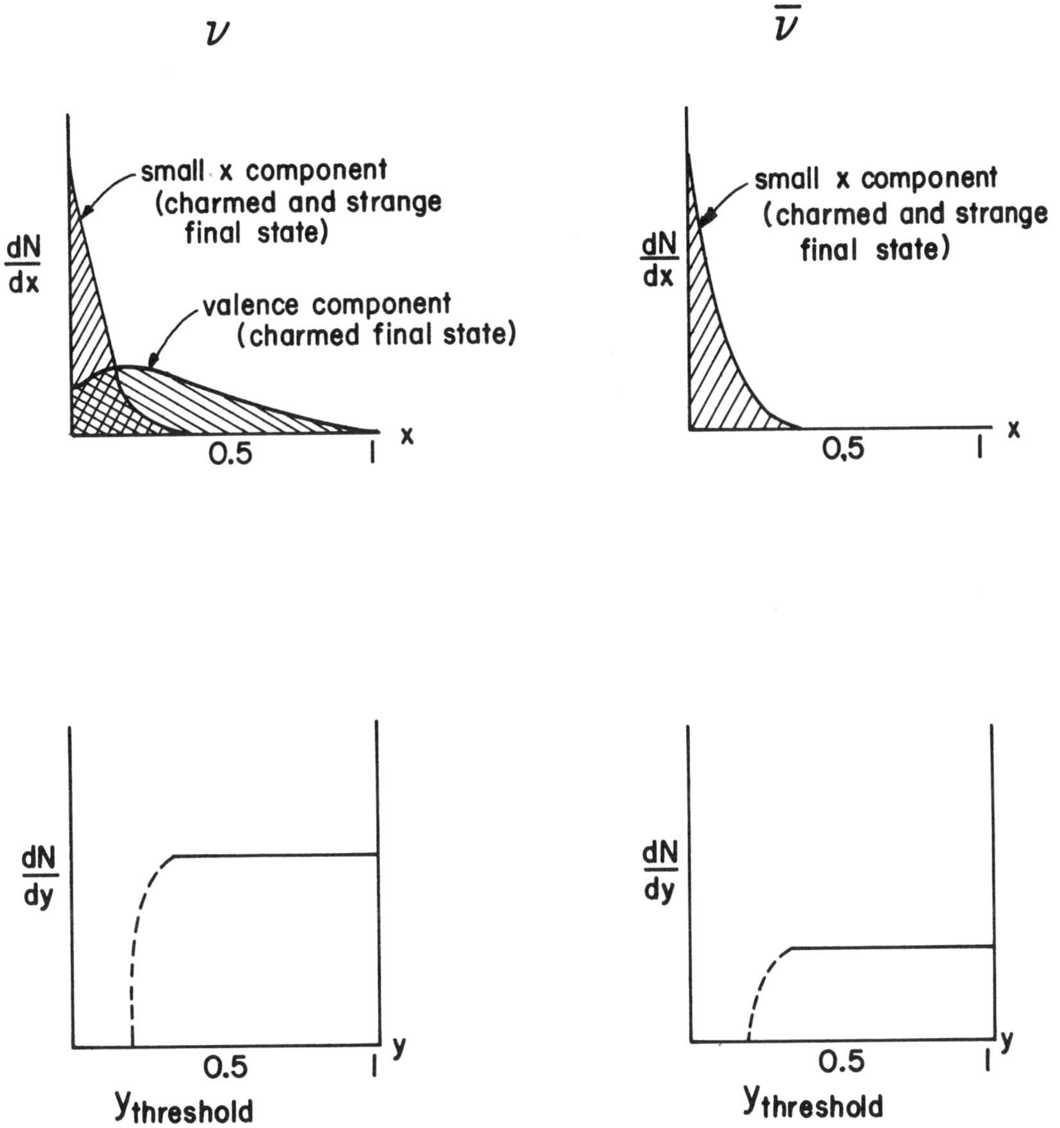

ν
ν̄
dN/dx
small x component
(charmed and strange
final state)
valence component
(charmed final state)
0.5
1
x
dN/dx
small x component
(charmed and strange
final state)
0,5
1
x
dN/dy
0.5
1
y
$y_{threshold}$
dN/dy
0.5
1
y
$y_{threshold}$

Fig. 7

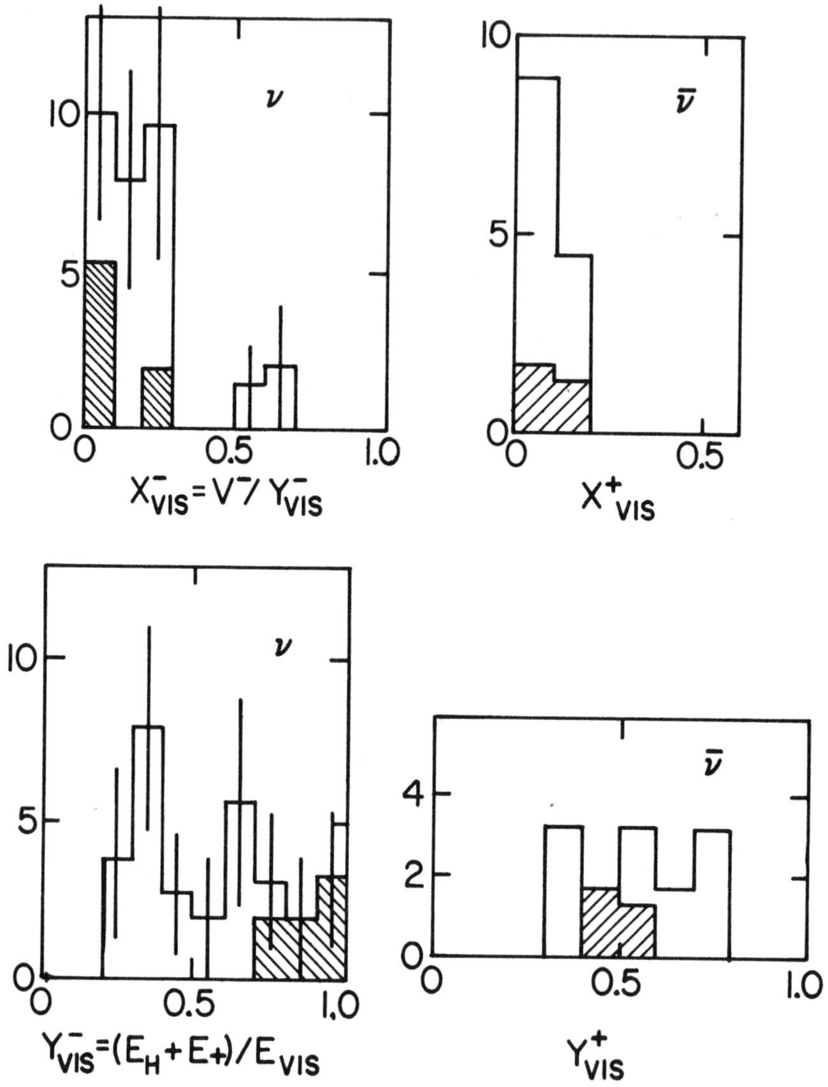
ν
10
5
0
0
0.5
1.0
$X^-_{VIS} = V^- / Y^-_{VIS}$
$\bar{\nu}$
10
5
0
0
0.5
X^+_{VIS}
ν
10
5
0
0
0.5
1.0
$Y^-_{VIS} = (E_H + E_+)/E_{VIS}$
$\bar{\nu}$
4
2
0
0
0.5
1.0
Y^+_{VIS}

Fig. 8

6. Exclusive Channels

There are several exclusive charm producing reactions for which we can make semiquantitative estimates. These processes are of interest in experiments where final state particles are detected and identified, such as in the 15-ft bubble chamber at Fermilab.

Single charmed baryon productions

$$\nu + p \rightarrow \mu^- + C_1^{++},$$

$$\nu + n \rightarrow \begin{Bmatrix} C_1^+ \\ C_0^+ \end{Bmatrix} + \mu^-$$

have been discussed elsewhere;[11] they are expected to be rather rare. One of these processes may have a bearing on the BNL event[15]

$$\nu + p \rightarrow \mu^- + \Lambda + \pi^+ + \pi^+ + \pi^+ + \pi^- .$$

Another class of processes for which one can make quantitative estimates is the charm-strangeness two-body associated productions of the type

$$\nu + p \rightarrow \mu^- + K^+ + C_{0,1}^+ .$$

Near threshold, barring the existence of resonances in the hadronic final states, the generalized Born approximation of Adler[16] and Shrock[17] should be fairly reliable. Shrock and I have considered this approach, and concluded[18] that the cross sections for these processes are about 10^{-41} cm^2, around $E_\nu = 8$ GeV. The total cross section for $\nu p \rightarrow \mu^- K^+ C_0^+$

is plotted in Fig. 9 (this is preliminary). Since the total νp cross section is about 5×10^{-38} cm^2 at these energies, detection of these processes would be very difficult.

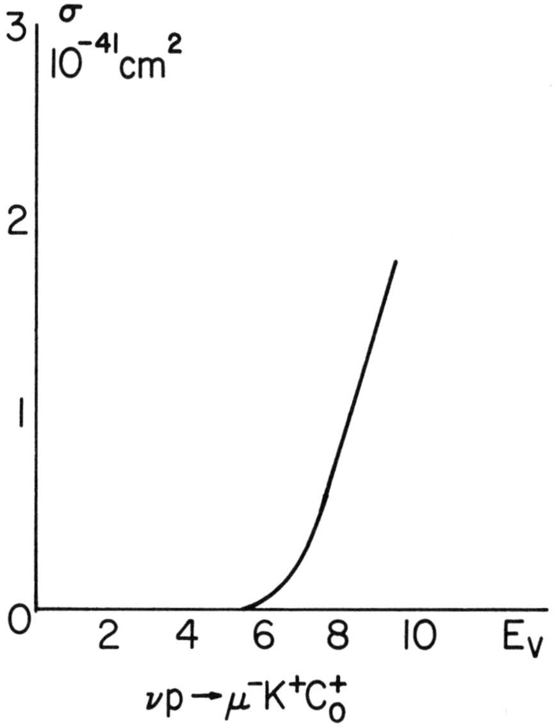

Fig. 9

A process of particular interest is the diffractive production of F^*, which is a 1^- $c\bar{s}$ bound state:

$$\nu + N \to \mu^- + F^{*+} + N$$

Many authors have commented on this process and computed its cross section.[11, 19-23] It is given by

$$\frac{d^2\sigma}{dxdy} = \frac{G_F M E_\nu}{\pi} \frac{\cos^2\theta_c}{\gamma_{F^*}^2} \left(\frac{\mu^2}{Q^2 + \mu^2} \right)^2$$

$$\frac{Q^2(1-x)}{2\pi} \left[y^2 \sigma_\perp + 2(\sigma_\perp + \sigma_L) \frac{1-y-Mxy/2E_\nu}{1+2xM/yE_\nu} \right]$$

where

M: mass of the nucleon,

μ: mass of F*,

and $\sigma_\perp$ and σ_L are the transverse and longitudinal F*N elastic cross sections, and γ_{F^*} is defined by

$$\left\langle F^{*+}(\epsilon) \,\middle|\, J_\mu(0) \,\middle|\, 0 \right\rangle = \frac{\mu^2}{\gamma_{F^*}} \epsilon_\mu .$$

In the following I shall simply assume $\sigma_L \simeq 0$, $\sigma_\perp \simeq \sigma_{el}$ ($F^*N \to F^*N$). However, there is one effect of extrapolating the initial F* off the mass shell which is likely to be quite important, viz., the minimal momentum transfer allowed. So we[23] multiply $\sigma_\perp$ by $\exp(b\,t_{min})$ where, in the Bjorken limit,

$$t_{min} \cong \frac{-M^2x^2}{1-x} \left[1 + \frac{1}{2ME_\nu y} \left(M^2 + \frac{\mu^2}{x} \right) \right],$$

and $b \simeq 4\ (\mathrm{GeV})^{-2}$.

In Fig. 10 I show a figure from the paper of Gaillard, Jackson, and Nanopoulos.[22] What is plotted is the diffractive vector and axial-vector boson production cross sections as fractions of the total neutrino cross section. Roughly, the ratio of the ρ and F* cross sections is given by[22]

$$\left[\frac{\sigma_{tot}(F^*N)}{\sigma_{tot}(\rho N)}\right]^2 \frac{\gamma_\rho^2}{\gamma_{F^*}^2} = \left(\frac{5\ \text{mb}}{26\ \text{mb}}\right)^2 \left(\frac{m_\rho}{\mu}\right).$$

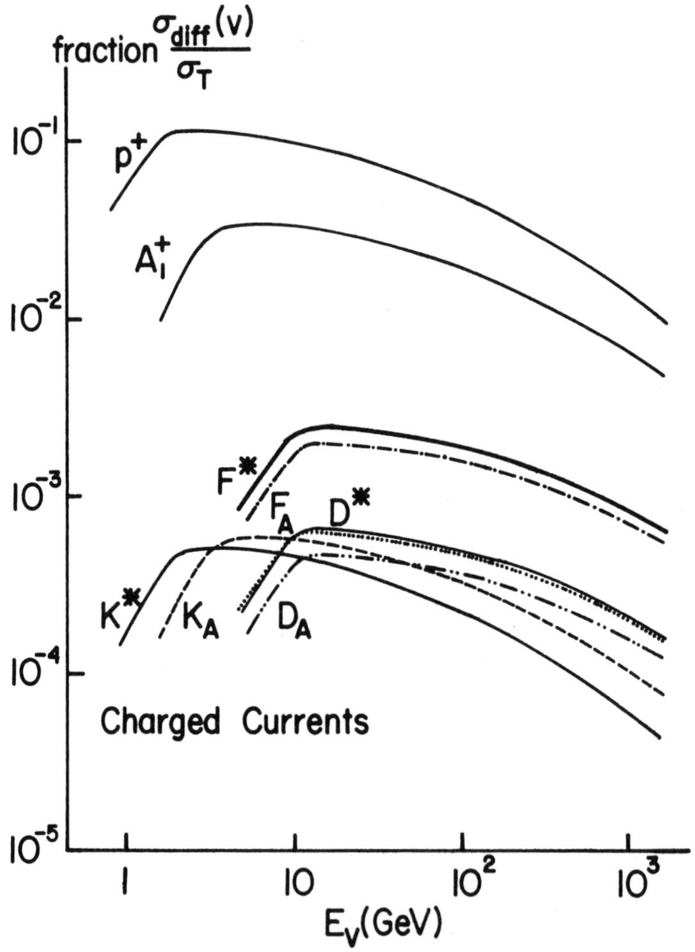

Fig. 10

Thus we expect that the F* diffractive production is about 2×10^{-3} of the total neutrino cross section. However, near the effective threshold of charmed particle production, i. e., at the energy range where deep inelastic, charmed particle production cross section begins to scale, the diffractive F* production may be an important, indeed dominant, source of charmed particles in the final state.

In Figs. 11 and 12 I have plotted the invariant hadronic mass squared (W^2) distribution and the x, y distribution of the diffractive F* production.

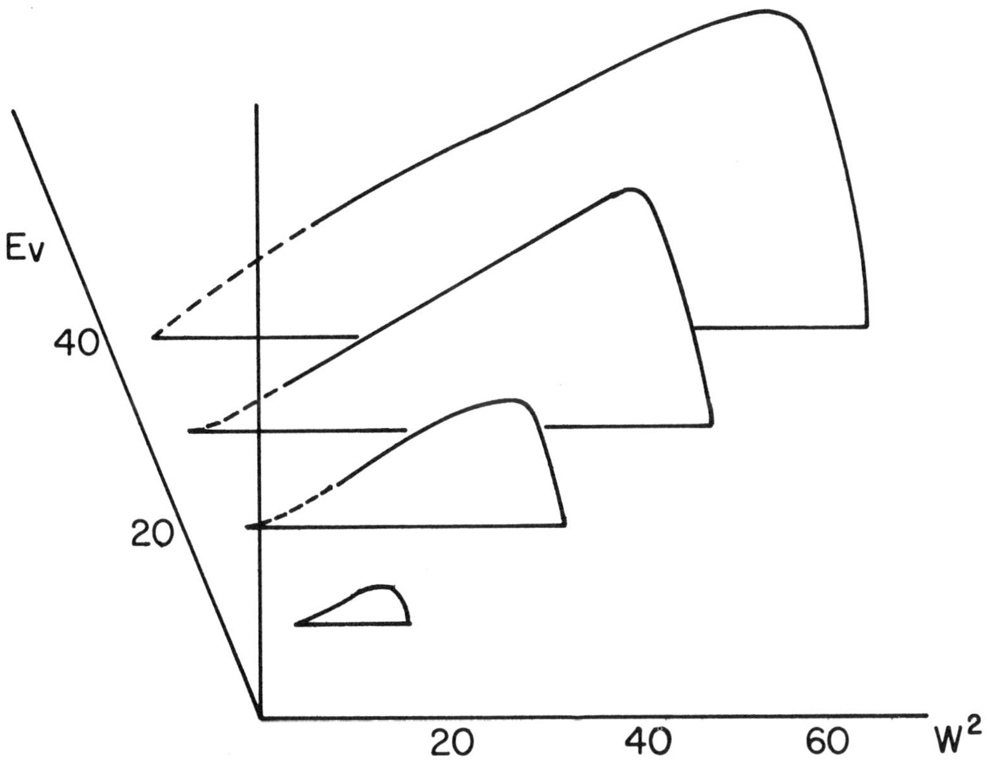

Fig. 11

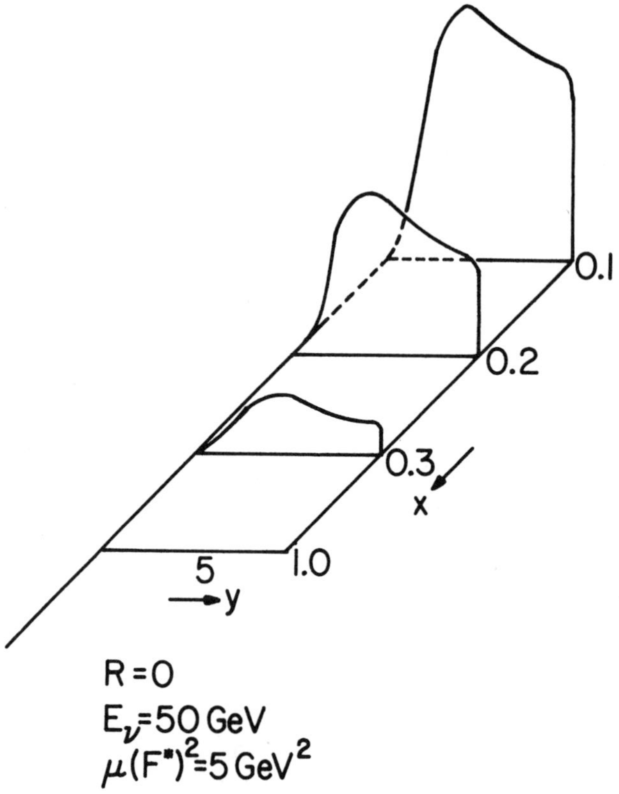

Fig. 12

We have already discussed signatures of F* production.[11] If F* is sufficiently heavier than F, then the decays

$$F^* \to F + \eta, \; D + K$$

may be dominant. If these processes are not energetically possible, the electromagnetic decay

$$F^* \to F + \gamma$$

is expected dominant. F^+ would then cascade:

$$F^+ \to \pi^-\pi^+\pi^+,\ K^-K^+\pi^+,\ ---$$
$$\eta \ell^+ \nu,\ X\ell^+ \nu,\ ---.$$

7. Dimuon Events of the Same Sign

In the minimal model, dimuon events of the same sign are explained in terms of associated production of a charmed pair:

$$\nu + N \to \mu^- + C + \overline{C} + ---$$
$$\hookrightarrow \mu^- + \nu + ---.$$

While we have no way of estimating charmed pair production in neutrino reactions, strange particle pair production in neutrino reactions is known to be substantial ($\sim 15\%$). To explain the ratio $\sigma(\nu \to \mu^-\mu^-)/\sigma(\nu \to \mu^-\mu^+)$ of about 0.1 we must assume that charmed pair production is about 1% of the total neutrino cross section above, say, 40 GeV.

An important corollary of this assumption is that the trimuon events of the type

$$\nu + N \to \mu^- + C + \overline{C} + ---$$
$$C \hookrightarrow \mu^+ + \nu + ---$$
$$\overline{C} \hookrightarrow \mu^- + \nu + ---$$

must exist at the level of 10^{-2} of the dimuon events of opposite sign.

Acknowledgment

I have benefitted greatly from discussions with D. Cline, W. Ford, M. Einhorn, M. K. Gaillard, T. Y. Ling, A. Mann, and S. Treiman on this subject.

References

[1] The interested reader may consult B. W. Lee "Lectures in Gauge Theories" in the forthcoming 1975 Les Houches Summer School Lectures to be published by North Holland Publishing Company, and references cited therein.

[2] A. Benvenuti et al., Phys. Rev. Lett. 34, 419 (1975); "Characteristics of Dimuons as Evidence for a New Quantum Number" to be published; "Dimuons Produced by Antineutrinos" to be published.

[3] B. C. Barish et al. "Neutrino Interactions with Two Muons in the Final States" in La Physique du Neutrino à Haute Energie CNRS (Paris, 1975), p. 131.

[4] R. W. Brown and J. Smith, Phys. Rev. D3, 207 (1971).

[5] L. N. Chang, E. Derman, and J. N. Ng, Phys. Rev. Lett. 35, 6 (1975).

[6] C. H. Albright, Phys. Rev. D12 (to be published).

[7] A. Pais and S. B. Treiman, "Natural Heavy Leptons as a Source of Dimuon Events: a Criterion," to be published.

[8] S. Weinberg, Phys. Rev. Lett. 19, 1264 (1967).

[9] A. Salam in Elementary Particle Physics, ed. by N. Svartholm (Stockholm, 1968), p. 367.

[10] S. L. Glashow, J. Iliopoulos, and L. Maiani, Phys. Rev. D2, 1285 (1970).

[11] M. K. Gaillard, B. W. Lee, and J. L. Rosner, Rev. Mod. Phys. 47, 277 (1975).

[12] A. Pais and S. B. Treiman, to be published.

[13] L. Wolfenstein, in Proceedings of the 1975 Lepton/Photon Symposium, to be published.

[14] C. Llewellyn-Smith, in Proceedings of the 1975 Lepton/Photon Symposium, to be published.

[15] E. G. Cazzoli,et al., Phys. Rev. Lett. 34, 1125 (1975).

[16] S. L. Adler, Ann. Phys. (N.Y.) 50, 189 (1968).

[17] R. Shrock, Thesis, Princeton University, 1975; Phys. Rev. (to be published).

[18] R. Shrock and B. W. Lee, to be published.

[19] B. A. Arbuzov, S. S. Gershtein, and V. H. Folomeshkin, Serpukhov preprints, IHEP 75-11 (1975); IHEP 75-25 (1975).

[20] V. Barger, T. Weiler, and R. J. N. Phillips, Wisconsin preprint COO-456 (1975).

[21] J. Pumplin and W. Repko, Michigan State University preprint (1975).

[22] M. K. Gaillard, S. A. Jackson, and D. V. Nanopoulos, CERN preprint, TH. 2049-CERN (1975).

[23] M. B. Einhorn and B. W. Lee, FERMILAB-Pub-75/56-THY (1975).

DISCUSSION

H. QUINN (Harvard)

Barish quotes a value of $\langle x \rangle \simeq .23$ for dimuon events. I know it's only a few events but do you feel this is consistent with the "naive charm" model?

B. LEE

Oh yes. Well, you see, the graph I showed you is based on at least 60 events and I think I'm free to throw away 4 events from other experiments. Incidentally, the definition of x and y measured in two experiments are different because in the Caltech experiment you use a dichromatic beam and you have some control over the incident energy.

D. LICHTENBERG (Indiana)

You discussed the production of charmed particles within the framework of the simple Weinberg-Salam model with the GIM mechanism. In this model charmed particles like to decay into strange particles. How do you reconcile this prediction with the fact that at SPEAR people do not see an anomalous number of K mesons above the charmed threshold?

B. LEE

Well I think you have to ask Harari that question. I would argue this way. Perhaps charmed particles have higher branching ratios into leptonic channels which cut down the hadronic branching ratios, and perhaps they have very small branching ratios into two body channels or three body channels, and furthermore maybe heavy leptons are being produced at SLAC which cuts down all the numbers. But I think we have to wait and see.

NEWLY DISCOVERED RESONANCES

Alvaro De Rújula*

This talk is a tour d'horizon at the charmonium picture of ψ-cholo-J, and its confrontation with experiment. I concentrate on the width of J/ψ, the systematics of meson masses in the standard model, the degree of naiveté of naive gluon counting, and charmonium spectroscopy with emphasis on what we will learn from paracharmonia.

*Physics Laboratories, Harvard University, Cambridge, Mass. 02138.
Work supported in part by the National Science Foundation, Grant MPS 73-05038 A 01.

INTRODUCTION

This is not a review of the theoretical or experimental situati concerning the new heavy mesons, but a written version of a brief and rather prejudiced talk[1]. I discuss the status of the "charmonium" picture in which the new mesons are interpreted as bound states of a heavy quark and its antiquark. Even in the best of circumstances, it will take a long time to establish whether or not these hypothetic constitutent quarks are "charmed" in the technical sense of carryin the quantum number used in the GIM scheme[2] to solve the problem of strangeness changing neutral weak currents. While the charm of charmonium is an entirely open issue, I will argue that the "onium" of charmonium is in a satisfactory shape.

In tables of particle properties one can find such heavy things as $\Delta(3250)$ whose discovery never made the newspapers. The excitemen that the codiscovery of the heavy meson $J/\psi(3095)$ produced was due t its spectacularly narrow width: fully three orders of magnitude smaller than what one would expect for a "normal" strongly interact resonance of the same mass. At first sight, the hypothesis that J/ψ is charmonium, and thus a hadron, seemed untenable. However, the possibility of heavy vector mesons (consisting mainly of a heavy quark-antiquark pair) being narrow had been entertained by theoretic prior to the experimental discovery. In their timely review of the search for charm, Gaillard, Lee and Rosner[3] predicted a total width 2MeV for a charmonium state of mass 2 GeV (for the actual mass they would have predicted $\Gamma = 3$ MeV, an overestimate of "only" a factor of 50). Appelquist and Politzer[4], on the other hand, were considerin a "Coulombic" picture in which the quark-quark binding forces would have become so weak that one could expect not only an extremely narrow resonance but a rapid succession of them with a hydrogen-like mass spectrum. Nature chose to sit halfway between these theoretical expectations.

THE LEPTON AND HADRON WIDTHS OF J/ψ

Let the ρ^o and the ω be mainly made of nonstrange quarks ($\bar{p}p\pm\bar{n}n$), the ϕ of a $\lambda\bar{\lambda}$ pair of strange quarks and J/ψ of a $\bar{p}'p'$ pair of charmed quarks (Why these assignments will be discussed in the next chapter). Let the quarks have the conventional fractional charges. Then, if the ρ,ω,ϕ and J/ψ all had the same mass, their leptonic widths would be in the ratio 9:1:2:8. Experimentally the widths are found to be in the ratio 9.4:1.2:2:7.3, where I have normalized to the ϕ meson. Impressive, but fallacious. What combination of widths and masses to use in a broken SU(4) - symmetric comparison is a very model-dependent question. We may anyway conclude that J/ψ has a "normal" leptonic width, not inconsistent with the most naive expectations.

In discussing the hadronic widths of the vector mesons I will consistently take the ratio to their ("normal") widths into e^+e^- pairs. I expect this to reduce the model-dependence of my considerations. I will further normalize the electron widths to equal quark charges. Thus, consider the following experimental information:

$$\gamma_\rho \equiv \frac{\Gamma(\rho \to 2\pi)}{\left\{\Gamma(\rho \to e^+e^-)/9\right\}} \sim (2.1 \pm .3)\ 10^5 \qquad (1a)$$

$$\gamma_\phi \equiv \frac{\Gamma(\phi \to 3\pi)}{\left\{\Gamma(\phi \to e^+e^-)/2\right\}} \sim 990 \pm 110 \qquad (1b)$$

$$\gamma_J \equiv \frac{\Gamma(J \to \text{hadrons})}{\left\{\Gamma(J \to e^+e^-)/8\right\}} \sim 98 \pm 26 \qquad (1c)$$

The first number is very large, not unexpected for a ratio of a strong process to a second order electromagnetic process. The relative smallness of the second number may be attributed to Zweig-Iizuka's rule[5]. In the decay to pions (nonstrange hadrons containing no λ quarks) the constitutent quarks in the ϕ must annihilate. This is not the case in ρ decay. $\gamma_J \neq \gamma_\rho$ because <u>different processes may have very different rates</u> (the rule becomes non-trivial only after other "allowed" and "forbidden" processes are compared). If J/ψ is below the threshold for decay into charmed pairs, its constituent charmed quarks must also annihilate in the decay. Thus γ_J should resemble γ_ϕ, not γ_ρ, since similar processes must have similar rates. (This rule is trivial enough not to have a name). In conclusion, when the hadronic width of J/ψ is adequately compared to its only conventional analog much of the mystery disappears and the new mesons can comfortably be allowed to join the evergrowing family of hadrons.

The above considerations are incomplete, in that I have not offered an explanation of the necessarily high degrees of "purity" in the $\bar{\lambda}\lambda$ composition of ϕ and in the $\bar{p}'p'$ composition of J/ψ. It would be interesting to have a more specific understanding of the relation between γ_J and γ_ϕ. I now proceed to describe the attempts to study these questions within colored gluon gauge theories of the strong interactions.[3,6,7]

MESON MASSES AND WIDTHS IN THE STANDARD MODEL

The "standard" model[8] is the conventional quark model with four quark flavours (p', p, n and λ). Each quark flavour comes in three colours, and is a triplet under color SU(3). The model is further restricted with the following assumptions:

a) Mesons are quark-antiquark color singlets, baryons are three-quark color singlets. This solves the statistics problem and reproduces the spectroscopic success of the quark model.

b) Quarks are strongly coupled to an octet of massless colored gluon fields G_a^μ via the color-SU(3) invariant Yang-Mills interaction $g_s\bar{\psi}\gamma_\mu\lambda^a\psi G_a^\mu$. This coupling is asymptotically free: g_s effectively becomes smaller at short distances. Quarks and gluons are colorful and not expected to be observable as free particles.

c) Quarks couple to photons as Dirac fermions with fractional charges. The interaction is pointlike before the strong gluon couplings are turned on. After they are turned on, asymptotic freedom insures that at short distances the interaction recovers its pointlike nature, with calculable corrections. The model thus explains approximate Bjorken scaling and predicts a specific pattern of deviations.

d) The weak interactions are introduced à la GIM[2], unifiable à la Weinberg-Salam[9] and perhaps superunifiable à la Georgi-Glashow[10]. I will not give further details on this point which is not directly relevant to my talk.

To accept J/ψ as a hadron we assumed with no justification that it is a very pure state of p' quarks, and that ϕ is a rather pure state of λ-quarks. In the standard model these assumptions find support in the systematics of the meson spectrum.[6] The $J^P=1^-$ mesons satisfy the equal spacing rules

$$\rho \simeq \omega \left[770 \text{ MeV} \simeq 784 \text{ MeV}\right] \tag{2a}$$

$$2K^*-\rho \simeq \phi \left[1014 \text{ MeV} \simeq 1019 \text{ MeV}\right] \tag{2b}$$

where particle names mean particle masses (not their squares). Take the naive point of view that the mass of a bound state equals the sum of the masses of the constituent quarks $\left[m(\lambda) > m(n) \sim m(p)\right]$ plus a binding energy which is constant in a given SU(3) multiplet. Then the equal spacing rules are satisfied provided ϕ is purely $\lambda\bar{\lambda}$, and ρ and ω are made of p and n quarks. This will be the case in any model with no significant interaction capable of annihilating a $\lambda\bar{\lambda}$ pair and recreating a light quark pair. The mass matrix is diagonal in quark labels and its eigenvalues are dictated by quark content.

Try to extend the above "success" to the $J^P = 0^-$ mesons. The result is rather unsatisfactory:

$$\pi = \eta \left[140 \text{ MeV} = 548 \text{ MeV}\right] \tag{3a}$$

$$2K - \pi = \eta' \left[854 \text{ MeV} = 958 \text{ MeV}\right] \tag{3b}$$

Something must complicate the mass matrix and allow for mixing of a strange quark pair with light quark pairs. In the standard model such a mechanism exists: the annihilation of one pair into two or more gluons that subsequently produce another pair. But, back into the 1^- mesons, such a mechanism also exists: the annihilation via three or more gluons. To explain the relative smallness of this effect one must assume that the coupling constant g_S is small enough at the vector meson masses for the three gluon effect to be small. These naive considerations become predictive if one considers yet another multiplet: the $J^P = 2^+$ mesons. These are P-wave bound states of quarks and annihilation, though it may proceed via two gluons, should be suppressed by the angular momentum barrier. Thus the equal spacing rules should work. They do:

$$A_2 = f \quad \left[1310 \text{ MeV} \simeq 1270 \text{ MeV}\right] \tag{4a}$$

$$2K_A^* - A_2 = f' \quad \left[1530 \text{ MeV} \simeq 1516 \text{ MeV}\right] \tag{4b}$$

The above discussion is relevant to a qualitative understanding of the decay width of ϕ into nonstrange hadrons. To lowest order in g_S this process involves the annihilation of the strange quark pair into three gluons. Three gluon annihilation must be small to explain the success of the equal spacing rule. Thus $\Gamma(\Phi \to 3\pi)$ must also be small. The "purity" of J/ψ as a $\bar{p}'p'$ bound state and its narrow width, for the same reason, become only one problem. It goes without saying that these considerations are more general than the gluon annihilation ansatz, but within the model we may go even further. The ratios γ_J and γ_ϕ of Eqs. (1b,c) can be explicitly computed in terms of the fine structure constant $\alpha = e^2/4\pi$ and its strong analog $\alpha_S = g_S^2/4\pi$. To leading order:

$$\gamma_J/8 \equiv \frac{\Gamma(J \to \text{hadrons})}{\Gamma(J \to e^+e^-)} =$$

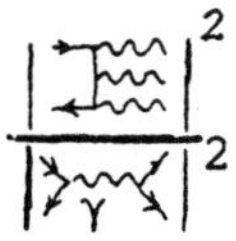

$$= \frac{5(\pi^2-9)\left\{\alpha_S\ (J)\right\}^3}{18\pi\alpha^2} \tag{5}$$

and similarly for γ_ϕ. From the experimental number for γ_ϕ one obtains $\alpha_S\ (\phi) \approx .5$. Optimistically assuming that ϕ is a sufficiently nonrelativistic bound state, that 1 GeV is large enough for asymptotic freedom to have set in and that $\alpha_S(\phi) \simeq .5$ is small enough for higher order corrections not to matter, we can predict[6,7] $\alpha_S(J)$. In the 12 quark model

$$\alpha_S(J) \sim \frac{\alpha_S(\phi)}{1 + \frac{25}{12\pi}\alpha_S(\phi)\ln\frac{m^2[J]}{m^2[\phi]}} \simeq .28 \tag{6}$$

Inserting this into Eq (5) we obtain a value for γ_J three times the observed one. Simply writing $\gamma_J \simeq \gamma_\rho$ would have been an over-estimate by a factor of 10. Admittedly, the ifs and buts are too many to conclude that we have completely explained the narrow width or that we have seen asymptotic freedom at work, but the picture is nice and consistent. I now proceed to discuss some of the ifs and buts

IS THE NAIVE PICTURE APPROXIMATELY CORRECT?

In this section I discuss some questions and a few answers concerning the consistency of the ideas presented so far.

1) For J/ψ to be narrow it must lie below charm threshold, or its decay would proceed fast and with no quark annihilation. In all theoretical estimates the lightest charmed particles are predicted to have a mass greater than $m_J/2$ and the condition is satisfied. In a naive quark model calculation [7] for instance, the nonstrange pseudoscalar D is lightest and $m_D \simeq 1.85$ GeV.

2) Bound state singularities should not spoil the perturbative calculation of γ_J. This problem has been energetically addressed by Appelquist and Politzer[11] who conclude that Coulomb and Yang-Mills singularities can be factorized into an amplitude (The "A" circle of Figure 1) that drops from ratios like γ_J or γ_ϕ. The "B" amplitude of the same figure, defined as two heavy quark irreducible in the annihilation channel has no bound state singularities. The width is proportional to $\Sigma\ |B|^2$, which has been checked to be finite to several orders of pertubation theory and not sensitive to eventually small parameters, like $m_J - 2m_{p'}$.

3) In the computation of quark-antiquark annihilation into an increasing number N of gluons (or gluons plus light quarks) to successive orders of renormalization group improved pertubation theory no small fractional momentum $\sim$ (mass of the decaying state)/N naturally shows up. Thus we may use a running coupling constant $\bar{\alpha}_S = \bar{\alpha}_S(Q^2 = M^2)$ with an argument coinciding with the decaying particle mass. This has also been checked to several orders of pertubation theory.[11]

4) The strong interaction "fudge factor": the effect of the amplitude for a three gluon state to evolve into the observed hadrons can be safely neglected. The idea is that when computing the width of J/ψ one trades the conventional hadron dynamics for the underlying field theory involving only quarks and gluons. The lowest order amplitude is then annihilation into three gluons. Indirect support for the viability of this procedure comes from the success of the quark model in inclusive electron and neutrino scattering and in the prediction of the e^+e^- cross section into hadrons below the region of new resonances. Here the evolution of the fundamental fields into hadrons is also ignored.

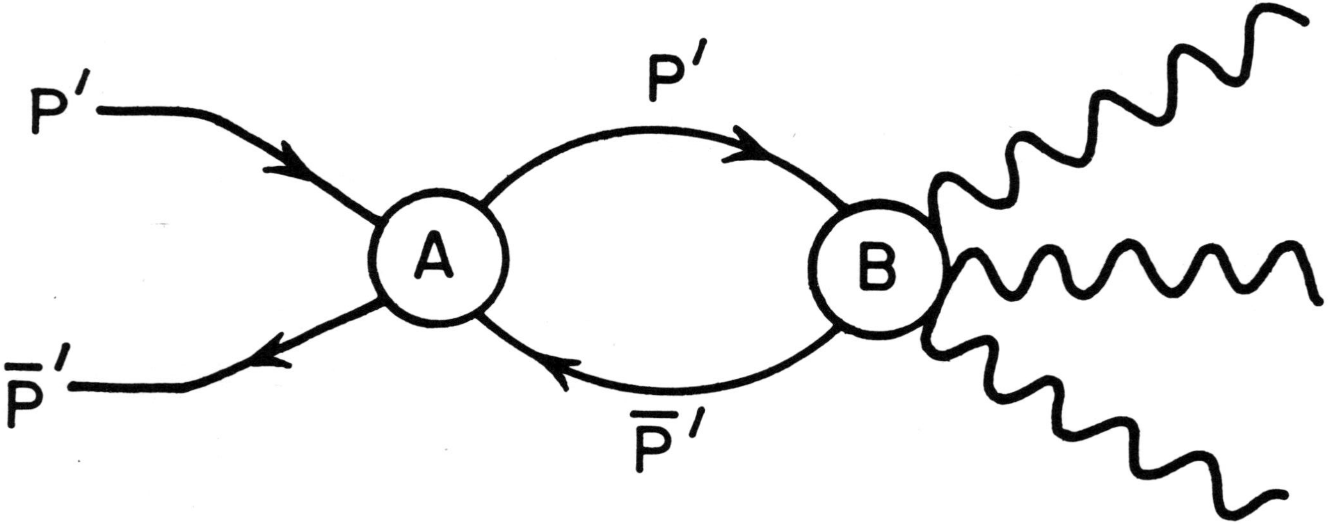

Figure 1

5) Quark-antiquark annihilation in the bound state must be a short distance phenomenon. This has been studied in models where the binding mechanism is explicit. For a potential linear in inter-quark distance Eichten et al.[12] conclude that the inverse of the charmed quark mass ($\sim$1.6 GeV) is roughly 1/5 of the classical turning point of the ground state of charmonium. Thus the model satisfies the condition to a satisfactory degree. Whether the strange quarks in the ϕ meson would also satisfy this condition is much more doubtful.

6) Surprises undetectable and untreatable in perturbation theory may affect the heavy quark annihilation process. Perhaps when two bound quarks are far apart their strong interactions, which by then are really strong, mediate the production of a light quark pair. The heavy quarks then approach and annihilate. The short distance three gluon ansatz would be wrong. Which picture (long or short distance) is closer to reality may be settled in the study of paracharmonium.

PARACHARMONIUM

If the $J^P = 1^-$ meson J/ψ is orthocharmonium (the S-wave bound state of a quark-antiquark pair in the triplet spin state) paracharmonium (the $J^P = 0^-$ singlet bound state) should exist and be nearby in mass. The particle of mass $\sim$ 2.75 GeV discovered in DESY among the γ ray decays of J/ψ may be paracharmonium. In the approach discussed in the preceeding chapter paracharmonium decay proceeds via two-gluon annihilation. The ratio of hadronic widths of para and orthocharmonium is then computable:

$$\frac{\Gamma_P}{\Gamma_O} \equiv \frac{\Gamma(\text{Paracharmonium} \rightarrow \text{hadrons})}{\Gamma(\text{Orthocharmonium} \rightarrow \text{hadrons})} \sim \frac{27\pi\left[\alpha_S[\sim 3\text{GeV}]\right]^{-1}}{5(\pi^2-9)} \sim 70 \qquad (7)$$

This large ratio is mainly two-body versus three body phase space. Should paracharmonium turn out to be approximately two orders of magnitude wider than orthocharmonium we would have a major success of the quark and gluon picture. Perhaps we could go as far as stating that we understand the strong interactions, at least those effects that we can hopefully treat in perturbation theory. Should quark-antiquark annihilation be a long distance nonperturbative process the ratio Γ_P/Γ_O would be roughly unity since now spin does not play a decisive role. Values of Γ_P/Γ_O well outside the range 1 to 70 would be very puzzling.

The cleanest determination of α_S at the mass of Paracharmonium would come[11] from the measurement of the ratio of the width of paracharmonium into two γ rays versus the width into hadrons (i.e., two gluons)[4,13].

$$\frac{\Gamma(\text{Para} \to 2\gamma)}{\Gamma(\text{Para} \to \text{hadrons})} = \frac{2\alpha^2}{\alpha_S^2} \sim 0.3\% \tag{8}$$

Unfortunately the present experimental situation is preliminary and we only have convoluted branching ratios. We know, for instance, that

$$\frac{\Gamma(3.1 \to \text{Para}\,\gamma)}{\Gamma(3.1 \to \text{all})}\ \frac{\Gamma(\text{Para} \to 2\gamma)}{\Gamma(\text{Para} \to \text{all})} \sim 2.10^{-4} \tag{9}$$

From the theoretical estimate Eq. (8) and the very rough preliminary experimental result Eq. (9) we may deduce $\Gamma(3.1 \to \text{Para}\gamma)/\Gamma(3.1 \to \text{all}) \sim 6\%$. This is an order of magnitude smaller than the most naive estimate based on the nonrelativistic quark model,[14] and a factor 2 or 3 smaller than more educated guesses.[15] This "failure" is not serious. In the estimate of this magnetic dipole transition a normal (g-2=0) magnetic moment of the quarks, has been assumed. Nothing in the standard model allows us to expect this to be a good approximation for on-shell photons. Similar calculations of $K^* \to K\gamma$ and other radiative transitions are also overestimates by a factor $\sim$ 3.

The ratio $\Gamma(\text{Para} \to 2\gamma)/\Gamma(\text{Para} \to \text{all})$ can also be guessed differently. Experimentally we know, with pathetic statistics, that $\Gamma(\text{Para} \to 2\gamma) \sim \Gamma(\text{Para} \to \bar{p}p)$. Theoretically we expect $\Gamma(\text{Ortho} \to \bar{p}p)/\ \Gamma(\text{Ortho} \to \text{all})$ (measured to be $\sim$ .2%) to be of the same order of magnitude as $\Gamma(\text{Para} \to \bar{p}p)/\Gamma(\text{Para} \to \text{all})$. Thus $\Gamma(\text{Para} \to 2\gamma)/\Gamma(\text{Para} \to \text{all}) \sim .2\%$ in rough agreement with Eq. (8).

CHARMONIUM SPECTROSCOPY

If J/ψ is a $J^{PC} = 1^{--}$ S-wave bound state of a charmed quark-antiquark pair one expects a full spectrum of radial and angular excitations. The narrow peak ψ' (3.695) may be the first S wave excitation of ψ, still below charmed threshold or very close to it. This identification led many people to predict narrow P wave states that would lie in mass between ψ' and ψ. Since the hadron width of ψ' is very small one expects γ-ray transitions between these states to have significant branching ratios[12,13]. In a Coulomb (1/r) potential the P wave states would be degenerate with ψ, while in a harmonic oscillator (r^2) potential they would be halfway between ψ and ψ'. In a gauge theory on a lattice one may argue for linear (r) potentials.[16,17] Thus one would expect[13] or predict[12,18] the baricenter of P wave states to be at about 3.5 GeV. In Figure 2 I have displayed the predictions of Appelquist et al.[13] and added an extensive list of synonyms.

Transitions $\psi' \to \psi\gamma\gamma$ via monochromatic γ rays where first seen at DESY. The product of branching ratios

$$\{B(\psi' \to P\gamma)\}\{B(P \to \psi\gamma)\} \equiv \left\{\frac{\Gamma(\psi' \to \text{P state} + \gamma)}{\Gamma(\psi' \to \text{all})}\right\}\left\{\frac{\Gamma(\text{P state} \to \psi + \gamma)}{\Gamma(\text{P state} \to \text{all})}\right\}$$

summed over P states is quoted to be in the range 4 - 12%. This is of the same order of magnitude as the upper limits for $B(\psi' \to P\gamma)$ from earlier SLAC data, implying that $B(P \to \psi\gamma)$ is of order 1. A large $B(P \to \psi\gamma)$ is a prediction of the naive charmonium approach: for P states the short distance annihilation of the quarks into gluons is suppressed by the angular momentum barrier.[13] These states have been named P_c by the DESY workers, apparently meaning positive charge conjugation , not P-wave charmonium.

States with mass $\sim$ 3.5 GeV have also been seen at SLAC in the decay sequence $\psi' \to$ P state + γ; P state $\to$ pions and/or kaons, and baptized "χ" to add to the general confusion. In the SLAC data one sees a narrow peak in the invariant mass of the final hadrons at 3.41 GeV and a broader peak (compatible with a superposition of two resonances) at 3.53 GeV. We expect[7] the lighter of this resonances

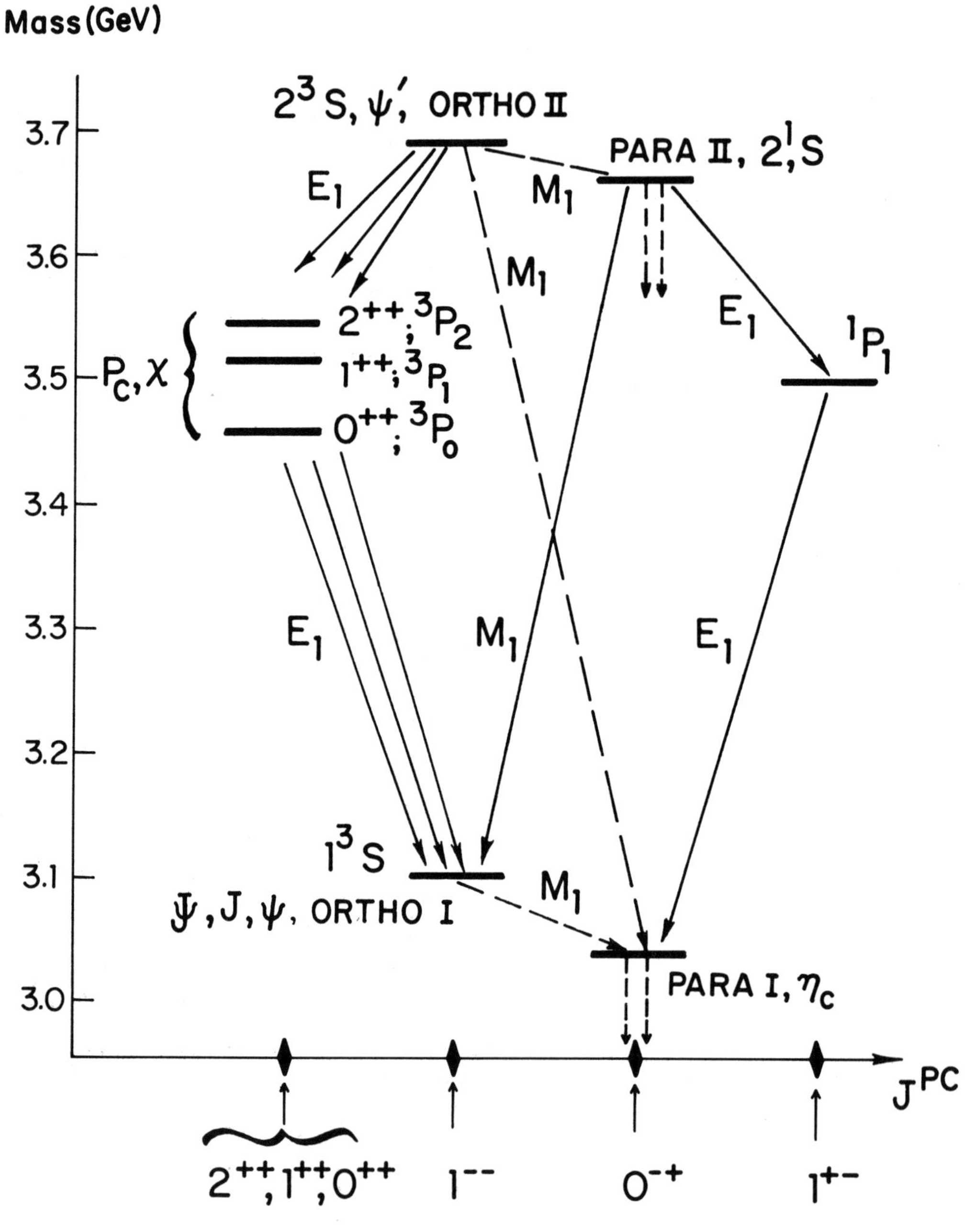

THE VOCABULARY AND SPECTRUM OF CHARMONIUM

Figure 2

to have $J^{PC} = 0^{++}$. This is compatible with the data, $\chi(3.41)$ is seen to decay into $\pi^+\pi^-$ and thus, has natural spin-parity. $\chi(3.53)$ may be a superposition of the P states 1^{++} and 2^{++}. Statistics are too poor to draw conclusions from the fact that it has not been seen to decay into $\pi^+\pi^-$.

The naive theoretical bounds[13] or estimates[12] of the widths $\Gamma(\psi' \to \text{P state} + \gamma)$ have turned out to be too large by an order of magnitude. Other than the naïveté of nonrelativistic quark model calculations one may find a reason for this in the proximity of ψ' to charm threshold. The proximity would cause the wave function of ψ' to contain a large admixture of virtual charmed meson pairs. This, in turn, would make ψ' "more orthogonal" to the P wave states. Predictions that are independent of wave function details (other than fine structure) should stand better the comparison with experiment. An example is the statistical rule

$$\frac{\Gamma(\psi' \to 2^{++}\gamma(k_2))}{5k_2^3} \sim \frac{\Gamma(\psi' \to 1^{++}\gamma(k_1))}{3k_1^3} \sim \frac{\Gamma(\psi' \to 0^{++}\gamma(k_0))}{1k_0^3}$$

which remains untested. The rule assumes that the E_1 transition dominates.

THE e^+e^- ANNIHILATION CROSS SECTION INTO HADRONS

Following convention, define $R = \sigma(e^+e^- \to \text{hadrons})/\sigma(e^+e^- \to \mu^+\mu^-)$ and $W = \sqrt{Q^2} = E(e^+) + E(e^-)$. The quantity R remains constant at $R \sim 2.5$ from $W \sim 2.5$ to ~ 3.7 GeV, except for the narrow resonances. Somewhere between W = 3.5 and 3.8 GeV, in a region where data is unfortunately sparse, R starts rising. Naive quark model advocates[6,7] would predict the rise associated with charm threshold to start at $W \sim 3.7$ Gev. With every new set of data the structure seen between 3.7 and 4.6 GeV becomes richer. At the moment, there are wide peaks at $W \simeq 4.1$ GeV and $W \simeq 4.4$ GeV, and perhaps a smaller peak at $W \sim 3.9$ GeV. Advocates of nonrelativistic linear potential guesses[18] had successfully predicted $J^{PC} = 1^{--}$ states at W = 4.15, 4.6, 5,... GeV. These states would be above charm threshold (broad) and the increase in phase space would make the higher mass ones too broad to be seen as peaks. Even the smaller peak at 3.9 GeV may have room in the naive picture. Tensor forces would inevitably mix S and D wave states and allow some of the "almost" D wave excitations to have nonvanishing wave functions of the origin and be significantly coupled to one photon.[12] The region between 3.6 and 4.2 GeV should in any case not be simple to study. We naively[7] expect six thresholds to be crossed at these energies: $D\bar{D}$, $D^*\bar{D}$, $D^*\bar{D}^*$, $F\bar{F}$, $F^*\bar{F}$ and $F^*\bar{F}^*$ (unstarred means 0^-, starred 1^-; D mesons are charmed, F mesons charmed and strange).

From $W \sim 5$ to $W = 7.8$ GeV the data are again compatible with a flat R at a level $R \sim 5.2$. The asymptotic value in a theory with the standard 12 fractionally charged quarks is $R = 10/3$. The production of a heavy lepton and its fast decay into hadrons would raise the expectation to "R" = 4 1/3, still below the data. Some have used this fact, with charming naïveté, to conclude that more than four quark flavours have been seen (tasted?) at SLAC energies. But our field theoretic understanding of R at large energies is based on asymptotically free field theories, where the heavy quark mass and the "scale parameter" Λ play a non-trivial role at any energy $W < \infty$. Moreover, it is clear that the theory cannot be directly compared with timelike data. The safest approach is to transform the data via a dispersion relation into spacelike information and compare it there to theoretical expectations. The conclusion of this analysis[19] is that e^+e^- and deep inelastic lepton scattering data are compatible with the standard model, for $m(p') \simeq 1.55$ GeV and $\Lambda \sim .5$ GeV, provided heavy leptons are being produced at SLAC energies, but not otherwise. The agreement is marginal and better data may rule out the model.

FINE AND HYPERFINE SPLITTINGS OF CHARMONIUM LEVELS

In lattice gauge theories[16,17] the spin-spin force between two distant static quarks vanishes exponentially with distance while the spin-independent force remains constant. At short distance, on the other hand, the quark-gluon coupling constant decreases and one-gluon exchange may be a good approximation to interquark forces, including spin-spin correlations. This idea has been used to relate "hyperfine" splittings between members of SU(6) multiplets of baryons and mesons, perhaps too naively, but very successfully.[7] Hyperfine splittings and mass differences between strange and nonstrange members of SU(6) multiplets are smaller than their mean masses and it is a consistent approach to assume SU(6) symmetry of the wave functions and perturb in the smaller hyperfine interactions and mass differences between constituent quarks. Unjustifiably extending this analysis to SU(8), one may predict the mass difference Δ between ortho and para-charmonium:[7]

$$\Delta = \text{Ortho-Para} \sim (\rho-\pi)\left(\frac{\rho}{\psi}\right)^2 \sim 30 \text{ MeV}$$

If the state at 2.75 GeV seen at DESY is paracharmonium, this is wrong by a factor of 10. One may doctor this result by "correcting" for the error incurred in assuming SU(8) symmetry of the wave functions at the origin $\psi(o)$. The leptonic widths of the vector mesons and the expectation values of one gluon exchange spin-spin interactions are both proportional to $|\psi(o)|^2$. Fudging with this fact, I get

$$\Delta = (\rho-\pi)\ \frac{9}{8}\ \frac{\Gamma(J \to e^+e^-)}{\Gamma(\rho \to e^+e^-)} \sim 400 \text{ MeV}$$

I interpret this as meaning that the SU(8) wave function symmetry was not a good guess, but the one-gluon exchange idea may be viable. Independent support for this point of view can be found in the work of Barbieri et al.[20] who assume SU(4) symmetry of the binding linear potential, compute wave functions and get a bigger $\Delta \sim 54$ MeV. The bag model with a bag constant modified to reproduce the $\psi'-\psi$ mass difference gives a nice value for $\Delta \sim 280$ MeV,[21] if hyperfine splittings are again attributed to one-gluon exchange.[7] The moral seems to be that Δ is very model dependent. The splittings between P-wave levels were also underestimated by a factor ~ 10 in the naive approach.[7]

PARACHARMONIUM II

Paracharmonium II, the 0^- comrade of ψ' should also exist and be nearby in mass. The mass difference between ψ' and Para II in the one-gluon exchange model is

$$\Delta' \equiv (\text{Ortho II - Para II}) \sim (\text{Ortho I - Para I})\ \frac{\Gamma(\psi' \to e^+e^-)}{\Gamma(\psi \to e^+e^-)} \sim 160 \text{ MeV}$$

Should this be a correct estimate, Para II would lie suspiciously close in mass to $\chi(3.53)$.

The smaller hyperfine splitting between the excited charmonium states diminishes the transition rate $\psi' \to$ PARA II + γ relative to $\psi \to$ PARA I + γ by a factor $\sim (\Delta'/\Delta)^3 \sim .1$. Moreover PARA II has a smaller $\gamma\gamma$ branching ratio than PARA I because of the extra hadronic decay PARA II $\to$ PARA I + 2π and its extra electromagnetic decays into ORTHO I and the 1^{+-} state at $\sim$ 3.5 GeV. Thus PARA II may be somewhat more difficult to find than PARA I, at least in the 3γ decays of ψ'. A nice signature for the detection of PARA II may be

$$\psi'(3.7) \to \text{PARA II} + \gamma$$
$$\hookrightarrow \text{Para I} + \pi^+\pi^- \ (\text{or } \pi^0\pi^0)$$
$$\hookrightarrow \gamma\gamma \ (\text{or } \bar{p}p)$$

Its cascade γ ray decays into ORTHO I and the 1^{+-} state would also be rather spectacular.

I am indebted to my colleagues at Harvard, particularly Tom Appelquist, Howard Georgi, Shelly Glashow, David Politzer and Howard Schnitzer for many heated discussions on the topic of this talk.

References

1) A more complete and detailed discussion of the same topics is contained in A. De Rujula "The Discreet Charm of the New Particles", to be published in the proceedings of the International Summer Institute on Current Induced Reactions, DESY, Hamburg, September 1975.

2) S. L. Glashow, J. Iliopoulos and L. Maiani, Phys. Rev. D2, 1285 (1970).

3) Mary K. Gaillard, Benjamin W. Lee and Jonathan L. Rosner, Revs. Mod. Phys. 47, 277 (1975).

4) T. W. Appelquist and H. D. Politzer, Phys. Rev. Letters 34, 43 (1975).

5) G. Zweig, unpublished. I. Iisuka, K. Okada and O. Shito. Progr. Theor. Phys. (Kyoto) 35, 1061 (1966).

6) A. De Rújula and S. L Glashow, Phys. Rev. Lett. 34, 46 (1975).

7) A. De Rújula, Howard Georgi and S. L. Glashow, Phys. Rev. D12, 147 (1975).

8) Too many people have contributed bits and pieces to the standard model for me to attempt to give a fair list of references at this point.

9) S. Weinberg, Phys. Rev. Lett. 19, 1264 (1967); A. Salam, in Nobel Symposium No. 8, edited by N. Svartholm (Almquist and Wiksell, Stockholm, 1968). p. 367.

10) Howard Georgi and S. L. Glashow, Phys. Rev. Lett. 32, 236 (1974); Howard Georgi, H. R. Quinn and S. Weinberg, ibid. 33, 451 (1974).

11) T. W. Appelquist and H. D. Politzer, to be published in Phys. Rev. D.

12) E. Eichten et al. Phys Rev. Lett. 34, 369 (1975).

13) T. Appelquist et al. Phys Rev. Lett. 34, 365 (1975). C. G. Callan et al., PHys Rev. Lett 34, 52 (1975).

14) J. M. Borenstein and R. Shankar, Phys. Rev. Lett. 34, 619 (1975).

15) G. Feinberg and J. Sucher "Relativistic Calculation of Radiative Ml Transitions in Charmonium". Submitted to this conference.

16) K. Wilson, Phys. Rev. D10, 2445 (1974).

17) J. Kogut and L. Susskind, Phys. Rev. D9, 3501 (1974) and to be published.

18) B. Harrington, S. Y. Park, A. Yildiz, Phys. Rev. Lett. 34, 168 (1975).

19) A. De Rújula and Howard Georgi "Counting Colored Quarks and Finding Fancy Flavors", to be published.

20) Barbieri et al., CERN preprint. 21) R. Jaffe, private comm.

DISCUSSION

O. GREENBERG (Maryland)

To follow on your last comment, how wide would a normal hadronic state at 2.4 GeV have to be?

DE RÚJULA

A few hundred MeV, is it narrower than that?

O. GREENBERG (Maryland)

Yes, it is.

DE RÚJULA

Perhaps a hundred.

O. GREENBERG (Maryland)

Less than a hundred.

P. O'DONNELL (Toronto)

I have two comments. The first is about your assumption of the normality of the leptonic width of the 3.1. About 7 years ago, van Royen and Weisskopf produced a fairly successful SU(3) quark model in which for some mysterious reason $|\psi(0)|^2 \sim m$ (which they later justified using SU(3) and Weinberg sum rules). If this ansatz is extended to SU(4) to include the ψ one can show that the "charmed" quark would have charge 4/3 and that the leptonic width of $\sim$ 6 keV is large! Also the K/π ratio at SPEAR and the HPWF y anomaly would be explicable.

DE RÚJULA

Yes, by extending a non-understood mystery which is this Weisskopf thing to an unexplored field you get to an unpleasant conclusion that the charge of the new quark is 4/3. Now you may do it if you wish, I don't wish to do it.

P. O'DONNELL (Toronto)

The second comment is that the discrepancy you have shown between the Feinberg-Sucher calculation and your estimate of the experimental radiative decay of the 3.1 GeV resonance is in the same direction and about the same magnitude as the discrepancy between recent experimental determinations of the $\rho \to \pi\gamma$, $K \to K\gamma$, $\varphi \to \eta\gamma$ and $\varphi \to \pi\gamma$ and the quark model of the old physics.

DE RÚJULA

Yes, they are also wrong by a factor of 3 in the same direction.

D. WEINGARTEN (Rochester)

I would just like to mention that I disagree with the speaker's remarks concerning theoretical predictions for the mass of the lowest pseudoscalar $c\bar{c}$ state. A correct prediction of ≈ 2.8 GeV actually was made. This was done by Borchardt, Mathur and Okubo (and perhaps by others of whom I'm unaware) by diagonalizing the mass-squared matrix for the pseudoscalar meson 16-plet assuming broken SU(4) and fixing the ratio of SU(4) to SU(3) breaking (i.e., the relative coefficients of T^{15} and T^{8}) to be the same as the result gotten from a mass-squared formula for the vector 16-plet.

DE RÚJULA

True, I had forgotten, true.

B. WARD (Purdue)

The ψ and ψ' appear as isolated singularities in the annihilation of an electron with its positron. In your description of these particles, you have ignored the corresponding singularities (driving terms) in the relevant renormalization group equations. Why is this legitimate?

DE RÚJULA

The singularity due to the fact that I have massless vector particles? Yes, those are Yang-Mills and Coulomb singularities that provide for the binding. Now, how that really actually works explicitly, we do not know because nobody understands the infrared problems. In any case the only way to study the value of R seriously is to convert it via a dispersional relation into a statement of what the photon propagator is in a space-like region and then compare it to the theory, because in a space-like region the theory has not these problems that we've talked about. If you do that you essentially get to the same conclusions that four quarks is a reasonable theory.

B. WARD (Purdue)

No, because it's well-known that as Professor Adler was pointing out in his lecture earlier whatever one has in the annihilation region if it is systematically carried to the space-like region it has to show up, and in particular in functions of the type that you are describing.

D. LICHTENBERG (Indiana)

How do the predicted gamma decay rates of the $\psi(3.7)$ into the P states of charmonium compare with the measured decay rates?

DE RÚJULA

The original predictions are wrong. They are an overestimate by one order of magnitude. This is not particularly surprising, among other things because the 3.7 is close or perhaps even above charmed threshold; therefore it's going to have a complicated wave function containing charm-anticharm mesons. That makes it more orthogonal to the wave functions of the P wave states and it is an effect that lowers the theoretical prediction. Also, these things are very model dependent because these are E_1 transitions which depend on the radius of the system and things like that, so an overestimate by a factor of ten should also not be a reason to abandon the theory.

J. SUCHER (Maryland)

I have two comments: The first is that one possibility for dealing with the apparently large hyperfine splitting in charmonium is to imagine that there is a contribution to the binding potential arising from axial-vector gluon exchange; Feinberg and I are exploring this.

The other is that today's speakers, as well as the session chairman, ought to be congratulated for the fact that we're running so closely on schedule.

GAUGE THEORIES: HOW DOES IT STAND?

Abraham Pais*

*Rockefeller University, New York, N.Y. 10021.
This chapter is a transcript of Session Chairman Pais's comments.
Work supported in part by the Energy Research and Development Administration, Contract E(11-1)-2232B.

I shall attempt to give a brief qualitative overview of the present status of gauge theories and will divide my comments in two main rubrics. I shall first touch on the "new phenomena". By this I mean neutral current processes as well as the recent indications for charm production in neutrino and in anti neutrino reactions. Thereafter, I shall turn to the "old phenomena", to be defined shortly. In regard to the new phenomena, it cannot be stressed enough that we should treat these early experimental results both with respect and with caution. We should never forget the lesson that it took about forty years after the discovery of β-radioactivity before we had fairly reliable β-spectra.

We face an odd situation.

Concerning the new phenomena and given the quite limited experimental facts, there seems nothing terribly wrong with the idea that we are dealing with the weak-electromagnetic gauge group SU(2) x U(1) and, more specifically, with the simplest imaginable quark content of this group, namely four quarks (modulo color) arranged in multiplets in the well known way, that is, two left-doublets and four right-singlets. In what follows I shall call this the standard SU(2) x U(1) model. On the other hand, in regard to certain old phenomena, this picture leaves the darkness so entirely unrelieved that it cannot possibly be the whole story.

As to the neutral current processes, we all know that the theoretical SU(2) x U(1) proposal provided a main impetus for the search of strangeness conserving neutral processes and that such reactions have been observed. It is far too early to state that the known facts confirm the standard SU(2) x U(1) scheme. But as we heard this morning from Professor Adler, for some reasonable value of the mixing angle experiments are not in conflict with this simple picture. However, as he pointed out, the important question of neutral current induced Δ-production is still far from resolved. Much more experimental information on this point is clearly needed, and the same is true for the purely leptonic scattering processes.

As was discussed by Professor B.W. Lee, the prompt dimuons produced by beams which contain preponderantly neutrinos[1] (as compared to $\bar{\nu}$) and by enriched antineutrinos beams[2] indicate that charmed particles are being produced at FNAL. The main indication[3] for this comes from the experimental value of the quantity $\langle p^-\rangle/\langle p^+\rangle$, where $\langle p^{\mp}\rangle$ is the average momentum of the negative (positive) muon, in the laboratory system. In what follows it will be assumed that these dimuons are indeed largely due to charm production. Here "charm" is used in a generic sense. It does not necessarily mean the charm associated with the four quark model mentioned earlier. Yet, upon a somewhat closer inspection, it would appear that in fact the standard SU(2) x U(1) model may well provide a first qualitative understanding of the dimuon phenomena, particularly if we accept the early indications that dimuon production by anti neutrinos is roughly of the same order of magnitude as the production by neutrinos.

What does this model say about charm production? In the context of the naive parton model, where valence quarks only are considered and the quark sea is neglected, one would of course have a mechanism for charm production by neutrinos but none such for anti neutrinos. In this model it has therefore to be assumed that the sea does play a role. In any event, its influence is not inhibited by Cabibbo factors (as is the aforementioned valence contribution). If the average branching ratio B for muonic decay of strongly stable charmed hadrons (weighted over the different kinds of such hadrons) is ~0.1, one could have a qualitative understanding of what goes on[4]. If this picture is correct then $\bar{\nu}$-produced charm provides a sensitive probe of the sea parton distributions. For example, as Professor Lee has pointed out, in such a picture charm production by antineutrinos may by anticipated to occur dominantly at small x. Note further that, in this picture, the like sign muons could be mainly due to associated production of charmed particles. If so, trimuon events should also be seen at a rate roughly ~0.1 of the like sign dimuon rate.[4]

Also in regard to the recently discovered radiative decays of the 3.7 and 3.1 Gev resonances, the standard model commends itself ("charmonium spectroscopy"[5]). Never mind that the positions and decay widths of the newly found levels are not precisely what was predicted. However, it cannot be said that these findings by themselves rule out more complicated options. The absence of direct evidence for charm production at SLAC does remain a puzzlement. But then, this puzzle persists also for alternative options.

What are these options? I have particularly in mind several recently proposed variants of SU(2) x U(1), and would like to comment on the way they fare in connection with the dimuon effects. As a simple example, consider the recent suggestion[6] that $(p',n)_R$ shall form an SU(2) x U(1) doublet (instead of two singlets as in the standard case). This provides a new mechanism for charm production by neutrinos off valence quarks, and in such a way that no Cabibbo suppression intervenes. In other words, it constitutes a major enhancement as compared to the standard picture. This, in itself, is acceptable as long as the above mentioned quantity B were correspondingly smaller. But no corresponding enhancement occurs for charm production by $\bar{\nu}$ and so there is a difficulty if indeed the ν and $\bar{\nu}$-productions of charm are comparable.

However, one might invent ways out by introducing more charm quantum numbers. For example let there be an additional quark n' with another charm and let $(p,n')_R$ form a second right-doublet. Then if the mass of n' is not much larger than that of p', one has a new way to enhance $\bar{\nu}$-producted charm to a level comparable to ν-production. But this raises two new difficulties.[4] First, if there is a second charm associated with a quark mass comparable to p', then why did the search (up to 7.6 Gev) at SLAC not reveal a second ψ-like family? Secondly consider the ratio

$$X = \frac{\sigma^{\nu}_{total}}{\sigma^{\bar{\nu}}_{total}} ,$$

where $\sigma^{\nu,\bar{\nu}}_{total}$ are the inclusive $\nu(\bar{\nu})$ cross sections off isoscalar targets. The naive parton picture predicts X=1/3 for the standard model. Experimentally $X \approx 0.3 \pm 0.1$ at $E \approx 80$ Gev, compatible with 1/3, though there is certainly room for play. However, if we are well over the charm thresholds (as the dimuon events seem to indicate) then the above two-charm example gives X=1, which does not look good. Of course this last argument is model dependent. Not only does it rest on the parton picture with its scaling implications but also on the naive version thereof in which X is grossly determined by valence contributions. Still, even allowing for modest contributions from the sea and for modest deviations from scaling, a value one for X looks implausible at the energies concerned.

Thus there are three types of arguments which all seem to point[4] to the standard model as the simplest viable picture proposed so far: 1) the relative magnitude of the dimuon effect induced by ν and by $\bar{\nu}$, 2) the absence of more than one ψ-family, 3) the magnitude of the ratio X. Further experimental developments may put such arguments in a different light. Nor do these arguments preclude the existence of very heavy quarks associated with not as yet excitable charm; several such models are current.[7] In any event, from the point of view of the new phenomena the standard model does not look unreasonable, if we may trust present experimental indications.

Contrast this with the situation in regard to some of the old phenomena.

As a first example of these, consider the proton-neutron mass difference, or rather the proton quark-neutron quark mass difference. In a satisfactory synthesis of weak-electromagnetic interactions this difference should be calculable.[8] Such calculability is absent not only in the standard SU(2)xU(1) model, but in any model based on this gauge group. (The smallest gauge group which allows for a calculable n-p mass difference is[9] SU(2)xU(1)xU(1).) If one regards the naturalness[8] of the n-p mass difference as a desideratum for an acceptable gauge theory, then one must look beyond SU(2)xU(1). This same desideratum should also apply to superunified theories in which strong, electromagnetic and weak interactions are synthesized. None of the theories of this kind studied to date incorporate a calculable n-p mass difference. More generally, a satisfactory theory which contains strong isospin as a natural symmetry has yet to be developed.

And what about this strange angle of about $13^{o}40'$? It seems reasonable to suppose that the Cabibbo angle should be calculable in a satisfactory theory. But in the standard SU(2)xU(1) model this angle is necessarily a phenomenological parameter.[10]

For these and other reasons, the study of larger structures (of which one may ask that they shall include SU(2)xU(1) in some appropriate broken symmetry limit) is imperative. This in part motivates the investigations which were discussed this morning by Professor Weinberg.

There are further questions which one may ask. For example, will the gauge theories finally give a rationale as to why parity-nonconservation is confined to the weak interactions? Will we get closer to an understanding of CP-violation? We must also ask if our criteria for building gauge models are free of prejudices. For example, is there really only one single charge raising (and one charge lowering) current for all $|\Delta Q|=1$ weak processes? Note that if $\Delta S=0$ and $|\Delta S|=1$ processes of this class were associated with distinct currents associated with vector mesons with respective masses M_1 and M_2, then if $M_2=2M_1$ were a natural mass relation, one would have the correct Cabibbo angle to within a few percent.

In any case, as said earlier we have a remarkable situation. There appear to be no flagrant objections if one attempts to accommodate the new phenomena in the standard SU(2)xU(1) model, with its single kind of charm and with its left-doublets. The existence of further not as yet excitable forms of charm is not excluded. On the other hand, in such a picture some important old questions remain as unsolved propositions. The view that this is intolerable is not as wide spread as I believe it should be. One way to alleviate the situation is to search for hierarchies of symmetry breaking in which SU(2)xU(1) emerges as a good approximation at presently available energies and momentum transfers. Such was the tenor of Professor Weinberg's talk this morning.

It seems clear that the subject of gauge theories is only in its early stages and that it will continue to give us much stimulus, demanding as it does that we know group theory, that we know field theory and that we know physics.

References

1. A. Benvenuti et al., Phys. Rev. Letters 35, 1199, 1203 (1975).
2. A. Benvenuti et al., Phys. Rev. Letters 35, 1249 (1975).
3. A. Pais and S. B. Treiman, Phys. Rev. Letters 35, 1206 (1975).
4. A. Pais and S. B. Treiman, Phys. Rev. Letters 35, 1556 (1975).
5. C. Callan et al., Phys. Rev. Letters 34, 52 (1975); T. Appelquist et al., ibid, 34, 365 (1975); E. Eichten et al., ibid, 34, 369, (1975); B. Harrington, ibid, 34, 706 (1975).
6. A. de Rujula, H. Georgi and S. L. Glashow, Phys. Rev. Letters 35, 69 (1975).
7. See ref. 4 for literature and further examples.
8. For the technical meaning of terms like "calculable", "naturalness", "phenomenological" see H. Georgi and A. Pais, Phys. Rev. D10, 1246 (1974), where references to earlier papers on these topics are also found.
9. A. Pais, unpublished. Along with this extended continuous group, an invariance is needed under the discrete conjugation $p_R \leftrightarrow n_R$.
10. See reference 8, Section IIB 1.

FUNDAMENTAL THEORY: NEW PARTICLES, NEW IDEAS

Sheldon L. Glashow*

*Harvard University, Cambridge, Mass. 02138.
This chapter is S. Glashow's talk given at the Conference banquet.

Truth, in elementary particle physics, is not determined by consensus. Ours is one of the few disciplines incorporating an absolute standard of truth: theoretical ideas contrary to experimental facts must be discarded. (There is an important proviso here: Remember the notorious wrong results concerning radium E, $\pi \rightarrow e\nu$, $K \rightarrow \bar{\mu}\mu$, $\Delta S = -\Delta Q$ decays, and the split A_2. Theorists must be very smart to know which experiments to believe). Moreover, our discipline also provides an absolute standard of relevance: theoretical ideas that make no statement about possible experiments are not part of our discipline, no matter they are interesting or rigorous. Nonetheless, if an election were held today, a certain picture of elementary particle physics could probably emerge victorious, winning a plurality if not a majority of physicists' votes. I would like today briefly to review this "standard picture" and to indicate the reasons for its wide following.

But first, it may be appropriate to comment on the emerging consensus. Few partisans of the standard model would agree that it is likely to be a correct theory: it is merely the most plausible model available at the present time. It is both theoretically incomplete, and experimentally imperfect. It does explain, or at least promises to explain, a wide variety of elementary particle phenomena, ranging from hadron spectroscopy to the detailed properties of weak interactions. Its existence serves to classify theoretical developments: one can contribute to the orthodox picture, attempt to destroy it, or try to develop an alternative heterodoxy. The evolution of a standard model poses no serious threat to the development of our discipline: the waning power of authority in Western society ensures that young researchers can hardly be expected to be overawed by our model. Let them destroy it if they can. But it just may be the correct model.

The standard model is in no sense "my model". Very many theorists and experimentalists have contributed to its development over at least the past 15 years, and they must all share the credit and the responsibility for it. I am not here to propagandize my work or theirs, but merely to point out that a more-or-less standard model now exists, if only to serve as a basis for future dialogue.

Let me list some of the essential aspects to and implications of the new orthodoxy. I will describe the standard model as restrictively as I can in order to isolate most precisely what it involves.

(1) We agree to abandon all hope, for the moment, to include gravity in the theory. Although an ultimate synthesis must correct this omission, we

assume that a reasonably coherent model of particle physics without gravity exists, and that it is accurately predictive at sufficiently low energy. Gravitational effects are expected to become appreciable at the vicinity of the Planck mass (about a microgram), and our picture will certainly fail at that point. But, not for the forseeable future will direct evidence at such energies be available.

(2) We put our faith in the validity of relativistic quantum mechanics, i.e., local renormalizable quantum-field theory. These dynamical systems appear to be more-or-less internally consistent. Moreover, quite unlike classical or quantum mechanics, or non-local quantum field theories (if such things exist), RQM is exceedingly restrictive: we are limited to fundamental fields with spins 0,½, or 1 and to couplings no more complex than quartic meson couplings, Yukawa interactions, or generalized Yang-Mills interactions. Thus, the standard picture excludes the notion that all (as opposed to none) of the hadrons are "elementary" in the sense that they correspond to quantum fields. It also excludes the existence of fundamental quantum fields corresponding to "strings" or "bags" or magnetic monopoles. Of course, such entities might emerge as epiphenomena, following from the underlying renormalizable local field theory. The study of such systems could prove fruitful and even predictive, but in the standard model they can only be regarded as phenomenological way-stations. Ultimately, ours is the search for the correct Lagrangian and the reason it is what it is, and the development and implementation of techniques to compute observable particle data in terms of the small number of parameters appearing in the Lagrangian.

(3) Matter is quarks and leptons. The leptons (including muons, electrons, neutrinos, and possibly other yet to be, or being, discovered) are truly elementary particles corresponding to fundamental fermion fields. The continuing validity of Quantum Electrodynamics supports this view. Not so for hadrons: these are to be composite systems made up of quarks. Not vague pointlike constitutents with unspecified properties called "partons", but spin ½ fractionally-charged quarks. Mesons are bound states of quark-antiquark while baryons are bound states of three quarks. Just two kinds of quarks suffice to build up ordinary matter. These have electrical charges 2/3 and -1/3 and have almost the same masses. To construct the strange particles, a third kind of quark is needed which is significantly heavier than the other two and carries electric charge -1/3. The hierarchy of phenomenological symmetry groups merely reflects the quark mass spectrum: Isotopic spin is good simply because the p and n quarks are nearly degenerate; insofar as chiral SU(2) is good it is because these quarks are light; approximate SU(3) is useful because none of the quark mass splittings is large compared to the level-splitting characteristic of the quark-binding mechanism, the Regge slope. The picture is more powerful that the eight-fold way, for it provides an explanation (however arbitrary, since we must yet explain why only $\bar{q}q$, qqq and $\overline{qqq}$ bind into hadrons) for the absence of exotic hadrons. Under the rules for hadron binding, mesons can appear only as 1's or 8's and baryons as 1's, 8's, and 10's. We also know that something else is a

conspicuous component of hadrons: deep inelastic electron scattering reveals that much of the proton's momentum resides on these "gluons", and that they must be electrically neutral. (This excludes the Han-Nambu hypothesis that the quarks are integrally charged). This view is confirmed in the comparison with deep inelastic (anti) neutrino scattering at energies not larger than 30 GeV: the neutrino results could have been correctly predicted from electron scattering data under the assumptions of (a) the quark model with electrically neutral gluons, and (b) scaling behavior.

The successes of the quark model in describing hadron spectroscopy are quite impressive. To obtain these results, it must be assumed that the dominant forces responsible for quark binding are spin independent. Thus we obtain a phenomological and very predictive approximate SU(6) symmetry. More importantly, we must treat the quarks as if they satisfied bose-einstein statistics: baryon wave functions must be symmetric under interchange of all known dynamical variables.

(4) Quarks come in colors and flavors. The solution to the statistics problem lies in the introduction of a new dynamical variable: quark color. It is assumed that the baryon wave function is antisymmetric under interchange of quark colors. This is most simply, if not uniquely, accomplished if each kind of quark comes in just three colors. The word flavor is used to distinguish the p, n and λ quarks. We now have nine quarks in each of three colors and three flavors. Does this not vastly increase the number of hadron species we can imagine? This problem is approached by introducing an SU (3) group acting on the color degree of freedom. It is assumed that color SU(3) is an exact symmetry, and that only color singlet states exist as observable states. Since all color singlets can be decomposed into baryons (qqq), anti baryons ($\overline{qqq}$), and mesons ($\bar{q}q$) (in correspondence with the three invariant tensors of SU(3)), not only have we eliminated 8/9 of all conceivable meson states and 26/27 of all conceivable baryon states, but we explain why no other quark configurations (e.g., q, qq, $qq\bar{q}$, etc.) correspond to hadrons. We have done away with fractionally charged hadrons, exotic hadrons, as well as unobserved but non-exotic things like S-wave 20-plets of hadrons in one stroke: Elegant, but quite arbitrary.

(5) Why are observable states necessarily color singlets? A possible reason can be found in a new model of strong interactions: an exact color SU(3) gauge theory. We demand that color SU(3) is an exact (not spontaneously broken) local symmetry. Arguments exist that an exact non-Abelian gauge theory behaves very differently from an exact Abelian gauge theory like QED. Lattice arguments suggest that the force between quarks does not dwindle with $1/r^2$, but remains constant. To separate a pair of quarks by one inch would thus require about as much energy as to lift me 30 feet off the ground. Long before this point is reached, a quark-antiquark pair is created. Thus, any attempt to isolate a quark merely results in the production of a hadron. Similar arguments apply to any other system of quarks and gluons that are not color singlets. They may "exist" within a hadron, but may never be isolated. The case for quark confinement by gauge color SU(3)

is not yet very firm, but proponents of the standard theory would argue that the possibility is too good not to be true.

Buying quark confinement by gauge gluons, we find several significant bonuses. Among possible field theories, only non-Abelian gauge theories may be asymptotically free. Although the coupling may increase at small momenta yielding confinement by infrared slavery, it must also decrease at large momenta, so that the confined quarks may appear free, or point-like, when probed at high momenta. Thus, it is alleged, may we understand the scaling behavior of deep-inelastic lepton scattering. Paradoxically, but quite consistently, quarks are bound by forces so strong that they cannot be isolated, yet so weak that the proton's constitutents may often seem free.

Another bonus is in the nature of the long-range quark-confining force. In lattice models, it is recognized to be primarily a spin-independent force, which is also independent of quark flavor. We thereby obtain the long-awaited justification for the successes of non-relativistic SU(6).

Splittings among SU(6) super multiplets find their origins in the (unexplained) splittings in mass among the quark flavors, in the small spin-dependent part of the qurak-confining force, and in the possibility that the constituent quarks of neutral mesons may annihilate. These mechanism seem adequate to explain the general features of the observed hadron intermultiplet splittings. The interpretation of the $\Sigma\Lambda$ mass difference as a hyperfine splitting is one minor triumph of the standard model. We have yet to successfully attack the problem of splittings among SU(6) multiplets, which clearly depend much more on the detailed dynamics.

In the standard model, the primary strong interaction is mediated by the exchange of massless, neutral, unobservable gluons between colored systems. Color singlet systems, like hadrons, cannot exchange single gluons. What are ordinarily regarded as strong interactions - the meson-nucleon or nucleon-nucleon force - are only the pale remnant of the primary force acting between neutral systems, like the Van der Waals force is the weak, short-range remnant of the electromagnetic force acting between uncharged systems.

Note the different roles played by color and flavor in the standard model. The strong interaction "sees" only color: gluons can change the color of a quark but not the flavor, and the consistency of the theory demands that the strength of the interaction is independent of quark flavor. On the other hand, the wondrous variety of hadron states reflects the variety of quark flavors. The proton is charged and the neutron not because different flavors have different electric charges. The strange particles are heavier because different flavors have different masses. Electromagnetic interactions, weak interactions, and the agency responsible for quark masses "see" flavor, but not color: the W may change the flavor of a quark but not the color, and its couplings to differently colored quarks are the same.

(6) It is expected that the interactions involving quark flavors are carried by a non-Abelian gauge theory, just as are the interactions involving quark colors. However, this gauge theory is spontaneously broken so that only electromagnetism appears gauge invariant, and only the photon is massless. The theory is renormalizable, and its two central predictions have been verified by experiments: Effects of the predicted neutral-current interactions at low energies have been seen, while effects of the predicted very heavy intermediate boson at low energies have not. Necessary for the renormalizability of this theory seems the essential commutativity of weak and strong interactions: the fact that the two gauge groups operate on different dynamical variables: color and flavor. Other possibilities are much more contrived, and may not work at all.

The fact that all interactions of elementary particles are mediated by gauge interactions suggests the appealing possibility that they are all (strong, weak, and electromagnetic) different manifestations of the same simple gauge theory. Such grand unifications seem inevitably to lead to the existence of other, weaker interactions not yet seen in experiment. These may lead to violations of baryon number $(P \rightarrow \pi^+\nu)$ or of lepton number $(n+n \rightarrow p+p+\bar{e}+\bar{e})$. However, there are both theoretical and experimental reasons that these interactions are very strongly suppressed. To accomplish such suppression, dimensional parameters must arise corresponding to masses well beyond the normal purview of elementary particle physics - masses on the order of the Planck mass where gravitational effects become important. Nonetheless, it seems an attractive possibility to expunge our theory of those conservation laws that do not correspond to local symmetries: baryon number and fermion number.

(7) How many flavors have quarks? Three flavors suffice from pre-1975 hadron spectroscopy. However, there are several arguments for the existence of at least one additional quark flavor. One argument involves the technical question of triangle anomalies, which must be absent for the theory to be renormalizable. With only nine quarks and the observed leptons, the anomalies do not cancel. A simple and elegant way to remove the anomalies is by the introduction of a fourth quark flavor with charge 2/3 which participates in weak interactions. With two weak doublets of quark flavors, just as there are two weak doublets of leptons, the theory certainly appears more symmetrical, and the anomalies cancel.

There are, of course, other ways to remove the anomalies. A less ambiguous argument for the fourth quark emerges from the nature of the weak interactions. Unified models of weak and electromagnetic interactions predict neutral currents, and with three quark flavors these neutral currents carry strangeness ±1 as well as zero. Although the strangeness conserving neutral currents have been seen, strangeness violating neutral currents are very strongly suppressed in nature. By a very wide margin, the simplest way to eliminate such phenomena to order G is by the introduction of a fourth quark with $Q = 2/3$ or $Q = -4/3$. The latter choice is no good, however, because the unwanted phenomena reappear in order αG, a circumstance also excluded by experiment. Thus, the standard model involves the existence of at least four quark flavors, and many suggestions have been made that there are even more.

With just four flavors and four leptons, the fundamental matter fields divide into two disjoint and equal sets: n, p, $\bar{e}$, ν on the one hand, which suffice to build the world; and $\lambda\, p'\bar{\mu}\nu'$ on the other, which seem relevant only to the professional particle physicist. Why nature did not make do with half as many building blocks is an intriguing puzzle. The old mystery "Why is the muon?" now seems inextricably bound to two other mysteries: "Why is there strangeness? Why is there charm?"

(8) The November revolution. Nothing said yet could not have been said before the discovery of J or ψ and its kin. But before then, the standard theory had many fewer advocates. We may see in the new particles many aspects of the standard model. They are supposed to be states of "charmonium" - mesons, made up of $\bar{q}q$ like all mesons, but finally revealing the existence of the fourth quark. Their narrow widths are explained by the notion of asymptotic freedom: they are narrow simply because heavy quarks have difficulty annihilating. A whole spectroscopy of new states has been predicted, and several predicted states have been found.

The new quark must bind to the old quark to produce a new class of "charmed" hadrons. Their existence, with predicted behavior under weak interactions, and with masses near 2 GeV, is the remaining severe test for the standard theory. At the moment, the evidence for such states is not convincing: one promising bubble chamber candidate at BNL, about a hundred unexplained dimuons at FNL, and not much else. If charmed hadrons do not exist, another explanation for ψ/J must be found: necessarily involving new fundamental hadrons, new interactions, and new weakly-decaying particles - certainly a less elegant explanation than the standard line. Without charm, our theory of weak interactions would have to be radically restructured, inevitably also requiring the existence of not-yet-observed degrees of freedom. The alternatives to charm seem less palatable than charm itself.

This is the standard model: it is an ambitious attempt to encompass all of elementary particle physics. It provides a natural end to the sequence of spectroscopies physics has encountered: Atomic to nuclear to hadronic to quark. For, how can we study the structure of a particle we cannot in principle produce. As I have failed adequately to stress, the standard model remains incomplete. Arguments for quark confinement are tentative, and there are many aspects of hadron-spectroscopy imperfectly understood. We have no idea how many leptons, or quark flavors, or other things remain to be discovered. Furthermore, the standard model seems epistemologically unsound, depending on hadronic constituents which cannot be isolated. Perhaps someday we will have a new standard model which does not need such crutches. Not yet.

THE STATE OF QUANTUM GRAVITY

Stanley Deser*

*Physics Department, Brandeis University, Waltham, Mass. 02154.
Work supported in part by the National Science Foundation.

I Introduction

Quantum gravity is beginning to diverge in rate of publication as well as in other ways, so I will have to skip many interesting developments and be rather telegraphic about those I do cover. The bibliography contains some recent reviews as well as a sample of the current literature, however. I have construed "quantum gravity" in the broad sense to include recent research in which there is a gravitational field and a quantized system, rather than just the quantized metric. My survey will be divided roughly as follows. First, we will consider the behavior of systems in which matter is quantized, but the metric is classical (whether fixed or dynamically determined). Included here will be the work initiated by Hawking on black hole radiation, where an evolving geometry affects quantized matter fields. This is the area with the most potential for observational consequences. On a more theoretical level in this context, we shall describe some gratifying general results concerning the renormalizability of quantized field systems in an arbitrary metric, whatever its nature. These considerations also show how the classical Einstein equations will be modified by matter vacuum polarization. Our next step will be at the semi-classical level for gravity, namely a discussion of the low frequency tree structure of gravitation and its essentially unique determination by special relativistic quantum requirements. These results justify the Einstein theory in the low frequency domain irrespective of the properties of the "ultimate" correct quantum model of gravity.

Quantum gravity proper -- the study of quantization of the metric field itself -- will be reviewed next. We will sketch the status of the renormalizability problem of conventional Einstein theory coupled to various matter systems, discuss possible improvements and some of the new calculations of radiative corrections as well as technical clarifications (e.g., anomalies). We will also mention the solution of the old canonical versus covariant quantization equivalence problem (Feynman integration measure, etc.), as well as a couple of classical relativity results which are relevant to the quantum side (field energy, null plane formulation). It seems superfluous at this conference to mention either supersymmetry or the possible meaning of the possible appearance of the Planck mass in broken higher symmetry models of the other interactions.

II Quantized Matter in Classical Geometries

(a) Particle Creation

Because it couples to the kinetic energy, an external gravitational field is much more complicated in its effects on quantized matter than say an external electromagnetic one. Even the particle interpretation of an otherwise free field becomes intractable because it is a "Fourier transform", global concept: arbitrary time-dependent metrics, or the changing role of the "time" coordinate in different parts of spaces (as in the Schwarzchild case) make it difficult to decompose. On top of these problems is that of vacuum polarization, i.e., how to regularize an appropriate $\langle T_{\mu\nu}\rangle_g$ in the presence of the metric. This is an essential point, amongst other places, in computing reaction back on the metric, and so on the sources producing it, due to the quantum field's alterations. There is now an extended literature dealing with these problems, both in "realistic" and model (e.g., two dimensional) situations. The main interest lies in radiation of particles (as seen at infinity) by a collapsing classical matter distribution through the time-dependent metric it generates. While there is still a great deal of uncertainty about both details, local interpretation and the wider implications of the process, there is complete agreement from independent calculations about the original results found by Hawking, namely that there is a steady flow of radiated quanta to spatial infinity, and that the spectrum is thermal uncorrelated radiation of black body form for each type of particle (boson/fermion),

$$\langle N \rangle = \Gamma\left[\exp(\omega/kT) \mp 1\right]^{-1}$$

with $kT = \hbar/8\pi GM$ in the simplest spherical neutral collapse. In the absence of back reaction, this process proceeds until the whole mass has exploded away (the lifetime is $\sim M^3$). Hawking has tried to draw some startling conclusions about a new type of indeterminacy for all of physics owing to loss of information down the black hole and its final "naked singularity" state, which would only allow a density matrix description of the whole process. Also, baryon number information will die with the collapsing object, since its eventual mass decrease will eventually trap them inside. Similar attempts have been made by Parker to get the 3° microwave spectrum as the accompaniment of the initial big bang. Clearly what is needed at this point is a way of computing corrections to these processes before extrapolating the consequences too far. For example, even aside from quantum gravity effects, and the quantum nature of the collapsing matter, the reaction from $\langle T_{\mu\nu}\rangle_g$ will generally involve effective vacuum polarization terms proportional to $\alpha R_{\mu\nu}{}^2 + \beta R^2$, etc. already at the classical level. This area will certainly be receiving intensive study.

(b) Renormalization in an External Metric

Before coming to the dramatic problems of quantum gravity itself, we look at a less general but very important question involving renormalizability, namely do flat-space renormalizable quantum field theories (e.g., φ^4 or QED or Yang-Mills) stay that way on earth (or in an earthquake, as the authors put it)? The answer -- yes -- agrees with our equivalence principle intuition, with a couple of provisos. In three recent papers, Freedman et al., concluded that the result holds to all orders in an external metric and to all powers of momentum transfer to the metric. The only new element lies in the need for introducing a non-minimal coupling term $\sim R\phi^2$ for scalar fields; its lowest order coefficient is the familiar 1/6 of conformal invariance, but it may become renormalized to higher orders, so there is one extra parameter (as with $\lambda\phi^4$). Let me sketch a derivation which Boulware and I used. One is interested in the general $n^{\underline{th}}$ derivative with respect to the external metric of the m-particle amplitude,

$$\delta^n/\delta g^n_{\mu\nu} \left\langle \psi_1 \cdots \psi_m \right\rangle_g \Big|_\eta \equiv \Gamma(q_1 \cdots q_n ; p_1 \cdots p_m)$$

Now the Ward identity tells us (this is just the extended equivalence principle of the next section) that Γ is fully determined to order q^2 in any one "graviton", and by symmetry therefore up to $\mathcal{O}(q_1{}^2 \, . \, . \, q_n{}^2)$. But each external graviton insertion preserves the order of a diagram since the vertex ($T^{\mu\nu}$) and extra denominator have cancelling powers of momentum. Since the maximal order of divergence without gravitons is quartic (vacuum diagrams), and logarithmic for all graphs with connected lines (except for scalar self-energy), only the $\mathcal{O}(q_1{}^2 \, q_2{}^2)$ part of a two graviton insertion (into a vacuum graph only) and the $\mathcal{O}(q^2)$ part of a one graviton insertion are dangerous.

The vacuum graph insertions were of course irrelevant in flat space, but now correspond to logarithmically divergent vacuum polarization contributions $\sim \alpha R_{\mu\nu}^{\ 2} + \beta R^2$ to whatever classical action (e.g., Einstein) determines the metric. [There can also be $\sim R$ and cosmological, $\sim \sqrt{-g}$ terms.] This is the reason we emphasized the unavoidable presence of such terms in classical gravitation. Note that the $(R_{\mu\nu}^{\ 2}, R^2)$ terms would not be seen at the single external graviton level since $R_{\mu\nu}^{\ 2}$ is quadratic in the metric (and R itself is a total divergence to linear order). Turning now to the more relevant graphs, those with external matter lines, we find one new term of $\mathcal{O}(q^2)$ in the scalar case $\sim R\phi^2$, the famous new improved non-minimal coupling, which thus requires one new parameter. [$RA_\mu^{\ 2}$, the only possibility in the vector case is, of course, excluded by gauge invariance]. All other graphs with external matter lines are fixed to be of the non-graviton form by the combined gravitational plus, e.g., Yang-Mills Ward identities (these can be shown to still hold), and there are therefore no alterations in the renormalizability properties of flat space models.

What about the cosmological term? Our argument shows that it comes from quartic $\mathcal{O}(q^0)$ graphs determined by the vacuum energy in flat space; but it can be cleverly set to zero to start with. There is another possible source of a cosmological term in theories with Higgs scalars, where the translation leaves a constant term in the flat space action. If it is taken seriously, one has a term $\sim \lambda \sqrt{-g}$ where the characteristic size of $\lambda \sim L^{-4}$ is a meson compton radius rather than a cosmological one ($\sim 10^{27}$cm). Either way, whether through the divergent diagrams or the Higgs mechanism, there is an enormous mismatch with observation, so a counter term which removes it exactly is required. It is not clear whether all this is trying to tell us something about the correct matter system. In this connection, there is one exception, supersymmetry models, for which Zumino has shown that the flat space vacuum energies of the fermions and bosons cancel exactly (in all theories the leading terms cancel between bosons and fermions). It is amusing, incidentally, that plane gravitational waves, like their counterparts in QED, are unaffected by vacuum polarization (to all orders), provided the cosmological term is absent.

Aside, then, from pure gravitational vacuum polarization terms, we conclude that the idealization of gravity to a classical external metric does not disturb our flat space renormalizability results.

III The Tree Level Structure of Gravity

The work I am reporting on here has recently been published, so I will be brief. About ten years ago, Feynman and Weinberg independently proposed a program to show that general relativity with its nonlinear structure is a consequence of the helicity 2 of the graviton; Weinberg obtained a number of basic results, and Boulware and I were able to complete the program. The idea is to begin with special relativistic quantum particle theory, in particular with the idea that all interactions are mediated by exchange of quanta, narrow the possibilities down to a massless spin 2 exchange as the primary carrier of gravitational forces and then discover the constraints on the "currents" coupled to a helicity 2 particle (no field theory proper need be invoked). The observational input is that there exist, between at least some macroscopic static systems, long range attractive forces, and that light is also affected. The macroscopic long range character excludes fermion exchange, while long range plus static excludes spin higher than 2. Spin 1 is excluded because it leads to repulsion between the "charges", while spin 0 can couple to light (e.g., via $\phi F_{\mu\nu}F^{\mu\nu}$), but not in anywhere near the observed way. Once spin 2 is singled out, one may use the Van Dam-Veltman argument to exclude any finite range (it would predict 75% of observed light bending). But Lorentz invariance then determines the matrix element $M_{\mu\nu}(p',p)$ which couples to the graviton polarization tensor $\epsilon_{\mu\nu}(p'-p)$ through first order in momentum transfer; this is the analog of the photon low energy theorems. The $M_{\mu\nu}$ is necessarily of the form of the matrix element of the usual stress tensor of the system, including the spin contribution. Most important here, the coupling must be universal (equivalence principle) which means that graviton-graviton vertices must exist as well. The latter consists of an infinite series of higher and higher vertices, whose sum $V = \sum_{n=3}^{\infty} V_n$ satisfies the Ward identity associated with the graviton propagator whose solution for S-matrix purposes is unique (apart from field relabellings in usual language). It can be shown to be generated by the cubic and higher parts of the Einstein action, together with minimal coupling to matter (except, as usual, for $R\varphi^2$). Thus, the total tree diagram generator for gravitational interactions is the usual Einstein action plus minimal coupling. The result holds to all orders in the gravitons and to quadratic order in their momenta. Note that this derivation is independent of whether the ultimate nature of gravity is as a fundamental field or e.g., a collective excitation. The low frequency effective action does not involve any quantum field considerations. Also, it is shown that there exist closed loop diagrams, namely ladders and crossed ladders which are responsible for the Kepler orbits and perihelion precession for example, which can be evaluated from this action without worrying about ultraviolet behavior.

Similar derivations of the self-coupling may be given at the corresponding classical field level, starting from an infinite range pure spin 2 field, but the quantum result is perhaps more compelling. Other indications that the helicity constraints are strong come the fact that the Born amplitude for graviton-graviton scattering is uniquely determined from kinematical constraints for helicity 2. This aspect, and the factorization properties of various Born amplitudes involving gravitons have been studied by a group at Brandeis.

IV The Quantized Gravitational Field

We come now to "real" quantum gravity, where the metric field is quantized. I will review the status of the renormalizability problem and possible remedies. We will also mention some examples of the many new explicit calculations of divergent and convergent processes involving gravitons and matter. Finally, we will touch on some interesting formal developments such as the old problem of equivalence of quantization and canonical quantization.

(a) Nonrenormalizability; explicit calculations

When the Einstein tree action is taken seriously as the basis for a quantized field theory, rather than as an effective action, dimensional considerations alone make it likely that closed loops will give higher and higher divergences in each order, because the Einstein coupling is through a vertex $T_{\mu\nu}$ which is quadratic/linear in momentum for bosons/fermions. So one-loop effects start off quartically divergent, while at two loops, one has $\sim \kappa^2 \Lambda^6$ cut-off dependence, etc., where κ^2 is essentially the Newtonian constant. The internal counting is the same for graviton-matter and graviton-graviton vertices. Therefore, only miraculous cancellations can save the theory from requiring an infinite series of pure gravity counter terms proportional to increasing powers of the curvature (and hence of derivatives!), as well as matter-gravity and pure matter terms (e.g., for QED $\sim \kappa^{2n} F^{2n+2}$). The latter in particular would "pollute" the pure matter sector, an important point to remember! Now, if miracles are to occur, they had better start at one-loop order or else as we shall see, there will be a radical change in the free-field part of the gravitational action itself.

Let me recall the results of 't Hooft and Veltman on pure gravity and gravity plus scalar coupling, together with those of van Nieuwenhuizen and myself on Brans-Dicke models, on Einstein-Maxwell, Einstein-Dirac and (with H.-S. Tsao) on non-abelian gauge fields plus gravity. In all cases but pure gravity, one obtained counter terms which were of the non-renormalizable type, i.e., did not vanish on shell and were not proportional to the original action. The exception of pure gravity must not be misunderstood, and I wish to emphasize that all dimensionally possible covariant counter terms were present in the calculation; there were no cancellations at all. Instead, it just happens to be impossible at the one-loop, and only at the one-loop, level to have a non-vanishing counter term. This is because the most general possible form for ΔL is

$$\Delta L \sim (1/n-4) [\alpha_1 R^2_{\mu\nu\alpha\beta} + \alpha_2 R^2_{\mu\nu} + \alpha_3 R^2].$$

[Since dimensional regularization kills tadpoles and vacuum graphs here, there are no $\sim$ R or $\sqrt{-g}$ counter terms]. Indeed, none of the α_i vanish by explicity calculation, but rather there is (in 4 dimensions) an identity reducing the a priori nonvanishing $R^2_{\mu\nu\alpha\rho}$ term to the other two. But there is no other identity left, and so possible 2-loop terms such as $\Delta L \sim \kappa^2 \, tr \, R^3_{\mu\nu\alpha\beta}$

and so on cannot be gotten rid of and there would have to be a real cancellation here and at higher orders. No one has yet done a 2-loop calculation (1 loop is discouraging enough), so of course it is conceivable that the miracle occurs. However, the point seems to be moot at best since as soon as coupling to matter is included, nonrernormalizable terms appear right away. With scalars, for example, all possible counter terms again appear, namely ΔL is a sum of $R_{\mu\nu}{}^2$, $\kappa^4 T_{\mu\nu}{}^2$ and $\kappa^2 R_{\mu\nu} T^{\mu\nu}$ type terms, which, when taken on the combined mass shell ($G_{\mu\nu} = \kappa^2 T_{\mu\nu}$ and $\Box\phi = 0$) no longer vanish. The hope that Brans-Dicke theory might be better because the dimensional coupling constant is replaced by a scalar field is also dashed because the one-loop corrections look like those in the scalar case above.

For coupling to spinors, one must introduce as basic gravitational variables the vierbeins, but it can be shown that quantization in terms of vierbein or metric is equivalent. One particular counter term which can be obtained explicitly and is not equivalent on shell to any others is proportional to the fourth power of the axial current, $\Delta L \sim \sqrt{g}\,[(\bar\psi\gamma^5\gamma^a\psi)\eta_{ab}(\bar\psi\gamma^5\gamma^b\psi)]^2$ where η is the local Minkowski metric. [There is nothing surprising about the γ_5's; the contributions of the separate Majorana pieces are just adding up here].

One might hope that coupling to the other long-range gauge fields, Maxwell or general nonabelian SU_n multiplets would look better. In the Maxwell case, one finds a sum of terms in the form $(R_{\mu\nu}^2,\ \kappa^2 R_{\mu\nu}T^{\mu\nu},\ \kappa^4 T_{\mu\nu}^2)$ which do not cancel on shell, one of the form $(D_\mu F^{\mu\nu})^2$ which does, but a number of apparently allowed other terms such as $\kappa^2 R_{\mu\nu\alpha\beta} F^{\mu\nu} F^{\alpha\beta}$, $\kappa^4 (F^2)^2$ are absent. Once again, further analysis shows that there is no cancellation involved, but that the missing terms are forbidden by duality invariance (rotation of E and B into each other) of Maxwell theory, since the only local invariant is $T_{\mu\nu}$ itself, and of course $(D_\mu F^{\mu\nu})^2$. This transformation can be shown to be locally implementable for Maxwell theory in flat or curved space and is therefore reflected (just like gauge invariance) in ΔL. [A recent attempt by Kallosh to link the absent terms to a hidden scale invariance of the combined system can be shown to be incorrect]. Although the general Yang-Mills case involves two extra parameters a priori, the number of components and the self-coupling constant of the SU_n field, it can be shown that its ΔL is of the same form as for Maxwell (even though Yang-Mills is <u>not</u> duality invariant), aside from the usual renormalizable $F^a_{\mu\nu} \cdot F^a_{\mu\nu}$ term.

There is as yet no full scale calculation of ΔL involving more than one matter field; however the various contributions to $T_{\mu\nu}{}^2$ and $R_{\mu\nu}{}^2$ all have (by unitarity) the same sign, and it is unlikely that some magic combination would give sufficient cancellation. In addition, there are some recent calculations by a Brandeis group involving both electromagnetic and gravitational interactions of charged fermions or scalars which give rise to ugly infinities. These include fermion-fermion scattering and graviton radiative corrections to the electromagnetic form factor in which $F_1(0)$ is happily finite but $F_1(q^2)$ is not. Indeed, things have reached the point where a calculation (Berends and Gastmans) which shows that $F_2(0)$ is finite, giving a (g-2) contribution of $7/2(GM^2/2\pi)$ is regarded as an anomaly! There is actually an interesting question regarding whether the form factors should include graviton-photon vertices or not; F_2 is infinite if not, and it is really physics that decides here.

Some other recent calculations which bear mentioning are explicit verifications of the theorem of Sec. II that external graviton insertions do not give rise to new infinities (photon radiative corrections to the graviton vertex, for example), and corrections to the bending of light. There have also been some earlier calculations of such things as graviton loop corrections to the long-range Newtonian potential, which are $\sim G/r^3$ (Radkowski). Finally, we mention the existence of trace anomalies (Capper and Duff) which will in general be present (though finite in one loop order) unless the relevant Ward identities are valid for all dimensions and do not involve the dimension explicitly. These have been seen (Berends-Gastmans) in the (finite) fermion triangle diagram with two external photons and one external graviton, whose trace over the graviton indices does not vanish as the fermion mass goes to zero. The effect of anomalies should be pursued, particularly with regard to models which start from a conformally invariant action.

b) "Improved" models

The specter of higher and higher powers of curvature in counter-terms from higher loop orders is distressing, and far worse than in other nonrenormalizable models. However, there is one way to turn defeat into (pyrrhic?) victory, as was first suggested by DeWitt and Utiyama long ago. The one-loop quadratic terms in $R_{\mu\nu}$ (and only these) have bilinear "free field" terms in the metric which go as p^4, and therefore if taken as part of the propagator rather than vertex terms, they give rise to graviton denominators which go as p^{-4} (as well as p^{+4} vertices from the higher powers in g expansion of $R_{\mu\nu}{}^2$). Power counting then shows that, because the coefficient of $R_{\mu\nu}{}^2$ is dimensionless, the overall effect is to restrict all divergences to $R_{\mu\nu}{}^2 + R^2$ form (more generally, with matter, there will also be R and $\sqrt{-g}$, but no other, contributions). The theory has now become renormalizable, not surprisingly, since we have basically added a Pauli-Villars or Lee-Wick regularization through the $R_{\mu\nu}{}^2$ terms. In arriving at these conclusions, care must be exercised not to pick the wrong combinations of $R_{\mu\nu}{}^2$ and R^2, to choose appropriate fourth order gauge breaking terms, to make sure the corresponding ghost sector is well-behaved, etc. Little is published on this, but some of the work has been done, and it seems to be O.K. The price of good behavior is of course a ghost when we decompose $(\alpha p^4 + \kappa^{-2} p^2)^{-1}$. For most people the price is too high, and they are probably right. I still feel personally that this line is worth pursuing, and that the ghosts will somehow become harmless.

A particularly appealing model is one in which the world is initially conformally invariant, namely

$$L = \sqrt{-g}\,[C^{\mu}{}_{\nu\alpha\rho}\, C^{\nu}{}_{\mu\beta\sigma}\, g^{\alpha\beta} g^{\rho\sigma} - \tfrac{1}{4} F_{\mu\nu}^{2} + \bar{\psi}\gamma D\psi]$$

where C is the Weyl (conformal) tensor, and there is no dimensional coefficient. It is very appealing because only in four dimensions can we have both the vector field conformally invariant and a conformally invariant purely geometrical (without compensating scalar fields) gravitational action which has a quadratic free field part (massless fermions are always invariant). In 2n dimensions one would have to use an action proportional to C^n because although C^2 itself is invariant, the $\sqrt{-g}$ times two $g^{\alpha\beta}$'s is not, and C^n is proportional to g^n in lowest order. The usual Einstein term would arise from conformal breakdown either through anomalies or by another mechanism. Of course, this action still has a (dipole) ghost, so there is no immediate improvement on this score. Incidentally, fourth derivative models appear from recent investigations to behave rather normally in other respects, i.e., both as canonical systems and with regard to Feynman rules.

What if nature does not choose the above models? It seems very difficult to restore perturbation renormalizability in any other way, which has led people to suggest that gravitation is not a basic field, and that its origin might be as a collective matter excitation. This is of course a modern version of Mach's principle, and is subject to the same problems, i.e., we don't even know how to get the Lorentz metric either classically or from quantum fluctuations. Earlier speculations by Sakharov and by Klein that fluctuations would give rise to terms like $\langle \bar{\psi}\psi \rangle R$ already presupposed a Riemann space and do not seem crazy enough. I would prefer not to have to relegate geometry to be an appendage of matter just yet, at least.

c) Related Developments

As is well known, it is possible to exhibit explicitly the equivalence of covariant quantization methods, based on path integration, gauge-breaking etc., with canonical quantization based on p's and q's. Faddeev and Popov have carried out the same program for gravitation, although the absence of an axial gauge analog (which linearizes Yang-Mills considerably) means that some heavy computation is involved to determine the weight function $\mu(g)$ which multiplies the product $\pi dg_{\mu\nu}(x)$ in the integration. The result is that $\mu(g)$ is simply $(\sqrt{-g})^{-5}$ which justifies some (and negates other) earlier work on this question. I suspect a simpler derivation is possible, but the important thing is that the measure and the equivalence are established. With dimensional regularization, the specific form of the weight turns out to be somewhat academic, because as Faddeev and Popov note, it corresponds to an addition to the action proportional to $\delta^4(0)$, which is set to zero anyway. If this is a consistent procedure, one could also settle old questions concerning whether the integration is to be carried out only over metrics with (+++-) signatures, etc.

Formally, this means a step function inclusion in the measure, to which the $\delta^4(0)$ remark applies as well when it is promoted to the action. Life seems almost too easy with dimensional regularization in this respect.

The next item is also known from vector gauge theories, namely the null plane formulation of classical Einstein theory (Aragone et al.; M. Kaku). Amazingly, one can actually give an explicit form of the action in terms only of the two unconstrained degrees of freedom, with no remaining constraints. The result is highly non-polynomial, which dampens the motivating hope that an interacting string renormalizable formulation would be possible, as that would require a finite number of interaction vertices. Still, the null form is very interesting for its own sake, and the null Hamiltonian in particular has a number of interesting properties.

The normal Hamiltonian of Einstein theory has also seen some progress in the sense that its positiveness properties have been explicitly proved in a number of cases by Geroch and his group and a full derivation may soon be given. This would be an important complement to the existing indirect derivations. I mention also that some apparently tachyonic modes of the Einstein field have been shown to be perfectly respectable. In view of the importance of positive energy in quantum theory, this property should be checked for any new model. A final formal point lies in the correspondence between background field and diagrammatic expansions in higher loops. This seems to be in order, from recent work.

We have unfortunately not had time to deal with other approaches to gravity, such as the elegant twistor calculus of Penrose, whose spirit is to keep light cones unfluttering but to quantize the points of space-time, in contradistinction to normal quantization. Nor have we mentioned the "Yang-Mills" approach to gravity, recently suggested by Yang, or the "gravity" approach to Yang-Mills in which a higher dimensional space includes both fields a la Kaluza-Klein. Finally, kinks, which were first suggested long ago in the context of relativity by Finkelstein and Misner, have been studied again in this context by Finkelstein and others at the classical level. A quantum treatment seems far off, however.

It is a pleasure to thank the Princeton University Physics Department for its hospitality while part of this paper was written.

References

I have attempted to collect this (highly incomplete) list under the same headings as in the text. Two very recent reviews with quite extensive references are:

"Quantum Gravity", C. J. Isham, R. Penrose, D. W. Sciama, eds., Oxford University Press (1975).

B. S. DeWitt, Phys. Reports, 19 No. 6 (1975).

Section II a):

S. W. Hawking, Comm. in Math. Phys. 43, 199 (1975).

M. Veltman, Phys. Rev. Letters 34, 777 (1975).

Some recent preprints include:

S. W. Hawking, OAP-412 (Cal Tech)

B. J. Carr, OAP-415 (Cal Tech)

D. Boulware, RLO 1388-693 (U. of Wash.)

R. Wald, EFI 75-13 (Chicago)

L. Parker, UWM 4867-75-8 (Wisconsin - Milwaukee)

W. G. Unruh (McMaster)

U. H. Gerlach (California)

B. S. Fulling and P. C. W. Davies (King's College, London)

B. S. Fulling, P. C. W. Davies and W. G. Unruh (King's College, London)

Section II b):

D. Z. Freedman, I. J. Muzinich, E. J. Weinberg, Ann. Phys. 87, 95 (1974).

D. Z. Freedman and E. J. Weinberg, Ann. Phys. 87, 354 (1974).

D. Z. Freedman and S. Y. Pi, Ann. Phys. 91, 442 (1975).

D. Boulware and S. Deser (unpublished)

M. Veltman, Phys. Lett. 50B

B. Zumino, Nucl. Phys. B89, 535 (1975).

G. W. Gibbons, Comm. in Math. Phys. (in press)

S. Deser, J. Phys. A (in press)

Section III:

R. P. Feynman, unpublished Cal Tech lectures (1962).

S. Weinberg, Phys. Rev. B135, 1049 (1964);
Phys. Rev. B138, 988 (1965).

D. Boulware and S. Deser, Ann. Phys. 89, 193 (1975).

H. Van Dam and M. Veltman, Nucl. Phys. B22, 397 (1970).

D. Boulware and S. Deser, Phys. Rev. D6, 3368 (1972).

S. Deser, J. Gen. Rel. Grav. 1, 9 (1970).

Section IV:

R. Utiyama and B. DeWitt, J. Math. Phys. 3, 608 (1962).

G. 't Hooft and M. Veltman, Ann. Inst. H. Poincare 20, 69 (1974).

S. Deser and P. van Nieuwenhuizen, Phys. Rev. D10, 401; 410 (1974).

S. Deser, P. van Nieuwenhuizen, H. S. Tsao, Phys. Rev. D10, 3337 (1974).

S. Deser, M. Grisaru, P. van Nieuwenhuizen, C. C. Wu, Phys. Lett. B58, 355 (1975).

T. T. Wu and C. N. Yang, ITP-SB-75/23, 31.

R. Kallosh, Phys. Lett. 55B, 321 (1975).

S. Deser and C. Teitelboim (Princeton preprint)

M. Grisaru, P. van Nieuwenhuizen, C. C. Wu (Brandeis preprints)

F. A. Berends and R. Gastmans, Phys. Lett. 55B, 311 (1975) and report in 1975 SLAC Conference Proceedings.

D. M. Capper and M. S. Duff, Nuovo Cimento 23A, 173 (1974).

A. F. Radkowski, Ann. Phys. 56, 314 (1970).

A. Gavrielides, T. K. Kuo, S. Y. Lee, Phys. Rev. (in press).

K. Stelle (unpublished).

C. Bernard and A. Duncan, Phys. Rev. D11, 848 (1975).

O. Klein, Physica Scripta 9, 69 (1974).

A. D. Sakharov, Sov. Phys. Doklady 12, 1040 (1968).

L. Halpern (F. S. U. preprint).

L. D. Faddeev and V. N. Popov, Sov. Phys. Usp. 16, 777 (1974).

C. Aragone and A. Restuccia, Phys. Rev. D15 (in press).

M. Keku, Nucl. Phys. B91, 99 (1975).

R. Geroch and P.-S. Jang (to be published).

S. Deser, Phys. Rev. D12, 943 (1975).

R. Penrose in "Quantum Gravity".

C. N. Yang, Phys. Rev. Lett. 33, 445 (1974).

Y. M. Cho and P. G. O. Freund, EFI 75-15 (Chicago preprint)

D. Finkelstein and G. McCollum, J. Math. Phys. 16, 2250 (1975).

DISCUSSION

J. BERNSTEIN (Stevens)

Might it be that gravitation is renormalizable only at special values of the coupling constants; i.e., an eigen-value condition?

DESER

Well the problem is that at the one loop level already the counter terms happen to be independent of the coupling constants so that in perturbative sense you're starting off on the wrong foot. An answer to your question has been provided by people like Salam who claim that if you do non perturbative calculations all would be well, but there's never been to my knowledge anything near this sort of nice possibility which exists in QED. And remember there you're going from renormalizable to finite and it's a deeper step here.

B. ZUMINO (CERN)

Your slide had all sorts of cases in which there was gravity plus something but never three things mixed together. Is it known what happens? And secondly, as long as I'm asking a question, some people have been saying that maybe in fact it's not even three things put together but you have to put in an infinite number, for instance, people who have been working with dual models claim to have things which in some way are renormalizable. Of course, they have an infinite series of particles but in the low energy approximation it looks exactly like gravity. So either by putting together let's say 15 or an infinite number can it possibly work?

DESER

Let me talk about an infinite number first. I had a slide somewhere here with which I was going to talk about the dual models. The point there is that indeed there were things that looked like pre-gravitation. However, if you wanted to make a real quantum field out of it you could only do it if that field could be expressable in a purely polynomial way because that's all that you can support with interacting strings and in fact a program was carried out by a number of people - Kaku who I think is here took general relativity and did it by expressing it in null plane coordinates in which case one can amazingly enough solve explicitly all the constraints in terms of the two dynamical variables. However, even with this breakthrough the thing is hopelessly non-polynomial so that at least he concludes that there is no manifest way of using the renormalizable string game. As far as I know that's the picture. To come back to 15 from infinity, I think that it's true that - I know of no calculation which simultaneously deals with two matter fields. However, it's hard to see how if the two matter fields are in any nonviolent i.e. renormalizable interaction with each other how that could happen because if you look at the way the contributions to $R_{\mu\nu}^2$ and $T_{\mu\nu}^2$ build up in the individual field cases the signs are always the same. That's not a proof of course for two reasons because there might be other cross terms that could win and because the interactions might generate up. But, if the interactions are relatively mild that's unlikely, and on the other hand I wouldn't want to try a non-renormalizable matter model in the hope that its troubles will just cancel the Einstein trouble. It is certainly something that should be done.

B. ZUMINO (CERN)

Has anyone put let's say the spinors and the Yang-Mills field in at the same time?

DESER

No, not that I know of but you can - I mean you really don't have to do that much, because you can of course recycle all the graphs you've already used and then you just have the additional ones and see if they contribute terms of the same form. Strangely enough that has not been done. We just never got around to it but our feeling was certainly that it was highly unlikely.

SUPERSYMMETRY

Bruno Zumino*

*CERN, Geneva, Switzerland.

1. Supersymmetry and Graded Lie Algebras

The word "Supersymmetry" was introduced in Ref. [1] to denote a particular kind of graded Lie algebras. Over the last year it has come to be used in the physics literature as a synonym for graded Lie algebra, and is now beginning to infiltrate the mathematical literature as well. The elements of a graded Lie algebra are graded, for instance one distinguishes between even and odd elements of the algebra (although finer gradings may sometimes be useful). The graded Lie algebra is specified by giving for any two elements their Lie bracket, which is again an element of the algebra. For two even elements, or for an even and an odd element the Lie bracket is odd (commutator); for two odd elements the Lie bracket is even (anticommutator).

As an example, let Q_a be a four-component Majorana spinor, $\bar{Q}_a = (\tilde{Q}\gamma^o)_a$ its Pauli adjoint. We may use the Majorana representation in which the Dirac gamma matrices have real elements. Let P_α be the four-momentum operator; we may consider the graded Lie algebra

$$\begin{aligned} \{Q_a, \bar{Q}_b\} &= -2(\gamma^\alpha)_{ab} P_\alpha \quad , \\ [P_\alpha, Q_a] = 0 \ , &\qquad [P_\alpha, P_\beta] = 0 \quad . \end{aligned} \tag{1}$$

Here, as usual, the curly bracket denotes the anticommutator. Q_a are the odd elements of the algebra, P_α the even elements. The matrix elements $(\gamma^\alpha)_{ab}$ of the Dirac matrices play the role of structure constants. One may enlarge this graded Lie algebra by adjoining the angular momentum operator $M_{\alpha\beta}$. Its commutation relations with Q_a and P_α are dictated by their property of being respectively a spinor and a four-vector. This enlarged graded Lie algebra is then an extension of the algebra of the Poincaré group. The elements $M_{\alpha\beta}$ are even; a finer grading would attribute to each element a grade equal to its dimension (in units of mass), which is 0 for $M_{\alpha\beta}$, $\frac{1}{2}$ for Q_a and 1 for P_α. In addition to Ref. [1], there are a few other papers which review the physical and the mathematical applications of graded Lie algebra [2-4].

During the last year both mathematicians and physicists have worked on the problem of classifying all semi-simple graded Lie algebras [5-7], which is the analogue of the problem solved by Cartan for Lie algebras. The classification will soon be complete. Notice however that in local quantum field theory the conserved generators must arise as space integrals of the line component of a conserved current

$$Q_a = \int J^o{}_a \, d^3x \ , \qquad \partial_\alpha J^\alpha{}_a = 0 \ . \tag{2}$$

Here α is a vector index. Let us, for the time being, leave unspecified the nature of the index a. If the Q_a are to be Fermi charges (odd elements), they must satisfy anticommutation relations, which implies that the current satisfies a locality condition

$$\{J_{\alpha a}(x), J_{\beta b}(x')\} = 0 \ , \qquad x, x' \text{ space-like} \ . \tag{3}$$

To satisfy the theorem on the connection between spin and statistics, the indices a and b must be spinorial and therefore the charges Q_a carry half-integral spin. This simple argument shows that supersymmetry must necessarily be intimately connected with the Poincaré group. In particular, one cannot take just any graded Lie algebra and use it as an internal symmetry, since this would require scalar Fermi charges. Arguments of this type show that the supersymmetries compatible with our usual field theoretic picture form a very restricted class [8].

An interesting exception is worth mentioning. It consists of the Slavnov transformations of Becchi, Rouet and Stora [9]. These are supersymmetry transformations, occurring in non-abelian gauge theories, which connect physical fields like the Yang-Mills fields with the Faddeyev-Popov ghosts. Invariance under the Slavnov transformations implies the Slavnov identities for the Green's functions, which are a very useful tool for proving the renormalizability and unitarity of non-abelian gauge theories. The generator of the Slavnov transformations is a scalar Fermi charge ρ and it satisfies the very simple supersymmetry algebra

$$\rho^2 = 0 \ . \tag{4}$$

Because of the occurrence of ghosts there is no contradiction with the spin-statistics theorem. We shall not discuss any further the Slavnov transformations here.

As described in Refs. [1-4] and in a number of papers quoted there, there exist local renormalizable field theories in four space-time dimensions [10] which admit the algebra (1), (2), (3). In these theories the particles arrange themselves in supermultiplets consisting of particles of both integral and half-integral spin. The conserved generators Q_a, applied to a one particle state give another one particle state differing in spin by one-half. When the supersymmetry is exact, the particles in a supermultiplet all have the same mass and their couplings to one another are related in a special way. These relations are preserved by renormalization. In general one finds that a supersymmetric theory is less divergent than the generic theory of its kind, due to compensation of divergences among diagrams involving different numbers of boson and fermion lines. However, no examples are known in which this compensation of divergences renders a renormalizable theory super-renormalizable, or a non-renormalizable theory renormalizable. Among the renormalizable supersymmetric field theories the supersymmetric Yang-Mills theories are of special interest [11]. They provide examples of Yang-Mills theories with scalars where the scalar couplings are expressible in terms of the Yang-Mills coupling constant and simplify the constructions of asymptotically free theories with scalars.

With the possible exceptions of the massless particles (photon, neutrino, graviton) no bosons and fermions having the same mass are known. Therefore, if supersymmetry is to be used in particle physics, it cannot be exact. Supersymmetry can be broken explicitly but softly [12], in such a way that the theory remains renormalizable and that the relations among masses and couplings are corrected by finite terms calculable in perturbation theory. The masses of the particles in a supermultiplet, instead of being equal, satisfy sum rules. There is no fundamental objection to this approach. Alternatively, supersymmetry can be spontaneously broken. Corresponding to the conservation law (2), this gives rise to a massless "Nambu-Goldstone" particle, which in our case is a spinor and a fermion. It has been suggested [13] that it should be identified with the (electron) neutrino. Unfortunately, this idea does not seem to be tenable. Bardeen [14] has pointed out that the conservation law (2) implies a low energy theorem which is not satisfied by the neutrinos emitted in beta decay. Furthermore, no anomalies seem to occur, which could violate the validity of the low energy theorem. A possible way out is a kind of Higgs effect, as it can occur in certain supersymmetric theories of gravitation [15], by which the Goldstone spinor disappears, absorbed into higher spin fields present in the theory.

Supersymmetry has been very successful in showing the existence of renormalizable field theories with special, previously unsuspected, properties. On the other hand, the search for realistic models has met with a number of difficulties. We mention a few here. Since bosons and fermions occur on equal footing in supermultiplets and the bosons known in nature are naturally described by real fields, one is led almost inevitably to the use of Majorana spinors and to spinors belonging to the adjoint representation of the internal symmetry group. This makes it difficult, even if not impossible, to construct models with quarks and models with fermion (baryon or lepton) number [16]. In view of the success of the quark model in its various forms, this is not a very satisfactory situation.

Up to this point we have imagined that an entire supermultiplet belongs to a given representation of the internal symmetry group, so that the algebra (1) and the internal symmetry algebra commute. One can envisage a less trivial way of introducing the internal symmetry, in which the spinor charges themselves are assigned to a representation of the internal symmetry group. For instance, the spinor charge Q_{ai} could carry, in addition to the old spinor index a, an index i which makes it into an SU(3) triplet, and possibly an additional triplet index j for color SU(3): Q_{aij}. Then the spinor charges have the quantum numbers of a quark field. Applying at rest the operator Q_{aij} an arbitrary number of times to a basic no-quark state, one generates a finite supermultiplet containing all states of the quark model, among them color singlets with the desired quantum numbers. However, many unwanted states (one-quark states, two-quark states, etc.) also belong to the supermultiplet and, if the supersymmetry is exact, have the same mass as the color singlets. One can imagine interactions which break the supersymmetry so as to give a lower mass to the color singlets. However, the mass splittings cannot be too large, if one wants to preserve some of the relations implied by the supersymmetry. For instance, the experimental lower bound for the mass of fractionally charged quarks, which is of the order of 10 GeV, is already much too large for this picture to make sense. For integrally charged quarks of the Han-Nambu variety the experimental lower bound on the mass is not as high and perhaps acceptable. At any rate, in this picture the quarks would have to be real, not confined. In addition to these difficulties concerning the spectrum, there is another serious difficulty with the use of spinor charges carrying internal symmetry quantum numbers. Nobody has succeeded, until now, in exhibiting a renormalizable local field theory with conserved spinor charges of this kind.

In concluding this section we must honestly admit that at the moment the prospects for applying supersymmetry to hadronic physics are not very good, at least within the usual field theoretic framework. On the other hand, geometric methods of the kind outlined in Section 3, may be very useful for the construction of realistic dual models with spin and internal symmetry. For leptonic physics, the supersymmetric Yang-Mills theories are perhaps still a hopeful starting point towards the construction of more realistic models. The most interesting application of the idea of supersymmetry at present is the possibility of constructing unified theories of gravitation and matter, along the lines described in Ref. [15] and in Sections 2 and 3. Such theories are expected to be less divergent than the traditional Einstein theory of gravitation [17] and represent perhaps a step towards the construction of a renormalizable theory.

2. Geometry of Superspace

The algebra (1) can be realized in terms of geometrical transformations in superspace. We call superspace a space whose points are labeled by four commuting coordinates x^α and by a number of additional totally anticommuting coordinates θ^a

$$\theta^a\theta^b + \theta^b\theta^a = 0 \quad , \tag{5}$$

which also commute with the x^α. This means that the x^α are even elements and the θ^a odd elements of a Grassmann algebra. We shall restrict ourselves to the simplest case, where the θ^a are the four components of a Majorana spinor, but in general they could consist of several spinors, belonging to some representation of an internal symmetry group. Consider the infinitesimal transformation in superspace

$$\begin{cases} \delta x^\alpha = i\bar{\zeta}\gamma^\alpha\theta \\ \delta\theta^a = \zeta^a \end{cases} \tag{6}$$

where ζ^a is an infinitesimal spinorial parameter, anticommuting with itself and with θ^a, but commuting with x^α. The product of two infinitesimal transformations like (6) gives

$$\begin{cases} \delta_2\delta_1 x^\alpha = i\bar{\zeta}_1\gamma^\alpha\zeta_2 \\ \delta_2\delta_1\theta^a = 0 \end{cases} \tag{7}$$

and the commutator

$$\begin{cases} (\delta_2\delta_1 - \delta_1\delta_2)x^\alpha = 2i\bar{\zeta}_1\gamma^\alpha\zeta_2 \\ (\delta_2\delta_1 - \delta_1\delta_2)\theta^a = 0 \ . \end{cases} \tag{8}$$

Observe that $2i\bar{\zeta}_1\gamma^\alpha\zeta_2$ is a real even element of the Grassmann algebra, so (8) is an infinitesimal translation. The algebra is the same as (1), which can be also written in terms of commutators

$$[\bar{\zeta}_1 Q, \bar{\zeta}_2 Q] = -2\bar{\zeta}_1 \gamma^\alpha \zeta_2 P_\alpha \quad , \tag{9}$$

provided one stipulates that the parameters ζ^a anticommute with Q^a and commute with P_α (a graded Lie algebra can always be transformed into a Lie algebra by using parameters belonging to a Grassmann algebra). Observe that the transformations (6) leave invariant the differential forms

$$\begin{cases} \omega^\alpha = dx^\alpha + i d\bar{\theta}\gamma^\alpha\theta \\ \omega^a = d\theta^a \end{cases} \tag{10}$$

in a very similar way as the translations leave invariant the differentials dx^α in four-space.

A superfield is a field in superspace $V(x^\alpha, \theta)$. Expanding it in power series of θ^a, the series terminate at some finite power, since the square of each component θ^a is zero from (5). For instance, if θ^a is a Majorana spinor, the highest power is four, when the four different components are multiplied together. Therefore a superfield corresponds to a finite supermultiplet of ordinary fields of various spins, two successive terms in the expansion containing fields which differ in spin by one-half. One may require that a superfield transform like a scalar in superspace

$$V'(x,\theta) = V(x',\theta') \tag{11}$$

under the transformation (6). This induces a rearrangement among the fields of the supermultiplet and their derivatives and gives rise to a representation of the supersymmetry algebra. A superfield can also carry Lorentz and internal symmetry indices which specify its transformation properties under those groups. In this case, all fields of the supermultiplet carry those indices.

The idea of superspace permits an interesting geometric unification of gravitation and (spinning) matter. Just as gravitation is introduced in

Einstein's theory by allowing space-time to be a curved Riemann space, instead of the flat Minkowski space, we can also imagine a curved superspace. The tensors describing the geometry of curved superspace will contain, in addition to the gravitational tensor, a number of other fields describing particles of various spins. Superspace will be curved inasmuch as it deviates from the flat superspace characterized by the group of motions (6) (plus translations and Lorentz transformations) or by the invariant differential forms (10). There are various possible ways to introduce a geometry in curved superspace. In analogy with Einstein's theory one may want to try a generalized Riemannian geometry. It is a remarkable fact that the generalization of Riemannian geometry to superspace can be completely carried out [18]. A point in Riemannian superspace is specified by its coordinates $z^M = (x^\mu, \theta^m)$ and the Riemannian geometry by a metric tensor $g_{MN}(z)$, with certain symmetry properties. Observe that the space-space elements $g_{\mu\nu}(z)$ of the metric tensor, as well as the spin-spin elements $g_{mn}(z)$, are commuting quantities, while the space-spin elements $g_{\mu n}(z)$ are anticommuting quantities, just like θ^m. The occurrence of anticommuting quantities requires special attention in defining correctly the Christoffel symbols, the curvature tensor, etc. However, up to signs, the definitions are essentially the usual ones. Densities can also be defined and an action principle formulated. The resulting equations

$$R_{MN} = 0 \tag{12}$$

are perfectly analogous to Einstein's equation. By expansion in θ^m the various elements of the metric tensor g_{MN} generate a large number of fields of various spins, of which the gravitational tensor $g_{\mu\nu}(x)$ (for $\theta^m = 0$) is but one. These fields satisfy a system of coupled equations which are obtained from (12) by equating to zero the coefficients of the various powers of θ^m. Gravitation and matter are unified and there is no need for a right hand side to the equations (12). Since (12) are invariant under general transformations of coordinates in superspace, they are also invariant under Einstein transformations involving only x^μ, which form a subgroup. Among the transformations involving θ^m one recognizes in particular gauge transformations on vector and other fields contained in the expansion of g_{MN}. Thus, Einstein and gauge transformations are also unified. Of the various earlier attempts to give a unified geometrical picture of gravitation and other fields, this resembles most closely the five-dimensional Klein-Kaluza theory [19]. However, the fact that the additional space-time dimensions in the present picture are fermionic and spinorial, instead of bosonic as in Klein-Kaluza, presents the combined advantages that physical space-time is obviously still four-dimensional and that half-integral spin fields arise very naturally. The theory of Riemannian superspace, however elegant and appealing, has certain shortcomings. The main problem is that the metric tensor of flat superspace, as naturally constructed from the differential forms (10), does not satisfy in general the equations (12), nor any simple covariant modification of them [20]. In the following we describe a different (non-Riemannian) geometry of curved superspace, which is tailored more closely to the flat superspace geometry.

Before deciding on the local properties of superspace which will replace for us those of Riemannian superspace, let us observe that the most general affine superspace is described by a manifold of points $z^M = (z^\mu, \theta^m)$ and by a set of canonical forms

$$\omega^A = dz^M E_M{}^A(z) \qquad (13)$$

and of connection forms

$$\varphi_A{}^B = dz^M \hat{\varphi}_{MA}{}^B(z) = \omega^C \varphi_{CA}{}^B(z) \quad . \qquad (14)$$

Here the index $A = (\alpha, a)$, where α is a space index and a is a spin index, similarly $B = (\beta, b)$. The superfields $E_M^{\ A}$ generalize the usual vierbein field. In flat space the canonical forms are given by (10) and we shall denote the corresponding vierbein field by a bar, so

$$\bar{E}_\mu^{\ \alpha} = \delta_\mu^{\ \alpha} \quad , \quad \bar{E}_\mu^{\ a} = 0 \quad ,$$
$$\bar{E}_m^{\ \alpha} = i(\gamma^\alpha)_{mn}\theta^n \quad , \quad \bar{E}_m^{\ a} = \delta_m^{\ a} \quad . \tag{15}$$

We denote by $E_A^{\ M}$ the matrix which is the inverse of $E_M^{\ A}$. In particular

$$\bar{E}_\alpha^{\ \mu} = \delta_\alpha^{\ \mu} \quad , \quad \bar{E}_\alpha^{\ m} = 0 \quad ,$$
$$\bar{E}_a^{\ \mu} = -i(\gamma^\mu)_{an}\theta^n \quad , \quad \bar{E}_a^{\ m} = \delta_a^{\ m} \quad . \tag{16}$$

Observe that $(\partial_M = \frac{\partial}{\partial z^M})$

$$\bar{E}_a^{\ M}\partial_M = \partial_a - i(\bar{\theta}\gamma^\alpha)_a\partial_\alpha = D_a \quad , \quad \bar{E}_\alpha^{\ M}\partial_M = \partial_\alpha \tag{17}$$

are exactly the covariant derivatives [1-4] used in flat superspace.

Differential forms in superspace are a generalization of the usual Cartan differential forms. Just as the differentials dx^μ are taken as anti-commuting, the differentials $d\theta^m$ commute with each other, while dx^μ and $d\theta^m$ anticommute. If the differentials are written on the left, as in (13) and (14), the operator of differentiation d starts operating from the right. One can easily figure out the appropriate generalization of Cartan's rules for differentiation and Poincaré's theorem is seen to be valid for any form

$$d(d\omega) = 0 \ . \tag{18}$$

Also, the inverse theorem is valid, in sufficiently small neighborhoods [21]. Just as in ordinary differential geometry, differential forms provide the most compact and convenient notation to describe the geometry of superspace [21]. So, the covariant differential of the canonical forms

$$D\omega^A = d\omega^A - \omega^B \varphi_B{}^A = \Omega^A = \tfrac{1}{2}\omega^C \omega^B \Omega_{BC}{}^A \tag{19}$$

defines the torsion forms Ω^A and the torsion tensor $\Omega_{BC}{}^A$, while the curvature forms $R_A{}^B$ and the curvature tensor $R_{CDA}{}^B$ are defined by

$$d\varphi_A{}^B - \varphi_A{}^C \varphi_C{}^B = R_A{}^B = \tfrac{1}{2}\omega^D \omega^C R_{CDA}{}^B \quad . \tag{20}$$

One finds, for instance, for the torsion tensor

$$\Omega_{BC}{}^A = (-1)^{bc+bm} E_C{}^M E_B{}^N \partial_N E_M{}^A - (-1)^{cm} E_B{}^M E_C{}^N \partial_N E_M{}^A - \varphi_{BC}{}^A + (-1)^{bc} \varphi_{CB}{}^A \tag{21}$$

where, as in Ref. [11], the small letters b, c, m in the exponent of (-1) are defined to take the value 0 if the corresponding capital letter B, C, M takes the space value β, γ, μ, while they take the value 1 if the corresponding capital letter takes the spin value b, c, m. Clearly

$$\Omega_{BC}{}^A = -(-1)^{bc} \Omega_{CB}{}^A \quad . \tag{22}$$

The Bianchi identities have the usual form

$$dR_A{}^B + \varphi_A{}^C R_C{}^B - R_A{}^C \varphi_C{}^B = 0 \quad . \tag{23}$$

Furthermore, generalizing the definition of covariant differential to any local vector $v^A(z)$

$$Dv^A \equiv dv^A - v^B \varphi_B{}^A \quad , \tag{24}$$

one finds that the second covariant differential does not vanish and that it involves the curvature forms

$$D^2 v^A = - v^B R_B{}^A \quad . \tag{25}$$

In order to give superspace a more precise geometric structure one must specify the geometry of the tangent spaces. In ordinary four-space this is done by taking the tangent spaces to be Minkowski spaces and by requiring that the geometry be independent of the particular choice of Lorentz frame specified locally. Here the analogues of the local Lorentz transformations will be a group of linear transformations on the canonical forms (13). The infinitesimal transformations can be written as

$$\delta\omega^A = \omega^B X_B{}^A(z) \quad , \tag{26}$$

where the matrices $X_B{}^A$, at each point of superspace, form a (graded) Lie algebra. The corresponding transformations of the connection forms (14) is

$$\delta\varphi_A{}^B = \varphi_A{}^C X_C{}^B - X_A{}^C \varphi_C{}^B + dX_A{}^B \tag{27}$$

so that (19) is indeed a covariant differential and the torsion forms transform exactly like the canonical forms. The matrix of forms $\varphi_A{}^B$ must belong to the same Lie algebra as $X_A{}^B$ and the transformation law (27) preserves this property.

The algebra of the matrices $X_A{}^B$ could be chosen to be the superspace generalization of the Lorentz transformations. This would mean assuming that the tangent space possesses a constant numerically invariant tensor η_{AB}, and its inverse η^{AB}, with the symmetry properties

$$\eta_{AB} = \eta_{BA}(-1)^{ab} \quad , \qquad \eta^{AB} = \eta^{BA}(-1)^{ab+a+b} \quad , \tag{28}$$

so that

$$X_A{}^C \eta_{CB} = - X_B{}^C \eta_{CA}(-1)^{ab} \quad , \tag{29}$$

which generalizes the antisymmetry property of infinitesimal Lorentz transformation. If one further assumes that the torsion forms (19) vanish, the geometry of superspace becomes exactly equivalent to the Riemannian geometry of Ref. [18], the only difference being that in Ref. [18] the generalization of the Ricci calculus was used, while here we are using vierbeins and differential forms. As already mentioned above we shall choose the geometry in the tangent space in a different way. For the sake of brevity and also in order to describe an interesting application of the formalism, we shall now restrict ourselves to the case of two space-time dimensions, so that the spinor θ^m also has two-components. However, there is no difficulty in developing the theory in four dimensional space-time.

3. Lagrangian for the Neveu-Schwarz Model

In this section we consider a "supersurface." The space indices μ, ν take the values 0 and 1, and the Majorana spinor θ^m has two components. With the appropriate definition of the Dirac matrices

$$\gamma^0 = \begin{pmatrix} 0 & 1 \\ -1 & 0 \end{pmatrix}, \quad \gamma^1 = \begin{pmatrix} 0 & 1 \\ 1 & 0 \end{pmatrix}, \quad "\gamma_5" = \begin{pmatrix} 1 & 0 \\ 0 & -1 \end{pmatrix}, \tag{30}$$

all the formulas of the previous sections apply. If, in the supersymmetry transformations (6), we let the spinor parameter ζ^a depend on x so that it satisfies the differential equations

$$(\gamma_\alpha \partial_\beta + \gamma_\beta \partial_\alpha - \eta_{\alpha\beta} \gamma \cdot \partial)\zeta = 0 \quad , \tag{31}$$

the commutator of two transformations (6) is no longer a translation, it is now a conformal transformation in two dimension. This larger "conformal" supersymmetry as a four-dimensional analogue. The forms (10) are no longer invariant under (6), instead they transform as

$$\begin{cases} \delta\omega^\alpha = -i\bar{\theta}\gamma \cdot \partial\zeta\omega^\alpha - i\bar{\theta}\gamma_5\gamma \cdot \partial\zeta\omega^\beta \epsilon_\beta{}^\alpha \\ \delta\omega^a = -\frac{i}{2}\bar{\theta}\gamma \cdot \partial\zeta\omega^a + \frac{i}{2}\bar{\theta}\gamma_5\gamma \cdot \partial\zeta\omega^b(\gamma_5)_b{}^a + \frac{1}{2}\omega^\beta(\gamma_\beta\gamma \cdot \partial\zeta)^a \end{cases} . \tag{32}$$

In other words they rescale by amounts appropriate to their respective dimension, they undergo related Lorentz transformations and in addition ω^a mixes in with ω^α, but not the converse.

This behavior of ω^A should be compared with that of dx^α under ordinary conformal transformation in flat space. In two space-time dimensions an infinitesimal conformal transformation

$$\delta x^\alpha = \xi^\alpha(x) \tag{33}$$

is characterized by the differential equations

$$\partial_\alpha \xi_\beta + \partial_\beta \xi_\alpha - \eta_{\alpha\beta} \partial \cdot \xi = 0 \tag{34}$$

(notice the analogy with (31)). Under (33) dx^α changes as

$$\delta(dx^\alpha) = dx^\alpha \frac{1}{4}\partial\cdot\xi + dx_\beta \frac{1}{2}(\partial^\beta\xi^\alpha - \partial^\alpha\xi^\beta) \quad , \tag{35}$$

namely it undergoes a rescaling and a Lorentz transformation, both x dependent. This behavior of dx^α in flat space determines the behavior of the vierbein field in curved space [22]: it undergoes local Lorentz transformations and local rescalings (Weyl Transformations). The local Lorentz transformations are taken as determining the geometry of the tangent spaces, while the Weyl transformations play a less fundamental role and invariance under them may or may not be demanded, depending on the particular dynamical system under consideration. In analogy with this, the transformation property (32) suggests that we specify the geometry of the tangent supersurface at each point by requiring the matrix $X_B{}^A$ to be of the form

$$X_B{}^A = \begin{pmatrix} -\ell\epsilon_\beta{}^\alpha & (\gamma_\beta\varphi)^a \\ 0 & \frac{1}{2}\ell(\gamma_5)_b{}^a \end{pmatrix} \tag{36}$$

where ℓ is the parameter of the local two-dimensional Lorentz transformation and the spinor φ a further (anticommuting) parameter, both dependent on x^μ and θ^m. Observe that the matrices (36) form a Lie algebra which is not the same as that specified by (29), nor is it a subalgebra (rather, (36) is a subalgebra of a contraction of (29)). The group given by (36) has several numerically invariant tensors, in particular

$$\eta_{AB} = \begin{pmatrix} \eta_{\alpha\beta} & 0 \\ 0 & 0 \end{pmatrix}, \qquad \eta^{AB} = \begin{pmatrix} 0 & 0 \\ 0 & \eta^{ab} \end{pmatrix} \tag{37}$$

where $\eta_{\alpha\beta}$ is the usual Minkowski metric and η^{ab} the corresponding metric for spinors (in the Majorana representation $\eta^{ab} = (\gamma^0)^{ab}$). Another invariant tensor is

$$\gamma_A{}^{BC} = (\gamma_\alpha)^{bc} \quad , \tag{38}$$

when the capital indices take the indicated values and zero otherwise, e.g., $\gamma_a{}^{bc} = 0$.

The simplest Lagrangian on the supersurface which is a density under general coordinate transformations and is invariant under the transformations (26), (36) is

$$L = (\tfrac{1}{2}V_a V^a - E_a{}^M \partial_M V V^a - if(-1)^m \gamma_\alpha{}^{bc} E_c{}^M E_b{}^N \partial_N E_M{}^\alpha)D \tag{39}$$

where

$$D = \det E_M{}^A \tag{40}$$

is the generalized determinant of the matrix $E_M{}^A$ defined as in the second of Refs. [18]. Observe that the last term of L is obtained by contracting the torsion tensor (21) with the numerical tensor $\gamma_A{}^{BC}$. The part containing the connection drops out. The (2 + 2)-dimensional superfields V_a and V are assumed to be in addition ordinary Minkowski four-vectors and all products are intended as four-dimensional inner products. The four-dimensional Lorentz group acts as an internal symmetry commuting with everything, so we need not indicate explicitly the four-dimensional indices.

Variation of V^a, V, $E_\alpha{}^M$ and $E_a{}^M$ gives the respective equations of motion

$$V_a = E_a{}^M \partial_M V \quad , \tag{41}$$

$$\partial_M(V^a E_a{}^M D)(-1)^m = 0 \quad , \tag{42}$$

$$LE_M{}^\alpha - if(-1)^{\ell+n+\ell n}\gamma_\beta{}^{bc}\partial_N(E_c{}^L E_b{}^N D)E_L{}^\alpha E_M{}^\beta = 0 \quad , \tag{43}$$

$$-LE_M{}^a - \partial_M V V^a D - if(-1)^m \gamma_\alpha{}^{ac} E_c{}^N \partial_N E_M{}^\alpha D + if(-1)^{m+mn}\gamma_\alpha{}^{ac} E_c{}^N \partial_M E_N{}^\alpha D$$
$$+ if(-1)^{m+m'}\gamma_\alpha{}^{bc}\partial_N(E_c{}^{M'} E_b{}^N D)E_{M'}{}^a E_M{}^\alpha = 0 \quad . \tag{44}$$

Multiplying (43) by $E_B{}^M$ one obtains a trivial identity for B = b, while for B = β one has

$$L\delta_\beta{}^\alpha - if(-1)^{m+n+mn}\gamma_\beta{}^{ac}\partial_N(E_c{}^M E_a{}^N D)E_M{}^\alpha = 0 \quad . \tag{45}$$

Similarly, multiplying (44) by $E_B{}^M$, one obtains for B = b

$$-L\delta_b{}^a - E_b{}^M\partial_M VV^a D - if(-1)^m\gamma_\alpha{}^{ac}E_b{}^M E_c{}^N\partial_N E_M{}^\alpha D$$
$$- if(-1)^n\gamma_\alpha{}^{ac}E_c{}^N E_b{}^M\partial_M E_N{}^\alpha D = 0 \quad , \tag{46}$$

and for B = β

$$-E_\beta{}^M\partial_M VV^a D - if(-1)^m\gamma_\alpha{}^{ac}E_\beta{}^M E_c{}^N\partial_N E_M{}^\alpha D + if\gamma_\alpha{}^{ac}E_c{}^N E_\beta{}^M\partial_M E_N{}^\alpha D$$
$$+ if(-1)^m\gamma_\beta{}^{bc}\partial_N(E_c{}^M E_b{}^N D)E_M{}^a = 0 \quad . \tag{47}$$

These complicated non-linear equations can be simplified by making the Ansatz (which could be further justified)

$$E_M{}^\alpha = \Lambda\bar{E}_M{}^\alpha \quad , \qquad E_M{}^a = \Lambda^{\frac{1}{2}}\bar{E}_M{}^a \quad , \qquad D = \Lambda \quad . \tag{48}$$

Here Λ is a superfield and $\bar{E}_M{}^A$ is the flat space vierbein given by (15). The fact that Λ occurs with different powers for different components of the vierbein is due to the fact that they have different dimensions. From (48) it follows, for the inverse matrix,

$$E_\alpha{}^M = \Lambda^{-1}\bar{E}_\alpha{}^M \quad , \qquad E_a{}^M = \Lambda^{-\frac{1}{2}}\bar{E}_a{}^M \quad . \tag{49}$$

Substituting (48) and (49) into (45) one finds

$$L\delta_\beta{}^\alpha + 2f\delta_\beta{}^\alpha\Lambda = 0 \tag{50}$$

or

$$L = -2f\Lambda \quad , \tag{51}$$

while the same substitution transforms (46) into

$$L\delta_b{}^a + \bar{E}_b{}^M\partial_M VV^a\Lambda^{\frac{1}{2}} + 4f\delta_b{}^a\Lambda = 0 \quad , \tag{52}$$

and (47) into

$$\partial_\beta V\bar{V}^a + if\gamma_\beta{}^{ac}\bar{E}_c{}^N\partial_N\Lambda\Lambda^{-\frac{1}{2}} = 0 \quad . \tag{53}$$

It is natural to introduce the superfield

$$\bar{V}_a = V_a D^{\frac{1}{2}} = V_a\Lambda^{\frac{1}{2}} \quad . \tag{54}$$

Eliminating Λ by means of (49) we find finally the equations of motion

$$\bar{V}_a = \bar{E}_a{}^M\partial_M V \ , \qquad \partial_M(\bar{V}^a\bar{E}_a{}^M)(-1)^m = 0 \quad , \tag{55}$$

and

$$\partial_\beta V\bar{V}^a - \frac{i}{2}\gamma_\beta{}^{ac}\bar{E}_c{}^N\partial_N L = 0 \ , \qquad L \equiv \frac{1}{2}\bar{V}_a\bar{V}^a \quad . \tag{56}$$

It is easy to see that (55) are the equations of motion, while (56) are the gauge and supergauge conditions of the Neveu-Schwarz-Ramond model [23]. Expanding the superfield V in terms of component field,

$$V(x,\theta) = A(x) + i\bar{\theta}\psi(x) + \frac{i}{2}\bar{\theta}\theta F(x) \quad , \tag{57}$$

the equations (55) become

$$\Box A = 0 \ , \qquad \gamma\cdot\partial\psi = 0 \ , \qquad F = 0 \quad , \tag{58}$$

while (56) give

$$T_{\alpha\beta} = 0 \ , \qquad \partial_\beta A\gamma^\beta\gamma^\alpha\psi = 0 \quad , \tag{59}$$

where $T_{\alpha\beta}$ is the total energy-momentum tensor of the fields A, ψ and F (the latter does not contribute, because of (58)).

We see that our Lagrangian (39) gives rise not only to the linear equations of motion (58) but also to the quadratic gauge and supergauge conditions (59). This is due to its invariance under the larger group of coordinate transformations involving both x^μ and θ^m. Previously proposed Lagrangians [24] were invariant under coordinate transformations involving only x^μ and therefore did not give rise to the supergauge conditions.

It is well-known that the gauge and supergauge conditions are important in eliminating the ghosts from the model. The method described here for deriving them suggests a natural procedure for finding generalizations of the model. For instance, in order to include internal symmetries, one can take a more complicated supersurface where the spinor θ belongs to a representation of the internal symmetry group.

The author wishes to acknowledge numerous helpful discussions with J. Wess on the subjects treated in this lecture.

References and Notes

[1] B. Zumino, in Proceedings of the XVII International Conference on High-Energy Physics, London, July 1974, published by The Science Research Council 1974.

[2] L. Corwin, Y. Ne'eman and S. Sternberg, Reviews of Modern Physics 47 (1975) 573.

[3] A. Salam and J. Strathdee, Phys. Rev. D11 (1975) 1521.

[4] S. Ferrara, to be published in Rivista del Nuovo Cimento.

[5] V. G. Kac, Functional Analysis and Its Applications 9 (1975) 91 (Russian).

[6] A. Pais and V. Rittenberg, Journal of Math. Phys. 10 (1975) 206.

[7] P. Freund and I. Kaplansky, University of Chicago preprint (1975), submitted to Journal of Math. Physics.

[8] R. Haag, J. T. Topuszanski and M. Sohnius, Nuclear Phys. B88 (1975) 257.

[9] C. Becchi, A. Rouet and R. Stora, Phys. Letts. 52B (1974) 344, and various University of Marseille preprints.

[10] The first authors to use the supersymmetry (1) in local field theory appear to have been Yu.A. Golfand and E. P. Likhtman, JETP Letters 13 (1971) 452 (Russian).

[11] S. Ferrara and B. Zumino, Nuclear Phys. B79 (1974) 413;
A. Salam and J. Strathdee, Phys. Letters B51 (1974) 353;
S. Ferrara and O. Piguet, Nuclear Phys. B93 (1975) 261;
A. A. Slavnov, Nuclear Phys. B97 (1975) 155.

[12] J. Iliopoulos and B. Zumino, Nuclear Phys. B76 (1974) 310.

[13] D. V. Volkov and V. P. Akulov, Phys. Letters 46B (1973) 109.

[14] W. Bardeen, to be published; B. de Wit and D. Z. Freedman, Stony Brook University preprint.

[15] D. V. Volkov and V. A. Soroka, JETP Letters 18 (1973) 529 (Russian).

[16] A. Salam and J. Strathdee, Nuclear Phys. B87 (1975) 85.

[17] B. Zumino, Nuclear Phys. B89 (1975) 535.

[18] P. Nath and R. Arnowitt, Phys. Letters 56B (1975) 177;
R. Arnowitt, P. Nath and B. Zumino, Phys. Letters 56B (1975) 81.
See also P. Nath's lecture in these proceedings.

[19] T. Kaluza, Sitzber. Preuss. Akad. Wiss. (1921) 966; O. Klein, Zs.f.Phys. 37 (1926) 895; Y. M. Cho and P. G. O. Freund, University of Chicago preprint, to be published in Phys. Rev.

[20] For a discussion of this point and for a possible solution of the difficulty in certain cases see the lecture of P. Nath in these proceedings.

[21] B. Zumino, in preparation. The operations of derivation and integration in Grassmann algebras are described very clearly in F. A. Berezin's book, The Mathematics of Second Quantization, Academic Press.

[22] B. Zumino, Lectures at the 1970 Brandeis University Summer Institute, Edited by Deser, Grisaru and Pendleton, MIT Press, 1970.

[23] A. Neveu and J. H. Schwarz, Nuclear Phys. B31 (1971) 86;
P. Ramond, Phys. Rev. D3 (1971) 2415;
J.-L. Gervais and B. Sakita, Nuclear Phys. B34 (1971) 633.

[24] Y. Iwasaki and K. Kikkawa, Phys. Rev. D8 (1973) 440;
L. N. Chang, K. Macrae and F. Mansouri, Phys. Letters 57B (1975) 59.

DISCUSSION

P. WAN NIEUWENHUIZEN (Stony Brook)

Is your definition of parallel transport equivalent to Cartan's definition whose connection is not restricted to be metric, but describes a rotation which is conformal?

ZUMINO

The answer is yes. It is a generalization of it. In fact what I use is exactly the Cartan's formulation with differential forms. Because now the co-ordinates are both x and θ, I have to first of all discover what to do with differential forms on a Grassmann algebra. It turns out to be rather easy. In fact, as you know differential forms ordinarily have the property dx, dy anticommute because x and y commute. Now $d\theta_1$, $d\theta_2$ commute because θ_1, θ_2 anti-commute. Further dx and $d\theta$ anti-commute because x and θ commute. Everything is constructed in such a way that the second differential of any form is zero automatically just like in the Cartan theory and also in the small, the inverse of Poincare's theorem is valid: If a form has a d equal zero, then it is the d of something. Then you can almost carry over completely the theory of an affine space as is done in books which use differential forms and in an affine space with torsion as well. However, this is not purely an affine space with torsion because it does have a local structure which although it is not Lorentzian, nevertheless it tells you something more so that it is not purely an affine space with torsion but is something in between a metric and an affine space. I believe for these spaces which have both x and θ the truly Riemannian structure is not what one needs. Now I am sure that Pran will convince you of the opposite. I hope he will because it would be better, it would be nicer, if it were Riemannian.

S. DESER (Brandeis)

I was wondering what the form would be for the pure gravitational part in two dimensions. It's clear that you couldn't have gotten gravitation with the Riemannian structure in two dimensions but would you say you have any correspondence with the Einstein theory without having a metric structure? What would the closest limit without matter, that is to say, what would be the correspondence of the purely gravitational part since you don't have a metric structure but affine plus something else?

ZUMINO

No, but you see I do have a metric structure in the purely "x" part because if you remember correctly those transformation properties for the purely spin and purely x part are Lorentzian, [See e.g., Eq. 36] and then it had something else where it mixes half integral and integral spin. So, as far as the purely x thing is concerned I had a metric like everybody else, and the special sub-group of this is the usual Einstein thing so my theory in particular has all the invariances of ordinary gravitation. It is just the way the spin is mixed with the rest which is different from the one of a purely Riemannian theory.

S. DESER (Brandeis)

That's why you have the torsion.

ZUMINO

And this is why I have the torsion. That's right.

P. FREUND (Chicago)

In the kind of theory with torsion that you have just described - which locally has a conformal structure - is the number of Fermi dimensions arbitrary or must it be equal to the number of Bose dimensions?

ZUMINO

I think it doesn't have to be equal. However, I must admit that so far I have studied in detail only the case in which there are four x's and four Majorana components θ (i.e. a single Majorana spinor). But I don't think it has to be just that way. However, since I have not actually done it maybe some difficulty comes up along the way.

P. FREUND (Chicago)

It is not obvious to me that it can be arbitrary.

ZUMINO

Well, O.K. I don't really know the answer to your question. The reason why I took exactly this case with the Majorana thing is because that one is the most difficult to do from the point of view of the purely Riemannian space. It is in fact one in which you cannot recall the flat space supersymmetry from the Riemannian space as I think Pran will explain, and because of this I was really lead into thinking that geometry was not Riemannian - super Riemannian.

S. Bludman (Penn)

Could you spell out difficulties with fermion number conservation. It would appear that a Majorana octet has just the right number of components to describe, not a quark triplet, but a quark quartet.

ZUMINO

Well the difficulties with fermion number conservations are a little bit involved technically. When you have an octet you know that it transforms in a certain way and you are not going to make it into a quartet or a triplet no matter how you combine the things. The fermion number conservation problem really very simply is this: If you have vectors which are real, the spinors are real. If you have scalars which are real, the spinors are real. If you want the spinors to be complex you have to take scalars or vectors which are complex. Then you are going to have scalars with fermion number just like the spinors have. I mean this is roughly speaking the basic difficulty. You will be led to mesons which are not like the mesons that we know. However, by combining things cleverly you can sometimes escape this. For instance one very simple model that Salam and Strathdee and also we did sometime ago, (actually Salam and Strathdee have studied this problem with the fermion number conservation in great detail) is to take two multiplets one with vectors and spinors which are real and another with scalars and spinors which are real, and then by being sufficiently clever, (super symmetry of course mixes these) the two majorana spinors can be put into a complex spinor and you can construct a Lagrangian which also has a phase invariance. Then the vectors and the scalars don't carry fermion numbers but the spinors do. Such things can be done but it's rather limited how far one can go. This is unfortunately the case because basic multiplets are naturally Majorana spinors.

SUPERSYMMETRY AND GAUGE SUPERSYMMETRY

Pran Nath*

*Department of Physics, Northeastern University, Boston, Mass. 02155.

Work supported in part by the National Science Foundation.

I. INTRODUCTION

In the usual formulation of gauge theories(1) the world is constituted of two types of objects: gauge fields and matter fields which are treated on different levels in the theory. For example, in non-abelian gauge theories the form of the self-interaction among the gauge mesons is uniquely determined by the constraints of gauge invariance itself. This does not hold, however, for the matter fields which constitute the sources of the gauge mesons. This distinction between gauge fields and matter fields is ancient. Thus in the most well-known of the gauge theories, i.e., the Einstein theory, whereas the self-gravitational interactions are themselves determined by the non-abelian nature of the Einstein gauge invariance, the sources of the Einstein equations (the stress tensor) have to be added in an ad hoc fashion. Thus one has

$$R_{\mu\nu} - \tfrac{1}{2}g_{\mu\nu}{}^{E}R = 8\pi G_E T_{\mu\nu} \quad , \tag{1}$$

where the only restriction on the source $T_{\mu\nu}$ is that it be a gauge tensor. A similar arbitrariness also exists in the modern gauge theory approaches. For example, the interacting Yang-Mills fields satisfy an equation

$$F_{\mu\ a,\lambda}^{\ \lambda} + gf_{abc}F_{\mu\lambda b}v^{\lambda}{}_{b} = J_{\mu a} \quad , \tag{2}$$

where $F_{\mu\nu a} = v_{\nu a,\mu} - v_{\mu a,\nu} + gf_{abc}v_{\mu b}v_{\nu c}$. Once again the totality of self-interactions of the gauge field $v_{\mu a}$ is determined by the non-abelian gauge invariance. One needs, however, additional theoretical ideas to determine the source such as the quark content and group representation.

Since non-abelian gauge invariance so uniquely determines the gauge interactions, one may wish to eliminate the arbitrariness arising from the sources by requiring that <u>all</u> fields be non-abelian gauge fields. This approach, however, poses an immediate problem: it is conventional wisdom that gauge fields are bosons but the sources involve both bosons and fermions. If one is to pursue the above approach further, it is then imperative that one treat bosons and fermions in a symmetrical way. This is where recent developments by Wess and Zumino, Volkov and Akulov(2) and Salam and Strathdee(3) involving concepts of fermi-bose supersymmetry and superfields assume a crucial role.

Supersymmetry, however, is a global symmetry and one would like to enmesh the ideas of non-abelian gauge invariance and supersymmetry to construct a theory where all fields, both bose and fermi, are gauge fields. This has been attempted by Richard Arnowitt and myself by extending supersymmetry into a local gauge invariance.(4) One is then led to develop field theories in higher dimensions (superspaces) where the first four coordinates are bose and the remaining coordinates are fermi, i.e., we define

$$z^A = (x^\mu, \theta^\alpha) \quad , \tag{3}$$

where x^μ are the bose coordinates ($\mu = 0, 1, 2, 3$) and θ^α are the fermi coordinates ($\alpha = 1, 2, \ldots, N$) satisfying

$$\{\theta^\alpha, \theta^\beta\} = 0 \quad . \tag{4}$$

Fields in superspaces (superfields) are then functions of 4 + N coordinates. An expansion of the superfield in powers of θ^α yields only a finite series since for any α one has $(\theta^\alpha)^2 = 0$. A superfield in general contains a mixture of different spins and statistics.

For the case of the ordinary supersymmetry(2) the fermi coordinates are Majorana spinors (i.e., N = 4) and supersymmetry transformations are

$$\begin{aligned} \theta^{\alpha\prime} &= \theta^\alpha + \delta^\alpha \quad , \\ x^{\mu\prime} &= x^\mu + \tfrac{i}{2}\bar{\delta}\gamma^\mu\theta \quad , \end{aligned} \tag{5}$$

with δ^{α} being the infinitesimal fermion transformation parameters. As already noted, supersymmetry is a global symmetry in that the parameters δ^{α} appearing in Eq. (5) are not functions of the coordinates. In attempting to make supersymmetry into a local gauge invariance we adopt the viewpoint that the linear transformations of ordinary supersymmetry Eq. (5) in superspace are analogous to the Poincare transformations of the 4-dimensional bose space. Then in analogy with the general covariance group in 4-dimensions the fundamental gauge group in the bose-fermi superspace is the group of arbitrary coordinates transformations

$$Z^{A'} = Z^{A'}(Z) \qquad (6)$$

which leave the line element

$$ds^2 = dZ^A g_{AB}(Z) dZ^B \qquad (7)$$

invariant. The superfield $g_{AB}(Z)$ then becomes the basic gauge field of the theory. Because the indices g_{AB} can be either bose or fermi, there are three independent sectors of g_{AB}. These are (1) bose-bose with $g_{\nu\mu}(Z) = g_{\mu\nu}(Z)$, (2) bose-fermi with $g_{\mu\alpha}(Z) = -g_{\alpha\mu}(Z)$ and (3) fermi-fermi with $g_{\alpha\beta}(Z) = -g_{\beta\alpha}(Z)$. The superfield expansion of the metric tensor $g_{AB}(Z)$ contains in addition to the gravitational potential $g_{\mu\nu}^{E}(x)$, a variety of other fields such as scalar, Maxwell, Dirac, etc., fields.

Before proceeding further to discuss the dynamics of the supergauge theory, we should first like to summarize some of the results that emerge from the above developments.

a. The supergauge group for spaces involving Grassmann (i.e., anticommuting) coordinates contains not only the Einstein gauge group but also the electromagnetic (and Yang-Mills) gauge groups. Further, the theory determines its internal symmetry group from the number of fermi dimension. For 8n fermi coordinates this group is O(2n).

b. A supergeometry in spaces with Grassmann coordinates may be formulated and provides a framework for unified gauge theories. The superfield equations of motion (described by the vanishing of the contracted Riemann curvature tensor, i.e., $R_{AB} = 0$) decompose into Einstein, Maxwell (or Yang-Mills), Dirac, etc., field equations. No arbitrariness to the sources of these field equations exists, and in this sense the theory is self-sourced. The supergeometry also produces a natural explanation for the existence of infinite order non-linearities in the Einstein theory of gravitation but non-linearities of only a finite order for other gauge theories (such as the Yang-Mills' theory).

c. The spontaneous symmetry breakdown of the supergauge group determines the allowed dimensionalities of the fermi space and hence its internal symmetry group. After the spontaneous symmetry breakdown has occurred, the Einstein and Maxwell gauge invariances are maintained so that the graviton and the photon remain massless. In the absence of the gravitational and electromagnetic interactions, the broken gauge group in the global limit becomes just the supersymmetry group. However, when gravitational and electromagnetic interactions are present the supersymmetry transformations hold only in a small geodesic region around a point. In this sense, the supersymmetry transformations are analogous to the Poincare transformations which are applicable also in a small local inertial region.

d. While the graviton and the photon remain massless after the spontaneous symmetry breakdown has occurred, there exists a self-contained Higg's mechanism responsible for mass-growth in the broken sectors of the supergauge group. As a consequence of the mass growth, the Einstein gravitational constant G_E and the electric charge e are related such that

$$G_E \approx e^2/M_G{}^2 \quad , \tag{8}$$

where $M_G \approx 10^{18}$ GeV. Such a relationship holds even though both the gravitational and the electromagnetic interactions are long range. Further, the very existence of charge itself depends on the breakdown of the supergauge group though its numerical value remains undetermined.

It has already been noted and perhaps should be restated once again that the field theories in higher dimensions where the additional dimensions are fermi have none of the problems concerning their observability which arise if these additional dimensions are bose.(5,6) This is so trivially in that the superfield expansion in the fermi coordinates must terminate after a finite expansion.

The geometry which provides the underlying framework for the developments discussed here is Riemannian. The structure of Riemannian geometry in spaces with Grassmann coordinates is a very direct and straightforward extension of the well-known geometrical structures of bose spaces. There also exists, however, the possibility of developing a different (non-Riemannian) geometry of the curved superspace for which I refer you to Bruno Zumino's lecture(7) in these proceedings.

II. SUPERGAUGE TRANSFORMATIONS

The superfield $g_{AB}(Z)$ which acts as the basic gauge fields of the theory satisfy the symmetry property

$$g_{AB}(Z) = \eta_{ab} g_{BA}(Z) \quad , \tag{9}$$

where $\eta_{ab} = (-1)^{a+b+ab}$ and $(-1)^a$ is the Grassmann parity of the coordinate Z^A with a = 0 or 1, when A refers to bose or fermi coordinates. The invariance of the line element of Eq. (7) for arbitrary coordinate transformations implies that the superfields $g_{AB}(Z)$ transform according to the rule

$$g_{AB}{}'(Z') = L_{A'}{}^{C} g_{CD}(Z) R^{D}{}_{B'} \quad , \tag{10}$$

where R and L refer to right and left derivatives, such that for an arbitrary transformation $Z^{A'} = Z^{A'}(Z)$ one has $dZ^{A'} = R^{A'}{}_{B} dZ^{B} = dZ^{B} L_{B}{}^{A'}$. We also note that

$$L_{A'}{}^{B} = (-1)^{a+ac} R^{B}{}_{A'} \quad . \tag{11}$$

The superfields $g_{AB}(Z)$ in general involve a large number of components. Some of these are real dynamical fields such as the gravitational potential $g_{\mu\nu}{}^{E}(x)$. However, not all the elements in the superfield expansion of $g_{AB}(Z)$ are necessarily dynamical since it is possible to eliminate many elements of the metric supertensor by supergauge transformations. To define which fields constitute non-dynamical gauge variables and to determine the physical significance of the dynamical fields, it is instructive to study their response under supergauge transformations. We thus proceed to examine Eq. (10) under infinitesimal coordinate transformations

$$Z^{A} = Z^{A'} + \xi^{A} \quad . \tag{12}$$

Equation (10) then assumes the form

$$g_{AB}{}'(Z) = g_{AB}(Z) + (-1)^{a+ac} \xi^{C}{}_{,A} g_{CB}(Z) + g_{AC} \xi^{C}{}_{,B} + g_{AB,C} \xi^{C} \quad , \tag{13}$$

where in Eq. (13) only the right derivatives are involved.

(i) Gravitational Gauge Group

The transformations of Eq. (13) contain the gravitational gauge group as may be seen by setting $x^{\mu'} = x^{\mu'}(x)$, $\theta^{\alpha'} = \theta^{\alpha}$ and transforming $g_{\mu\nu}(Z)$.

(ii) Electromagnetic (or Yang-Mills) Gauge Group

It is a remarkable fact that in addition to the gravitational gauge group, the supergauge transformations also contain the electromagnetic (or Yang-Mills) gauge group(8) when the fermi coordinates are chosen to be Dirac spinors (or an n-tuplet of Dirac spinors). We discuss here the more general case of the Yang-Mills gauge group.

First, we introduce Dirac spinor coordinates $\theta^{\alpha q}$ by adding a charge index q = 1, 2 to the Majorana spinor θ^{α}. (The combination $\theta^{\alpha_1} - i\theta^{\alpha_2}$ then describes a Dirac spinor coordinate.) We shall show next that the <u>mere existence</u> of the n-tuplet of spinor coordinates $\theta^{\alpha q a}$, a = 1, 2, ..., n, generates automatically a Yang-Mills invariance. To exhibit this phenomenon, we consider infinitesimal coordinate transformations in the superspace, such that $\xi^A = (\xi^{\mu}, \xi^{\alpha q a})$ is given by

$$\xi^{\mu} = 0 \,, \qquad \xi^{\alpha q a} = \Lambda^A(x)(\mu_A(x)\theta)^{\alpha q a} \,, \tag{14}$$

where $\Lambda^A(x)$ are the gauge functions and μ_A are the real antisymmetric matrices in the Majorana spinor representation(9) which satisfy the relation

$$[\mu_A, \mu_B] = -f_{ABC}\mu_C \,, \tag{15}$$

and f_{ABC} are the structure constants of SU(n). One may deduce the transformation properties of various elements of the fields appearing in $g_{AB}(Z)$ from Eqs. (13) and (14). Expanding $g_{AB}(Z)$ but exhibiting only a few of the lowest terms in powers of θ we have (suppressing the charge and the SU(n) indices),

$$g_{\mu\nu}(Z) = g_{\mu\nu}{}^{o}(x) + (\bar{\theta}\epsilon\mu_A\theta)P_{\mu\nu}{}^{A}(x) + \ldots \quad ,$$

$$g_{\mu\alpha}(Z) = (\bar{\theta}\mu_A)_\alpha B_\mu{}^{A}(x) + i(\bar{\theta}\mu_A)_\alpha(\bar{\psi}\gamma_\mu\theta) + \ldots \quad ,$$

$$g_{\alpha\beta}(Z) = \eta_{\alpha\beta} + (\bar{\theta}\mu_A)_\alpha\bar{\chi}_\beta{}^{A} - (\bar{\theta}\mu_A)_\beta\bar{\chi}_\alpha{}^{A} + \ldots \quad . \tag{16}$$

Under the coordinate transformations of Eqs. (13) and (14) (the gravitational potential) $g_{\mu\nu}{}^{o}$ transforms as a Yang-Mills scalar, i.e., $p_{\mu\nu}{}^{A}$ and $B_\mu{}^{A}$ transform as

$$B_\mu{}^{A'}(x) = B_\mu{}^{A}(x) - \Lambda^{A}{}_{,\mu} + f_{ABC}\Lambda^{B}B_\mu{}^{C} \quad ,$$

$$P_{\mu\nu}{}^{A'} = P_{\mu\nu}{}^{A} + f_{ABC}\Lambda^{B}P_{\mu\nu}{}^{C} \quad , \tag{17}$$

where $P_{\mu\nu}{}^{A}$ is related to $p_{\mu\nu}{}^{A}$ by $P_{\mu\nu}{}^{A} = p_{\mu\nu}{}^{A} + \frac{1}{2}d_{ABC}B_\mu{}^{B}B_\nu{}^{C}$. From Eq. (17) we learn that whereas the gravity field is a Yang-Mills scalar, $B_\mu{}^{A}$ transforms as the Yang-Mills gauge meson and $P_{\mu\nu}{}^{A}$ transforms like the regular representation of SU_n. Analogously the spinor fields ψ_μ and χ^A of Eqs. (16) transform as

$$\psi'(x) = (1 - i\Lambda^{A}\mu_A)\psi \quad ,$$

$$\chi^{A'}(x) = (1 - i\Lambda^{B}\mu_B)\chi^{A}(x) + f_{ABC}\Lambda^{B}\chi^{C}(x) \quad . \tag{18}$$

Here we find that ψ transforms as the fundamental (quark) representation for SU_n and χ^A transform according to the fundamental x the adjoint representation (e.g., 3 x 8 for SU_3).

One may simply limit the preceding consideration to the situation where one has only a doublet of Majorana's. Equations (14) then acquire the form $\xi^\mu = 0$, $\xi^{\alpha q} = \Lambda(x)(\epsilon\theta)^\alpha$. These transformations generate the electromagnetic gauge group. We observe then that the coordinate transformations in superspace induce both the Einstein and the electromagnetic (or the Yang-Mills) gauge groups. There arises then a synthesis of the general coordinate group of space-time and internal symmetry transformations of elementary particle physics in a unified framework in the bose-fermi superspace.

It is a striking fact (as noted earlier) that the mere specification of the fermi coordinates determines the nature of the internal symmetry group. The most general internal symmetry group allowed for a set of 8n fermion (or n Dirac) coordinates is O(2n) which contains SU_n.[10]

III. SUPERSYMMETRY SPACE GEOMETRY

The field equations that the gauge fields $g_{AB}(Z)$ satisfy must also obey the gauge transformations of Eq. (10). Since the field equations are a natural outcome of the geometrical structure of the supersymmetry space we here outline briefly their construction. We assumed a Riemannian geometry of the curved supersymmetry space for two reasons. First, such a construction is both a simple and a straightforward generalization of the ordinary Riemannian geometry that holds in bose spaces. Second, the supergauge symmetry is not an exact symmetry of nature and one must eventually discuss how such a symmetry may break. We shall see later (in Sec. IV) that the Riemannian superspace geometry restricts severely the form of its own breakdown, lessening the need for extra ad hoc symmetry breaking assumptions.

We introduce now the essential geometrical notions in superspace. A contravariant superfield vector $P^A(Z)$ is one which transforms according to

$$P^{A'}(Z') = R^{A'}{}_B P^B(Z) = P^B L_B{}^{A'} \tag{19}$$

under the coordinate group $Z^{A'} = Z^{A'}(Z)$. There exist, however, <u>two</u> types of <u>covariant</u> vectors

$$Q_{A'} = L_{A'}{}^{B} Q_B \ ; \qquad \hat{Q}_{A'} = \hat{Q}_B R^B{}_{A'} \quad . \tag{20}$$

A random contraction of a super- and a sub-index does not necessarily maintain the tensor property, e.g., $P^A Q_A = \hat{Q}_A P^A$ is a scalar superfield whereas $Q_A P^A$ and $P^A \hat{Q}_A$ are not.

We define the affinity $\Gamma_{AB}{}^{C}$ through the right covariant differentiation of a contravariant vector by parallel transport

$$V^A{}_{;B} \equiv V^A{}_{,B} + V^C \Gamma_{CB}{}^{A} \quad . \tag{21}$$

Next assuming that the space is Riemannian (i.e., $\Gamma_{BC}{}^{A}$ depends only on g_{AB}) and making the assuptions of (a) Leibnitz rule,[11] (b) $S_{;B} = S_{,B}$ for a scalar superfield and (c) $g_{AB;C} = 0$, one may determine the affinity $\Gamma_{AB}{}^{C}$ to be

$$\Gamma_{AB}{}^{C} = (-1)^{bc} \tfrac{1}{2}[(-1)^{bd} g_{AD,B} + \eta_{ab}(-1)^{ad} g_{BD,A} - g_{AB,D}] g^{DC} \quad , \tag{22}$$

where $\eta_{ab} = (-1)^{a+b+ab}$ and g^{AB} is the inverse metric

$$g^{AC} g_{CB} = \delta^A{}_B \quad . \tag{23}$$

The parallel transport of a vector around an infinitesimal closed loop allows one to determine the curvature tensor. One finds

$$R^D{}_{ABC} = -\Gamma^D{}_{AC,B} + (-1)^{bc} \Gamma^D{}_{AB,C} - (-1)^{c(d+e)} \Gamma_{AC}{}^{E} \Gamma_{EB}{}^{D}$$
$$+ (-1)^{b(c+d+e)} \Gamma^E{}_{AC} \Gamma^D{}_{EB} \quad . \tag{24}$$

As remarked earlier, one must be careful in the definition of a contraction. The only allowed contractions are

$$R_{AB} \equiv (-1)^c R^C{}_{ABC} \ , \qquad R \equiv (-1)^b g^{BA} R_{AB} \quad , \tag{25}$$

where R_{AB} is the contracted curvature and R the curvature scalar.

We note here that R_{AB} computed from Eq. (24) is similar to the usual bose space form except for the additional sign factors. These are the signs that allow one to have gravitational and electromagnetic (or Yang-Mills) phenomena in a common formalism.

The most general second order differential field equations are

$$R_{AB} = \lambda g_{AB} \quad . \tag{26}$$

One may also derive Eq. (26) from an invariant action in superspace[12] which has the form

$$A = \int d^8 Z \, (\sqrt{-g}\, R + 2\lambda \sqrt{-g}) \quad , \tag{27}$$

where $\sqrt{-g} \equiv [-\det g_{AB}]^{\frac{1}{2}}$ and the definition of the determinant in superspace is given in Ref. 12

IV. SPONTANEOUS BREAKDOWN OF SUPERGAUGE

Supergauge symmetry is not an exact symmetry of nature and we next discuss the possible mechanisms for its breakdown. We shall see that the supergauge symmetry may break in a spontaneous fashion.[13] Before, discussing our postulate of symmetry breakdown, we state an identity concerning the ordinary Wess-Zumino-Volkov-Akulov supersymmetry transformations which correspond to the invariance of the line element for a single Majorana spinor.

$$ds_o^{\,2} = (dx^\mu - i\beta\bar{\theta}\gamma^\mu d\theta)^2 + d\bar{\theta}d\theta \quad . \tag{28}$$

We denote the metric associated with Eq. (28) by $g^o{}_{AB}$ which takes the form

$$\begin{aligned} g^o{}_{\mu\nu} &= \eta_{\mu\nu} \quad , \\ g^o{}_{\mu\alpha} &= -i\beta(\bar{\theta}\gamma_\mu)_i \quad , \\ g^o{}_{\alpha\beta} &= \eta_{\alpha\beta} + \beta^2(\bar{\theta}\gamma_\mu)_\alpha(\bar{\theta}\gamma^\mu)_\beta \quad , \end{aligned} \tag{29}$$

where $\eta_{\mu\nu}$ is the Lorentz metric and $\eta_{\alpha\beta} = -(C^{-1})_{\alpha\beta}$ where C is the charge conjugation matrix.[14] One may extend the transformations of Eq. (29) to the case of N fermi coordinates (Eq. (29) corresponds of N = 4). The calculation of the contracted curvature tensor R^o_{AB} corresponding to g^o_{AB} then gives[15]

$$R^o_{AB} \equiv R_{AB}[g^o_{CD}] = \tfrac{1}{4}N\beta^2 g^o_{AB} + \tfrac{1}{4}(8 - N)\beta^2 \eta_{\alpha\beta}\delta_{A\alpha}\delta_{B\beta} \quad . \tag{30}$$

To discuss our postulate of symmetry breakdown, we consider the supergauge invariant field equations

$$R_{AB}[g_{CD}] = \lambda g_{AB} \tag{31}$$

and induce the spontaneous breakdown by the requirement that the vacuum metric reduce to the conventional supersymmetry metric, i.e.,

$$\langle 0|g_{AB}(Z)|0\rangle = g^o_{AB}(Z) \quad , \tag{32}$$

where $g^o_{AB}(Z)$ is the supersymmetry metric of Eq. (29). The vacuum expectation value of the field equations (32) then gives in the tree approximation

$$R_{AB}[g^o_{CD}] = \lambda g^o_{AB} \quad . \tag{33}$$

Comparing Eq. (33) with Eq. (30) we find that when $\lambda \neq 0$ the spontaneous symmetry breakdown of the supergauge group may occur provided N = 8 and $\lambda = 2\beta^2$. [The more general coordinate condition determined by the symmetry breakdown is $N_{Fermi} = 2N_{Bose}$.] We shall show later that spontaneous breakdown of the supergauge group may also occur with a vanishing λ in Eq. (31) but with a fermi dimension N > 8 and a more subtle form for the vacuum metric g^o_{AB}. This latter form of the breakdown must be adopted if one were to discuss not only the gravitational and the electromagnetic interactions but the weak interactions as well. We shall return to a brief discussion of this subject in Sec. V. Here we discuss spontaneous breakdown for the case when the fermi dimension N = 8 and Eq. (33) hold.

a. Gauge Group of the Vacuum Metric:

Before the symmetry breakdown, the metric g_{AB} possesses the Einstein and Maxwell gauge invariances. We wish to determine which, if any, of these invariances will be maintained after the spontaneous symmetry breakdown has occurred. Now these invariances of the full dynamical metric g_{AB} will remain intact provided the metric $g^{o}{}_{AB}$ for the vacuum state, i.e., the state of vanishing gravitational and electromagnetic field, possesses Einstein and electromagnetic gauge invariances. It is a straightforward task to exhibit the gauge invariance of the vacuum metric under the Einstein and the electromagnetic gauge transformations. We accomplish this by exhibiting the vacuum metric of Eq. (29) (for the case N = 8) in arbitrary Einstein and Maxwell gauges (since Eq. (29) actually holds in a specific Einstein and Maxwell gauge). We have

$$\begin{aligned} g^{o}{}_{\mu\nu}(Z) &= \eta^{o}{}_{\mu\nu}(x) + i\beta[\bar{\theta}\epsilon\gamma^{(o)}{}_{\mu}(x)\theta U_{,\mu} + \bar{\theta}\epsilon\gamma^{(o)}{}_{\nu}(x)\theta U_{,\nu}] \\ &\quad + \bar{\theta}\theta U_{,\mu}U_{,\nu} - \beta^{2}U_{,\mu}U_{,\nu}(\bar{\theta}\epsilon\gamma_{a}\theta)(\bar{\theta}\epsilon\gamma^{a}\theta) \quad , \\ g^{o}{}_{\mu\alpha}(Z) &= -i\beta(\bar{\theta}\gamma^{(o)}{}_{\mu}(x))_{\alpha} + (\bar{\theta}\epsilon)_{\alpha}U_{,\mu} + \beta^{2}U_{,\mu}(\bar{\theta}\epsilon\gamma_{a}\theta)(\bar{\theta}\gamma^{a})_{\alpha} \quad , \\ g^{o}{}_{\alpha\beta}(Z) &= \eta_{\alpha\beta} + \beta^{2}(\bar{\theta}\gamma_{a})_{\alpha}(\bar{\theta}\gamma^{a})_{\beta} \quad . \end{aligned} \tag{34}$$

In Eq. (34), $\eta^{o}{}_{\mu\nu}(x)$ is the Einstein metric for a gravitationally flat vacuum state so that

$$\eta^{o}{}_{\mu\nu}(x) = \langle 0|g_{\mu\nu}{}^{E}(x)|0\rangle \quad . \tag{35}$$

For a vacuum gravitational state, there always exists a global inertial frame, y^{a}. Then

$$\eta^{o}_{\mu\nu}(x) = e^{o}_{a\mu}(x)\ \eta^{ab}\ e^{o}_{b\nu}(x) \quad , \tag{36}$$

where η^{ab} is the constant numerical metric of the Minkowski space and $e^{o}_{a\mu}(x) = \partial y^{a}/\partial x^{\mu}$ is the Vierbein of the vacuum state. γ^{a} are the usual constant Dirac matrices and[16]

$$\gamma^{(o)}_{\mu}(x) = \gamma^{a}\ e^{o}_{a\mu}(x) \quad . \tag{37}$$

The quantity U(x) in Eq. (34) defines the arbitrary gauge freedom of the electromagnetic field for the vacuum state, i.e., $\langle 0|A_{\mu}(x)|0\rangle = U(x)_{,\mu}$ where $A_{\mu}(x)$ is the electromagnetic potential. Equation (34) is both Einstein and Maxwell gauge invariant and reduces to Eq. (29) in the special gauge $y^{a} = x^{a}$, $u(x) = 0$.

b. Domain of Supersymmetry Transformations:

The vacuum metric of Eq. (34) no longer satisfies the supersymmetry transformations of Eq. (5) (appropriately extended to apply to the case with Dirac spinor coordinates). This is so because Eqs. (5) hold only in a local frame. One may, however, exhibit the supersymmetry transformations for the case of zero gravitational and zero electromagnetic interactions in a global frame, under which the vacuum metric is invariant. One has for the vacuum case

$$\begin{aligned} \theta^{\alpha\prime} &= \theta^{\alpha} + \delta^{\alpha}(x) + i\beta\bar{\delta}(x)\gamma^{a}\theta\ e^{o}_{a\lambda}U_{,}{}^{\lambda}(\epsilon\theta)^{\alpha} \quad , \\ x^{\mu\prime} &= x^{\mu} + i\beta(\bar{\delta}(x)\gamma^{a}\theta)\ e^{o}{}_{a}{}^{\mu}(x) \quad , \end{aligned} \tag{38}$$

where $\delta(x) = \exp(\epsilon U(x))$. The ordinary supersymmetry form corresponds to choosing a local frame where $U = 0$ and $e^{o}{}_{a}{}^{\mu} = \delta_{a}{}^{\mu}$.

In the presence of real gravitational and electromagnetic interactions a single form for the supersymmetry transformations in the global frame such as Eq. (38), under which the full metric $g_{AB}(Z)$ is invariant, no longer holds. In fact, in the presence of interactions the supersymmetry transformations would hold only in a small geodesic region around a point in much the same way that Poincare transformations hold only in a local inertial frame. A similar observation in the context of combined supersymmetric and gauge invariant field theory models has been made by de Wit and Freedman.[17]

c. Massless Nature of the Graviton and the Photon in Broken Supergauge:

Since the gravitational and the Maxwell gauge invariances for the metric g_{AB} are maintained after the spontaneous breakdown of the supergauge symmetry, the broken symmetry situation preserves the massless nature of the graviton and the photon. This remarkable result may also be checked by an explicit calculation of the graviton and the photon equations of motion from the superfield equations Eq. (26). We consider the electromagnetic sector first. The electromagnetic gauge complete metric (retaining only the electromagnetic potential) corresponding to the guage complete vacuum metric of Eq. (34) is

$$g_{\mu\alpha}(Z) = -i\beta(\bar{\theta}\gamma_\mu)_\alpha + (\bar{\theta}\epsilon)_\alpha A_\mu + \beta^2(\bar{\theta}\epsilon\gamma_\lambda\theta)(\bar{\theta}\gamma^\lambda)_\alpha A_\mu \quad ,$$

$$g_{\mu\nu}(Z) = \eta_{\mu\nu} + i\beta(\bar{\theta}\epsilon\gamma_\mu\theta A_\nu + \bar{\theta}\epsilon\gamma_\nu\theta A_\mu) + \bar{\theta}\theta A_\mu A_\nu$$
$$- \beta^2 A_\mu A_\nu(\bar{\theta}\epsilon\gamma_\lambda\theta)(\bar{\theta}\epsilon\gamma^\lambda\theta) \quad , \qquad (39)$$

and $g_{\alpha\beta}(Z) = g^o{}_{\alpha\beta}(Z)$. Equations (39) represent the form of the metric <u>after</u> spontaneous breaking and the limit of no spontaneous breaking corresponds to $\beta = 0$ in which case the electromagnetic potential appears only in one term in the superfield expansion of $g_{\mu\alpha}$. In the absence of symmetry breaking the electromagnetic field equations arise as the coefficient of $(\bar{\theta}\epsilon)_\alpha$ term in $R_{\mu\alpha} = 0$. After spontaneous breaking, the field equations take the form

$R_{AB} = \lambda g_{AB}$. However, one may show that due to the existence of fermi derivatives, R_{AB} may develop terms linear in the electromagnetic potential and with no spatial derivatives. Gauge invariance then dictates these terms to be precisely those in R_{AB}, i.e., one has

$$\text{Terms with no spatial derivatives in } R_{AB}(Z) = \lambda g_{AB} \quad . \tag{40}$$

Equation (40) then implies the massless nature of the photon after the symmetry breakdown has occurred.

A similar situation holds for the graviton. Here not only does the graviton remain massless but there is in addition no generation of any cosmological term due to the breakdown of the supergauge symmetry. To exhibit this phenomenon we display the metric $g_{AB}(Z)$ with gravitational interactions (but suppressing all other interactions) after spontaneous symmetry breakdown has occurred. We have

$$g_{\mu\nu}(Z) = g_{\mu\nu}{}^{E}(x) \; , \quad g_{\mu\alpha}(Z) = -i\beta(\bar{\theta}\gamma^{a})_{\alpha}\, e_{a\mu}(x) \; , \quad g_{\alpha\beta}(Z) = g^{o}{}_{\alpha\beta} \; , \tag{41}$$

where $g_{\mu\nu}{}^{E}(x) = e_{a\mu}\eta^{ab}\, e_{b\nu}$. Initially, the fields in the Einstein sector appear to possess a cosmological term, i.e., expanding

$$R_{\mu\nu}(Z) = R^{(o)}{}_{\mu\nu}(x) + (\bar{\theta}\theta)R^{(1)}{}_{\mu\nu}(x) + \ldots$$

one has

$$R^{(o)}{}_{\mu\nu}(x) = \lambda g_{\mu\nu}{}^{E}(x) \quad . \tag{42}$$

However, $R^{(o)}{}_{\mu\nu}(x)$ is not the Einstein contracted curvature but involves an additional terms from the fermionic space. One finds

$$R^{(o)}{}_{\mu\nu}(x) = R_{\mu\nu}{}^{E}(x) + \Gamma_{\mu\alpha}{}^{\beta}\Gamma_{\beta\nu}{}^{\alpha} \quad ,$$

where

$$\Gamma_{\mu\alpha}{}^{\beta}\Gamma_{\beta\nu}{}^{\alpha} = -\frac{1}{8}\lambda\ \mathrm{Tr}(\gamma_\mu\gamma_\nu) = \lambda g_{\mu\nu}{}^{E}(x) \quad . \tag{43}$$

We find then that the apparent cosmological term in the gravitational sector is cancelled identically producing the familiar Einstein field equations.

d. Spontaneous Symmetry Breakdown and Higg's Phenomenon:

We have seen in the preceding discussion that after the supergauge symmetry breaks spontaneously, the graviton and the photon maintain their massless nature. This is not necessarily the case in other sectors of the supergauge multiplet. In these sectors there may be a mass growth after the supergauge symmetry breakdown involving an analog of the Higg's phenomenon whereby some elements of the supergauge multiplet are absorbed as longitudinal modes of a higher spin field which then acquires a mass.

We illustrate this phenomenon by considering specific elements in the supergauge multiplet which are connected by the supergauge transformation

$$\xi^{\alpha}(Z) = \sum_n (\epsilon\theta)^{\alpha}(\bar{\theta}\theta)^n\Lambda_n(x) \quad . \tag{44}$$

Consider then the set of vector fields $V_{n\mu}(x)$ and scalar field $\Pi_n(x)$ defined below

$$g_{\mu\alpha}(Z) = -i\varphi(x)(\bar{\theta}\gamma_\mu)_\alpha + \sum_n (\bar{\theta}\epsilon)_\alpha(\bar{\theta}\theta)^n V_{n\mu}(x) + \ldots \quad ,$$

$$g_{\alpha\beta}(Z) = \eta_{\alpha\beta} + \varphi'(x)(\bar{\theta}\gamma_\mu)_\alpha(\bar{\theta}\gamma^\mu)_\beta - \{(\bar{\theta}\epsilon)_\alpha\bar{\theta}_\beta - (\bar{\theta}\epsilon)_\beta\bar{\theta}_\alpha\}\sum_n \Pi_n(x)(\bar{\theta}\theta)^{n-1} \ . \tag{45}$$

Clearly $V_{0\mu}$ ($\equiv A_\mu$) is the electromagnetic potential, and $V_{1\mu}(x)$, etc., are vector fields in the higher θ sectors which are connected to A_μ through the field equations. Under the supergauge transformations of Eq. (44), $V_{n\mu}$ transform according to

$$V_{n\mu}'(x) = V_{n\mu}(x) - \Lambda_{n,\mu}(x) \tag{46}$$

whereas Π_n transform as

$$\Pi_n(x)' = \Pi_n(x) - 2n\Lambda_n(x) \tag{47}$$

where n = 0, ..., 3. The spontaneous breakdown of the supergauge is induced when $\varphi(x)$, $\varphi'(x)$ in Eq. (45) acquire vacuum expectation values such that $\langle 0|\varphi(x)|0\rangle = \beta$ and $\langle 0|\varphi'(x)|0\rangle = \beta^2$ so that the vacuum part of Eq. (45) reduces to Eq. (29). It is then possible to identify the modes $\Pi_n(x)$ for n = 1, 2, 3 as the fictitious Goldstone bosons which are absorbed by the set of vector fields $V_{n\mu}$ (n = 1, 2, 3) to produce the guage invariant combinations $\tilde{V}_{n\mu} = V_{n\mu} - 2n\Pi_{n,\mu}$. The fields Π_n (n = 1, 2, 3) then disappear as longitudinal modes for the vector fields $\tilde{V}_{n\mu}$ in the usual fashion, and the latter fields grow masses.

V. UNIFICATION OF ELEMENTARY INTERACTIONS

We have seen in the preceding discussion that the supergauge group even after a spontaneous symmetry breakdown preserves the exact invariances of the gravitational and the electromagnetic gauge groups. Since the process of symmetry breakdown involves a self-contained Higgs phenomenon, a relationship would be imposed between the observed gravitational and the electromagnetic coupling strengths and a unification of these two fundamental interactions of nature would emerge. In a more ambitious framework one would wish to tie together not only the gravitational and the electromagnetic but the weak interactions (and ultimately the strong interactions) as well. Here, we first discuss the confluence of gravitational and electromagnetic interaction and then briefly remark on the possibility of including weak interactions in the formalism.

a. Coupled Maxwell-Gravity:

For the present discussion it is sufficient to exhibit in the superfield expansion of $g_{AB}(Z)$ the terms

$$g_{\mu\nu}(Z) = g_{\mu\nu}^{E}(x) + \bar{\theta}\theta p_{\mu\nu}(x) + \ldots \quad ,$$

$$g_{\mu\alpha}(Z) = (\bar{\theta}\epsilon)_{\alpha}\, e\, A_{\mu}(x) + \ldots \quad ,$$

$$g_{\alpha\beta}(Z) = \eta_{\alpha\beta} + (\bar{\theta}\epsilon)_{\alpha}(\bar{\theta}\epsilon)_{\beta}\, f(x) + \ldots \quad . \tag{48}$$

Before the spontaneous symmetry breaking one obtains from $R_{AB} = 0$ the field equations

$$G_{\mu\nu}[g_{\lambda\rho}^{E}(x)] = -8(p_{\mu\nu} - \frac{1}{2}g_{\mu\nu}^{(E)}p^{\lambda}{}_{\lambda}) + \ldots \quad ,$$

$$K_{\mu\nu}(p_{\lambda\rho}) = -\frac{1}{2}f_{;\mu\nu} + \frac{e^2}{2}F_{\mu}{}^{\lambda}F_{\nu\lambda} + \ldots \quad ,$$

$$-\Box^2 f = \frac{e^2}{2}F^{\lambda\rho}F_{\lambda\rho} + \ldots \quad , \tag{49}$$

where $K_{\mu\nu}$ is the massless spin 2 differential operator. From Eqs. (49) we find that contrary to the usual expectation the Maxwell stress tensor does not appear as the source of the Einstein field. What he have in Eq. (49) is a field-current identity where $p_{\mu\nu}$ acts as the source of the field $g_{\mu\nu}^{E}$ while the Maxwell field appears in the sources of $p_{\mu\nu}(x)$ and $f(x)$. For Eqs. (49) to be physically meaningful there must be a superheavy spontaneous (or dynamical) mass growth M for $p_{\mu\nu}$ which then satisfies the equation

$$K_{\mu\nu}(p_{\lambda\rho}) - M^2 p_{\mu\nu} = -\frac{1}{2}f_{\mu\nu} + \frac{e^2}{2}F_{\mu}{}^{\lambda}F_{\nu\lambda} \quad . \tag{50}$$

Solving for $p_{\mu\nu}$ from Eq. (50) and using it as the source in the $G_{\mu\nu}^{E}$ equation one has

$$G_{\mu\nu}[g_{\lambda\rho}^{(E)}] = 8\pi G_E T_{\mu\nu}^{Maxwell} + \text{super-potential terms} \quad , \tag{51}$$

where $G_E = e^2/(2\pi M^2)$ and $M \approx 10^{18}$ GeV. (A similar superheavy mass also appears in the renormalization group analyses of the breakdown of SU(5).[18]) The above analysis also shows how two long range forces such as gravitation and electromagnetism (but with a huge disparity of coupling strengths) may be unified in a spontaneously broken symmetry scheme. In the analogous situation of the Weinberg-Salam model, the weakness of the coupling strength of the weak interactions was correlated with their short range nature. Here the structure of a field current identity gives to gravity both an infinite range part from $g_{\mu\nu}^E$ and an extremely short range part due to $p_{\mu\nu}(x)$. It is precisely the very short range nature of $p_{\mu\nu}$ which accounts for the extreme weakness of the gravitational interactions.

b. SU(2) x U(1) Model:

The preceding discussion involved an analysis of the gauge supersymmetry using four bose and eight fermi (or one Dirac) coordinates. This system was seen to include not only the gravitational but the electromagnetic gauge group as well. Further, the supergauge group allowed for a spontaneous symmetry breakdown, unifying gravitational and electromagnetic phenomenon. We recall, however, that the choice of 8 fermi (or 1 Dirac) coordinates was necessitated by $N_{Fermi} = 2N_{Bose}$. This condition was determined by the constraints of spontaneous symmetry breakdown allowed by the superfield equations $R_{AB} = \lambda g_{AB}$, $\lambda \neq 0$. However, if one assumes field equations to have the form $R_{AB} = 0$, spontaneous symmetry breakdown may occur with the number of fermi dimensions greater than 8. For this situation there is a remarkable theorem[13] that spontaneous symmetry breaking also predicts parity violations. We do not wish to prove this result here but shall consider spontaneous symmetry breakdown only within the framework of an SU(2) x U(1) invariance, involving a doublet of Dirac coordinates $\theta^{\alpha q a} \equiv (\theta^{\alpha q}{}_{\nu e}, \theta^{\alpha q}{}_{e})$, a = 1, 2 with the hope that the weak interactions along with the electromagnetic and gravitational interactions would arise from a common supergauge invariance.

Let μ_γ (γ = 1, 2, 3) be the SU(2) real, antisymmetric matrices.(9) We denote the generators of SU(2) x U(1) by T_A (A = 0, 1, 2, 3)

$$T_A = \{\tfrac{1}{2}\epsilon - \mu_3 P_+, \mu_\gamma P_-\} \quad , \tag{52}$$

where $P_\pm = \frac{1}{2}(1 \pm i\epsilon\gamma^5)$ and $\gamma^{5\dagger} = \gamma^5$. Here P_+ (P_-) is the right (left) chiral projection operator. To discuss spontaneous symmetry breaking we consider the following ansatz for the vacuum metric:

$$\begin{aligned} g^{o}{}_{\mu\nu} &= \eta_{\mu\nu} \quad , \\ g^{o}{}_{\mu\alpha} &= i\beta_A(\bar{\theta}\gamma_\mu \epsilon T_A)_\alpha \quad , \\ g^{o}{}_{\alpha\beta} &= \eta_{\alpha\beta} + \beta_A\beta_B(\bar{\theta}\gamma_a\epsilon T_A)_\alpha(\bar{\theta}\gamma^a\epsilon T_B)_\beta \quad . \end{aligned} \tag{53}$$

The contracted curvature $R^{o}{}_{AB}$ corresponding to Eq. (53) is

$$R^{o}{}_{AB} = 4(\beta_0{}^2 - \beta_0\beta_3)\left\{g^{o}{}_{AB} - \left(\eta\frac{1+\tau_3}{2}\right)_{\alpha\beta}\delta_{A\alpha}\delta_{B\beta}\right\} \quad , \tag{54}$$

where $\beta_\gamma = \beta_3\delta_{\gamma 3}$. Thus spontaneous symmetry breakdown of the field equations $R_{AB} = 0$ occurs provided

$$\beta_0{}^2 - \beta_0\beta_3 = 0 \quad . \tag{55}$$

Of the two symmetry breaking solutions (1) $\beta_0 = \beta_3$ or (2) $\beta_0 = 0$ only in the second is $g^{o}{}_{AB}$ invariant under SU(2) x U(1) gauge transformation of β^A. For $\beta_0 = 0$ one expands

$$g_{\mu\alpha} = g^o_{\mu\alpha} + (\bar{\theta}T_A)_\alpha B_\mu{}^A(x) + \ldots \quad ,$$

$$g_{\alpha\beta} = g^o_{\alpha\beta} + (\eta V_A)_{\alpha\beta}\varphi^A(x) + \ldots \quad , \tag{56}$$

where $V_A = T_{\tilde{A}} - T_A$ and $T_{\tilde{A}}$ is T_A with $P_+ \Leftrightarrow P_-$. Under SU(2) x U(1) gauge transformations

$$\xi^\alpha = \Lambda^A(x)(T_{\tilde{A}}\theta)^\alpha \quad , \tag{57}$$

one finds that

$$W_\mu{}^{1,2} \equiv B_\mu{}^{1,2} + \partial_\mu\varphi^{1,2} \quad ,$$

$$Z_\mu \equiv \tfrac{1}{2}(B_\mu{}^o + B_\mu{}^3) + \partial_\mu(\varphi^o + \varphi^3) \tag{58}$$

are gauge invariant and so can grow masses whereas the remaining combination

$$A_\mu \equiv \tfrac{1}{2}(B_\mu{}^o - B_\mu{}^3) \tag{59}$$

has $\delta A_\mu = -\partial_\mu\Lambda(x)$ and represents the photon and remains massless.

VI. CONCLUDING REMARKS

As our discussion in the preceding sections shows, gauge supersymmetry does indeed possess gauge invariances such as those of Einstein and of Maxwell (and when extended also of the electromagnetic and weak group such as SU(2) x U(1)) in a unified framework. A unification of the interactions comes about when the supergauge group undergoes a spontaneous breakdown giving rise to a relationship between the Einstein constant G_E and the fine structure constant α. Also the mere assumption that the equations of motion have the form $R_{AB} = 0$ demands parity violations for the weak interaction due to the fact that the equations of motion must generate their own spontaneous breakdown. To see this more clearly, one may compare the situation to the conventional unified gauge theories where one must always add to the Lagrangian an additional potential function of scalar fields $V(\varphi)$ to guarantee that the spontaneous breakdown occurs. In the present scheme all interactions are uniquely determined by the non-abelian gauge invariance of the theory, and such additional ad hoc assumptions are not required.

To develop more fully the significance of the theoretical scheme discussed here one must determine the correct dimensionality of the fermi space and consequently the nature of the internal symmetry group generated by the theory. Another problem concerns the construction of an "effective Lagrangian"[19] after spontaneous breakdown occurs and a superheavy mass appears in the theory. For low energy phenomenon (i.e., at energies small compared to the superheavy mass), modes associated with the superheavy mass which are not excited until at very high energies ($\approx M$) may effectively decouple leaving behind a relatively simple effective Lagrangian. This is particularly desirable in view of the rather complicated nature due to its size of the superfield multiplet g_{AB}. One must also raise the inevitable question concerning the renormalizability of the theory. This is obviously an enormously important issue in view of the non-renormalizability of the conventional matter coupled-gravity.[20] We hope that supergauge symmetry would have a bearing on this issue but have nothing to say about it at the moment.

Finally, due to space limitation we unfortunately are not able to discuss in detail some applications, e.g., by G. Woo[21] and P. P. Srivastava,[22] of the curved geometric superspace. These applications are made however in a different spirit that the approach presented here, i.e., to extract the dynamics of the conventional supersymmetry. Other works[10,22] which also we have not discussed here concern the algebraic structures in superspace and extension of the Ogievetsky theorem[23] to superspace.

References and Notes

1. See, e.g., S. Weinberg, Rev. Mod. Physics 46, 255 (1974).

2. J. Wess and B. Zumino, Nucl. Phys. B70, 39 (1974); D. V. Volkov and V. P. Akulov, Phys. Lett. 46B, 109 (1973); B. Zumino, Proc. 17th Int. Conf. on High Energy Physics, London, 1974.

3. A. Salam and J. Strathdee, Nucl. Phys. B76, (1974).

4. P. Nath and R. Arnowitt, Phys. Lett. 56B, 171 (1975).
R. Arnowitt and P. Nath, Invited talk at the Conference, The Riddle of Gravitation, Syracuse, New York, March 1975, NUB #2261. P. Nath and R. Arnowitt, Invited talk at the French Physical Society Meeting at Dijon, France, July 1975, NUB #2259.

5. Th. Kaluza, Sitzungsber. Preuss. Akad. Wiss. (1921), 966; O. Klein, Zs. f. Phys. 37, 895 (1926).

6. Y. M. Cho and P. G. O. Freund, University of Chicago Preprint, EFI 75-15 (1975).

7. See B. Zumino's lecture titled "Supersymmetry" in these proceedings.

8. Scale invariance group is also contained in the transformations of Eq. (13) and is generated by the $x^{\mu\prime} = x^{\mu}$, $\theta^{\alpha} = [\Lambda(x)]^{\frac{1}{2}}\theta^{\alpha\prime}$.

9. μ_A are $\{\frac{1}{2}i\lambda_A^{(a)}, \frac{1}{2}\epsilon\lambda_A^{(s)}\}$ where $\lambda_A^{(s,a)}$ are the n x n (symmetric, antisymmetric) Gell-Mann matrices and $\epsilon_{qq'}$ is the antisymmetric charge matrix with $\epsilon_{12} = +1$ ($\epsilon^2 = -1$). The μ_A then satisfy the relation $\mu_A\mu_B = -\frac{1}{2}f_{ABC}\mu_C + \frac{1}{2}\epsilon\, d_{ABC}\mu_C$.

10. This result has also been noted by P. G. O. Freund, University of Chicago Preprint, EFI 75/51.

11. The generalized Leibnitz rule for two tensors Q_1, Q_2 is $(Q_1Q_2)_{;B} = Q_1Q_{2;B} + (-1)^{bq_2}Q_{1;B}Q_2$.

12. R. Arnowitt, P. Nath and B. Zumino, Phys. Letters 56B, 1 (1975), 81.

13. R. Arnowitt and P. Nath, under preparation.

14. For a Majorana spinor $\bar{\theta}_{\alpha} = \theta^{\beta}\eta_{\beta\alpha}$.

15. This result was established in collaboration with S. S. Chang

16. $\gamma^{(o)\mu}$ is given by $\gamma^{(o)\mu} \equiv \gamma^{a} e^{o}{}_{a}{}^{\mu}$ where $e^{o\mu}_{a} \equiv \partial x^{\mu}/\partial y^{a}$ and $e^{o}{}_{a}{}^{\mu} e^{o}{}_{b\mu} = \eta_{ab}$. γ^{a} obey the condition $\gamma^{a}\gamma^{b} + \gamma^{b}\gamma^{a} = -2\eta^{ab}$.

17. B. de Wit and D. Z. Freedman, Stony Brook Preprint ITP-SB-24-75, June (1975).

18. H. Georgi, H. R. Quinn and S. Weinberg, Phys. Rev. Lett. 33, 451 (1974).

19. In the sense of T. Applequist and Carrazone, Phys. Rev. D11, 2856 (1975).

20. S. Deser and P. Van Niewenhuizen, Phys. Rev. D10 (1974), 401 and 411. Also see S. Deser in these proceedings.

21. G. Woo, Nuovo Cimento Letters 13, 546 (1975).

22. P. P. Srivastava, Nuovo Cimento Letters 13, 657 (1975) and Preprints from Instituto de Fisica, UFRJ, Rio de Janeiro, Brasil.

23. V. I. Ogievetsky, Lett. Nuovo Cimento 8 (1973) 988.

DISCUSSION

G. WOO (MIT)

Have you any insight into the fact that the Wess-Zumino supermetric is singular in your Riemannian geometry?

NATH

As you well know, under the ordinary supersymmetry transformations, the quantities $(dx_\mu - id\bar{\theta}\gamma_\mu\theta)^2$ and $d\bar{\theta}d\theta$ are separately invariant and the general form for the invariant line element is $(dx_\mu - id\bar{\theta}\gamma_\mu\theta)^2 + kd\bar{\theta}d\theta$, where k is an arbitrary constant. The presence of the $kd\bar{\theta}d\theta$ term is perfectly consistent with the Wess-Zumino supersymmetry. It is also precisely this term that allows us to invert the metric and develop in a simple and straightforward fashion the geometry in superspace. What you get of course is a Riemannian geometry. Now one may conceive of taking the $k \to 0$ limit to get back to the non-Riemannian situation. However, Bruno tells me that even in this limit the group that you get by taking the limit of the Riemannian geometry is not quite the one that he talked about in his lecture; it is somewhat bigger. So they are not quite identical even in that limit.

VOICE FROM THE AUDIENCE

What is the superspace Ogievestsky theorem?

NATH

Well, first the Ogievestsky theorem in Bose Minkowski space is that the closure of the conformal and the special linear groups is the general covariance group. In its superspace extension (see ref. 22), the superspace general covariance group is the closure of three subgroups: superconformal, special linear and an additional supergauge group. So that's the slight modification there is in there.

B. ZUMINO (CERN)

First of all, I want to say something to Gordon Woo's question. You said it correctly but let me restate it. Gordon Woo himself pointed out that if one takes a Riemannian superspace and one writes for instance the equations of a scalar superfield in Riemannian superspace and then one tries to go to the flat limit one does not get the correct Lagrangian of flat space supersymmetry and, in fact, he's the first one who pointed out that you have to go to some limit in order to recover the correct flat space supersymmetry, and that was in fact in a sense the first indication that the space may not be Riemannian. Now, in fact, if you do start from a Riemannian space and go to the limit, you don't get exactly the geometric structure I was trying to describe. You'll get the structure which contains it but still is not sufficiently specific so it's not just the question of a Riemannian limit.

The second thing I wanted to say was to that point concerning renormalization about which you said you have nothing to say. I have nothing to say either. However, I just want to make a remark which is due to Steve Weinberg. May be he doesn't remember it, but he once asked me if it is possible that at some point a gravitational theory will come out which is renormalizable, but in order to find out he wanted to know what are these cancellations which are present in ordinary supersymmetric theories. And in fact if you look at it you will see immediately that these cancellations are not enough. I mean roughly speaking the argument goes as follows: In a non-renormalizable theory you have an infinite number of arbitrary constants. For instance, already if you take the simplest non-renormalizable supersymmetric model which was studied by Lang and Wess you have an infinite number of arbitrary constants. All you have, on the other hand, is one additional conservation law or four or whatever, I mean a finite number of additional conservation laws and these are not going to make a non-renormalizable flat space supersymmetry theory renormalizable because they are just a finite number and you have an infinite number to get rid of. On the other hand, in gravitation one can never tell because there one has the infinite gauge groups coming back again and already in ordinary gravitation there appear cancellations which are not fully understood, as far as I know. May be you will explain how to understand them, but there are things which at first sight don't seem reasonable and then they happen anyway. This is what I remember of Steve Weinberg's argument. Now, you agree?

S. WEINBERG (Harvard)

I don't remember asking the question but I remember giving the answer.

USES OF SOLID STATE ANALOGIES IN ELEMENTARY PARTICLE THEORY

Philip W. Anderson*

The solid state background of some of the modern ideas of field theory is reviewed, and additional examples of model situations in solid state or many-body theory which may have relevance to fundamental theories of elementary particles are adduced.

*Bell Laboratories, Murray Hill, N.J. 07974.

As I have been trying during the past weeks to catch up with what you all have been doing in the past decade or more since last I took a detailed look, I find that more and more ideas from my field are showing up in your field. I thought it might perhaps be interesting for you to hear some discussion of the actual many-body theoretic context of these ideas, not only for purely historical reasons but in order to stimulate the drawing of yet more analogies in your field to the strange phenomena we solid staters now and then encounter in ours.

Another thing that I have found on the relatively rare occasions when I do discuss these things with E. P. theorists is that very often ideas which seem to them to be absolutely trivial and obvious, and to go without saying, are to me the most esoteric and difficult results; but also of course almost everything I say to them which seems very easy and simple to me they find abstruse, and difficult or impossible to understand. This is by way of pointing out that you and I have a very difficult problem of communication here, but I do think we have things to say to each other, so I hope you will bear with me.

I will first indulge myself in some ancient history about the many-body theorems which correspond to Goldstone and Higgs, trying to place these theorems which are so familiar to you into their original context. Next I'd like to continue in the same vein with some ideas about broken symmetry which are only just now appearing to be relevant to the E. P. field: phase transitions, and order parameter singularities. Finally, I'd like to use any time remaining to me to say a few things about how modern asymptotically free quark-gluon theories of strong interactions look to a many-body theorist, and why I think they behave like the Kondo effect, which may serve as rather a model of how such theories can behave.

Let me make the basic nature of the quantum many-body to E. P. analogy clear, if I can. So far as I know Nambu deserves the credit for this very fruitful idea. The ground state of a many-body quantum system is compared to the vacuum state of a field theory. The fact that many-body Hamiltonians very often have ground states which do not have the same symmetry as the Hamiltonian itself is the broken symmetry phenomenon; and many, if not most of the field theories you discuss here have this same feature in common. In general, you break the symmetry by introducing a field - usually scalar - which is taken not to have zero expectation value: this is the analog of what we call an "order parameter": a physical variable, defined locally, which describes the distortion of our system from its unperturbed ground state. In our theories symmetry-breaking is always "dynamical", at least in some sense. Such order parameters are usually, but not always, not constants of the motion - i.e. do not commute with the Hamiltonian; this is the typical milieu of the Goldstone-Higgs phenomenon.

The first conscious use of a Goldstone boson was in the problem of the ground state of an antiferromagnet in 1952.[1] Of course, the idea of phonons in solids as well as Landau's and Bogolyubov's theories of phonons in liquid helium already existed, but there was no concept of the role of these excitations in a consistent description of the ground state and of the quantum theory of broken symmetry systems.

The problem in the case of antiferromagnetism was the incompatibility of a theorem of Hulthen's,[2] that the ground state of the antiferromagnetic Heisenberg Hamiltonian

$$\sum_{ij} J_{ij} S_i \cdot S_j$$

had to be a singlet, with the obviously non-singlet character of the accepted Neel hypothesis for the ground state

$$= \prod_{A \text{ subl.}} \chi_{\uparrow}(i \text{ on } A) \prod_{B \text{ subl.}} \chi_{\downarrow}(i \text{ on } B).$$

The solution of this problem of incompatibility was first suggested, at least to me, by Conyers Herring; that the true ground state had to be in some sense a symmetric linear combination of all the apparently orientationally degenerate Neel states pointing in different directions. But that this symmetry argument had dynamical consequences both for the energy level spectrum and for the long-wavelength fluctuations of the order parameter - that is, the sublattice magnetization - was, I believe, first discussed in ref. 1. The fact is that the broken symmetry requires a Goldstone boson, the anti-ferromagnetic spin wave, which in turn has a zero-point fluctuation which diverges in the long-wavelength limit in just such a way that the exact ground state is rotationally invariant. But infinitely close to the exact ground state, in the $N \to \infty$ limit, is an infinitely degenerate manifold of states which can be recombined to give the broken symmetry ground state. In '58, I thought the same phenomenon might be responsible for the gauge invariance difficulties of the BCS theory which preoccupied people at the time. But in the superconducting case, a new feature turned out to be there.(3) The long-range nature of the electromagnetic forces destroys the argument from continuity in Q which requires the existence of the Goldstone boson which would be there in a neutral B.C.S. system - in fact has now been demonstrated as 4th sound in the B.C.S. anisotropic superfluid He_3.

In fact, the Goldstone boson is raised to a finite frequency and joins the two components of the photon in a triad of excitations which, at least at $Q = 0$, mimic the 3 components of a massive vector boson. The three components are required by rotational symmetry: in the many-body system, where the order parameter and the underlying many-body systems carry conserved quantum numbers such as baryon number, mass and charge, the ground state is not even Galilean invariant, much less relativistic, and analogies to field-theoretic ideas of broken symmetry must be handled with great care. Nonetheless the Higgs mechanism is very nicely demonstrated, in the example of the charged Bose gas, or the equivalent charged B.C.S. pair condensation, where one makes up the charged scalar field which breaks the symmetry out of a Fermion pair field. All of these phenomena were described in a paper in 1962[(4)] which is usually <u>not</u> the one of mine quoted in this context, for reasons I don't understand.

In the course recently, of writing on the idea of broken symmetry, I had occasion to try to put together in my own mind all of the existing general ideas and rules about broken symmetry in the many-body sense, and it was of course then an obvious thing to do to ask which of these things might have significance in the field-theoretic context. It happens that two of them have actually appeared in the literature within the past year or two.

Long before any dynamical ideas about broken symmetry existed, Landau seems to have been the first to emphasize the importance of symmetry classification of phases of macroscopic systems.[(5)] Landau, in fact, is responsible for the first meaningful - if trivial - theorem of broken symmetry: that a symmetry element can never be discarded gradually, so that a symmetry change always requires a sharp thermodynamic phase transition. Usually - but not always - the higher symmetry phase allows more fluctuations and has therefore more entropy, and occurs at higher temperature, while the low symmetry phase occurs at low temperature. Realizing then that modern unified theories involve broken symmetry in the vacuum - ground-state, one is immediately tempted to ask whether in the original cosmic fireball of the big bang theory, the temperature could have been so high as to restore full symmetry and equate, for instance, the weak and electromagnetic interactions in a single global gauge symmetry. Kirshnitz and Linde,[5] and later and in more detail Weinberg[7] as well as Jackiw[8] have answered this question in the affirmative.

A much more recent development in the discussion of broken symmetry in the many-body and solid-state context is the idea of classification of singularities of the broken symmetry, which to my knowledge was initiated by deGennes only a few years ago. To describe what I mean by a singularity it is probably best simply to give examples, which I do in Table 1. In general, when a solid state type of system condenses into a broken symmetry state it has to satisfy a number of external constraints - it will always have boundaries, for instance, where there will be some constraint on the order parameter (symmetry-breaking field,

Table 1

Order Parameter Singularities in Broken Symmetry

"Dim" of Order Parameter	Broken Group	Example	Types of. Singularities
1 (real scalar)	point (finite)	"Ising Model" anisotropic ferromagnet	domain boundary 
2 complex scalar, 2-dim vector	continuous Abelian	superfluid, supercond., x-y model	" + line (vortex)  $\oint d\varphi = 2\pi$
3 or more	non-Abelian continuous	crystal liquid xtal He_3	" + point : vacancy, Brinkman point in He_3 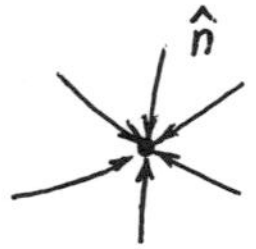

't Hooft opole? →

what else?

that is) or it will simply initiate its broken symmetry differently in different places, which are incompatible when they meet; or weaker forces such as dipole forces in a ferromagnet will be left over and enforce certain conditions. As a result in general the order parameter is not uniform in space. If Weinberg's phase transition occurs such a variation may be expected. But this non-uniformity is not continuous but achieved in such a way that ψ has its usual bulk value almost everywhere, but the order is seriously disturbed in restricted regions of a certain fixed geometry: planes, lines or points. Examples are given in Table 1. The key intuition here is geometrical: one must ask how or whether it is possible to get <u>continuously</u> from one stable value of the order parameter to another. If the possible routes can be placed in one-to-one correspondence with one-dimensional paths (an Abelian group) lines are necessary - essentially the "strings" of usual monopole theories. But in the non-Abelian theories, the possible routes can correspond to a 3-dimensional continuum: point singularities are possible. Such a singularity is a vacancy in a crystal; others are certain point singularities of liquid crystals or helium 3 - the latter are sometimes called "Brinkman spheres". In each case there is a quantization condition equivalent to the idea that a vector order parameter around a point must span 4π of solid angle, or that a vacancy in a crystal must be a whole vacancy, not a piece of one. Often the larger singularity as well as the smaller remains: dislocations in crystals, vortex lines in He_3; but I suspect boundaries are not possible in the Weinberg type of theory. Such extended singularities would be very peculiar astrophysical objects.

There is a precise analogy of the point singularity to the 't Hooft magnetic monopole(9) - the "'t Hooft opole" - in a unified gauge theory. I hope no one will construe this comment as an endorsement of any experimental result, but it would be delightful if such objects existed.

In ordinary many-body theory "condensates" - the matter exhibiting the order parameter - often carry conserved currents: baryon number or mass for helium, charge for a superconductor, mass for a crystal, etc. This is the appropriate milieu for what I have called "generalized rigidity" - such behavior as rigidity, superfluidity, etc. - which clearly must not be the case in a field theory context: your condensates must essentially carry only the quantum numbers of the vacuum.

In my final few minutes I'd like to make some abbreviated remarks about what one might call the "Theory of Unbroken Symmetry". This is a type of theory which we have only just in the past few years been realizing exists in ordinary low-energy examples, and which I suspect is very much a model for the currently favored color gauge theories of strong interactions. One point of analogy is that the kind of logarithmic approach to asymptotic freedom which is characteristic of their renormalization group behavior is fairly common in ours: it occurs, among other cases, in the B.C.S. theory, in the Kondo effect, and in one-dimensional quantum systems. It signifies two things: first, that the high-energy behavior is going to be harmless; and second, that something spectacular will happen at the infrared end of things. I don't think elementary particle physicists have quite appreciated the strength of the first statement: that an asymptotically free theory is not only renormalizable, it can be instantly renormalized without further nonsense by the simple process of inventing a "model" - like the B.C.S. or Kondo "models", both of which are terribly simple Hamiltonians which in fact are merely models which correctly describe the infrared behavior of complex systems. The idea is to introduce a real physical cutoff and coupling constant, which are to a great extent arbitrary and can be chosen more or less at will, so long as they are satisfactorily into the asymptotically free regime. Asymptotic freedom from this point of view is merely the statement that nothing above such a cutoff has any relevance to low energy physics.

One way of understanding this point is to realize that there are actually two renormalization group schemes, and it seems to be a source of confusion to the quantum field theoretic community that these two schemes go by the same name, and to a great extent satisfy the same equations.

The "classical" RNG of Goldberger, etc. is a useful sequence of statements about renormalizable field theories which follow from the arbitrariness of our choice of scale in such theories. An entirely <u>different</u> scheme is the one which we low-energy people, especially Ken Wilson, and myself on a smaller scale, have been applying to our various phase transition problems. This is the kind of thing Wilson already mentioned - the Block Spin technique. In this system, the scaling equations are not statements about mathematical solutions of the same problem, they are descriptions of equivalence of the results of measurements on <u>different physical</u> systems. The Callan-Szymancik equation,

$$\frac{dg}{(d(\ln\Lambda)} = Cg^3 + ..$$

in this description can be understood as connecting <u>different physical</u> problems with different high-energy cutoffs Λ, which need not necessarily be handled in a relativistically invariant way, but all of which correctly describe the same physics. It may not even need to have perfect gauge symmetry.

Asymptotic freedom in this context of course implies infrared troubles: necessarily, the perturbative scaling equation breaks down at low temperatures and the nature of the new physical problem is in doubt. In fact, we have in low-energy physics two types of exact solutions of such problems, each of which is interesting in the field-theoretic context. The first is the broken symmetry case of B.C.S. or many other phase transitions: the system has an infrared-unstable fixed point below which its behavior changes radically, and in particular it breaks its original symmetries: one makes the kind of sudden transition to a wholly different symmetry I talked about earlier. In this case the high energy RNG has little or no relevance to what really happens and is only a convenience: it may be much simpler, as in the B.C.S. case, to deal with the cutoff "model" problem than with the real one.

A second type of behavior has been found lately in two other types of systems.[10,11] In these - in both of which the precise character of the solutions is known through solution of a soluble case - there is no phase transition and no change from the aboriginal symmetry, which remains unbroken. In these systems it seems to be possible to squeeze down the high-energy cutoff to ordinary physical energies without changing the essential nature of the model - although it develops new parameters which affect numerical results. The essential nature of the system is that as $\Lambda \to 0$ $g \to \infty$: the scaling equations carry right through to a fixed point with infinite coupling constant.

In the Appendix I shall try to take you through an infinitely rapid course on this process in the case of the Kondo effect, where, to me, the formal analogy to what may be the true behavior of color gauge theories is very striking. I have in fact written a note suggesting that the $g \rightarrow \infty$ case can be converted to a reasonable sort of lattice model for elementary particles. What is particularly striking about such models is that in every case the actual low-energy spectrum, while perfectly exhibiting the symmetry of the original theory, has changed qualitatively in that certain types of excitation are missing. Thus this is a type of model in which quark confinement without recourse to broken symmetry seems perfectly possible.

I have one last point to make about this kind of system: that there is one way to avoid the infrared divergences which can also be useful. This lies in the suggestion by Collins and Perry[(12)] that a very high density of baryons will see only very weak forces - essentially the quarks will become asymptotically free if their Fermi energy is driven up into the asymptotic region. I, for one, believe this is so - and I also believe that one of the most promising ways to study how the quarks and gluons confine themselves and make real baryons is to do so in a series in 1/density or in $\frac{1}{E_F}$: the reason being that I am quite sure that all infrared divergences are eliminated in such a system by Fermi exclusion factors and by screening.

I thus conclude not with a many-body analogy but with a many-body problem which I think may be a very important task of elementary particle physics: the real theory of high density nuclear matter.

I hope I have at least said enough here to indicate to you that it is high time our two fields reestablished regular contact.

APPENDIX: The Kondo Problem

The Kondo problem is a very much over-simplified model of a magnetic impurity in a metal. The magnetic impurity is represented by a local spin $S = 1/2$ interacting via an exchange integral J with a Fermi sea of free, otherwise non-interacting electrons. The appropriate Hamiltonian is

$$H = \sum_{k\sigma}^{\varepsilon_k < E_c} \varepsilon_k n_{k\sigma} + J S \cdot s(0)$$

($s(0)$ being the spin density at the origin of space coordinates). The J term can cause spin-flip scattering of the free electrons, which is enhanced by the non-spin flip attractive elastic scattering. The result of the interaction of these two effects is a characteristic infrared divergence of the spin-flip scattering cross section as calculated in second Born approximation, $\sim \ln T/E_c$.

The density of states $\rho(\varepsilon_k)$ and the energy J are dependent on the normalization volume. The actual physical parameters of H are (a) the cutoff energy E_c (dimension energy) (b) a dimensionless coupling constant $j = J/E_c$, which is $\ll 1$ in the canonical Kondo Problem.

Recently[10] it has been realized that the renormalization group leads to an understanding of the Kondo H. Progressive reduction in the cutoff E_c leads to modification of j according to

$$dj\left(\frac{1}{j^2} - \frac{1}{2j} + \ldots\right) = -d\ln E_c .$$

Below T_K, one has a behavior first guessed at in detail by Mattis (although qualitatively suggested by Kondo and many others): the coupling constant becomes so large as to bind one "free" electron to S with opposite spin, with a constant binding energy T_K, and in all ways the resulting system scatters like a non-magnetic impurity; essentially, in a very real sense the spin S is "confined". A beautifully precise computer calculation by Wilson verifies the essentials of this result, for instance that the multiplicity of the levels in a finite model of the system is opposite to that required by S = 1/2.

As Wilson and Nozieres have shown, the result is equivalent to a renormalized normal non-magnetic impurity. Luther and Emery[(11)] have demonstrated a similar "confinement" phenomenon in the triplet excitations of a one-dimensional electron gas, where the essential mathematics has strong similarities to the Kondo problem.

REFERENCES

1. P. W. Anderson, Phys. Rev. 86, 694 (1952).
2. L. Hulthen, Thesis, Uppsala, 1938.
3. P. W. Anderson, Phys. Rev. 110, 837, 985 (1958).
4. P. W. Anderson, Phys. Rev. 130, 439 (1963).
5. L. D. Landau and E. M. Lifshitz, Statistical Physics, #79, 134, Pergamon, London, 1958.
6. D. A. Kirshnitz and A. D. Linde, Phys. Lett. 42B, 471 (1972).
7. S. Weinberg, Phys. Rev. 9D, 3357 (1974).
8. L. Dolan and R. Jackiw, Phys. Rev. 9D, 3320 (1974).
9. G. 't Hooft, Nucl. Phys. B79, 276 (1974).
10. Kondo theory: P. W. Anderson, G. Yuval, D. R. Hamann, Phys. Rev. B1, 4464 (1970); K. G. Wilson, Revs. Mod. Phys. 47, 773 (1975).
11. One-dimensional Electron Systems: A. Luther and V. J. Emery, Phys. Rev. Lett. 33, 589 (1974); A. Luther and L. Peschel, Phys. Rev. B9, 2911 (1975); S. T. Chui and P. A. Lee, Phys. Rev. Lett. 35, 315 (1975).
12. J. C. Collins and M. J. Perry, Phys. Rev. Lett. 34, 1353 (1975).

DISCUSSION

B. ZUMINO (CERN)

You spoke about the "Goldstone" and "Higgs" mechanisms. In relativistic field theory one now has examples of Goldstone spinors (which are fermions) and of Higgs-like phenomena in which Goldstone spinor of spin ½ disappears to give mass to a previously massless gauge field of spin 3/2. Is anything like this known in many body theories? Also, in relativistic local field theory there cannot be Goldstone particles of spin 1 or higher (in four dimensional space time). Are there Goldstone particles of spin larger than zero in non relativistic many body physics?

ANDERSON

I tried to think about fermions whether there are any fermion many body "Goldstone-on" and I don't think there are any. There are none that I know of. As far as the other theorem is concerned I think Bruno is trying to get me to mention one thing that I mentioned to him namely that probably there's at least one tensor "Goldstone-on" or "Higg-on" in many body physics. Namely, if you have a large enough solid body its elastic modes interact with relativistic modes and the resulting mess is at least of tensor symmetry if not more complicated.

S. WEINBERG (Harvard)

This is a statement but the question is whether you agree with it. I was trying to understand in the last few months why - although there are these two uses of the renormalization group that you've described - nevertheless it is possible to reproduce a lot of the second use of the renormalization group, that is the going from one effective Hamiltonian to another effective Hamiltonian, by methods which were really designed for the first application which was to discuss the scale invariance of Green's functions at very large momentum. And as far as I can understand it, the reason is that (I'm thinking particularly of the work of Brezin and his group) if you define a renormalization point and define coupling constants by giving the values of Green's functions of that renormalization point, then whether the theory is renormalizable or non-renormalizable - of course if it's nonrenormalizable you'll have to do it an infinite number of times - then all the integrals are effectively cut off at a momentum of the order of the renormalization point. When the momentum in the Feynman graph gets large compared to the value of the renormalization point, you see that the integral begins very rapidly to converge. So that in effect by introducing a variable renormalization point even though it looks like what you're doing is eliminating ultra-violet divergences, at the same time you're providing a method of having a floating cut off just like Wilson, and Kadanoff, and you do in applying the renormalization group to the study of effective Hamiltonians.

ANDERSON

I think I will agree with that, yes. Well in fact I will agree with that in the sense that I've really never thought that hard about it but it sounds right. Besides Steve Weinberg said it.

S. BLUDMAN (Penn)

Can you spell out the necessity or virtue of 't Hooft monopoles in removing cosmological order parameter singularities?

ANDERSON

I'm not saying that they're all that necessary. You know they might have all annihilated against each other if they came in equal numbers or perhaps the cosmos was small enough so that it all went the same way. So it is not an essential thing that some singularity be there, but it certainly is a possible thing that there could have been singularities. What's nice from my point of view is that one doesn't have to have such horrible things as boundaries. Whether one has to have lines I think depends on the groups and the broken symmetries and what has actually happened in that global decrease in the size of the areas that Weinberg put on that last slide of his [e.g. in the group symmetries remaining after successive symmetry breakdowns].

S. BLUDMAN (Penn)

Well I understand in the abelian case why you have to have Dirac strings. What I would like you to spell out is the role of the point singularities in the nonabelian case.

ANDERSON

I can only give you the many body example that I know which is if one had a tube of liquid He_3, it happens to have a vector order parameter called $\vec{n}$ and occasionally, and in the case of a tube, the boundaries I think make $\vec{n}$ want to be parallel to the boundaries. But one is stuck with the situation that - on the top it may have condensed with $\vec{n}$ pointing up the tube and on the bottom with $\vec{n}$ point down. The best way to solve that problem is obviously to stick a monopole in between. Supposing now one had that on some kind of cosmic scale, with a big boundary in between, and the boundary can be condensed perhaps into a sequence of these monopoles and the monopole density would give you the total amount of order parameter. I guess I really don't know enough about gauge theories to know whether this analogy is a solid one. This kind of thing incidently occurs in crystals. In crystals you have grain boundaries. In fact if you look at a grain boundary, you often find that it has condensed itself into a sequence of dislocations. There is a theorem that a grain boundary at long range is equivalent to a sufficient density of dislocations which are the line singularities of that theory.

K. WILSON (Cornell)

I know that I don't like the symptom that you read letters in the newspaper and they are always answering other letters, but I think I must defend myself against Steve Weinberg. I would like to stress the importance of the idea of the model Hamiltonian (or a completely new Hamiltonian) for describing the infra-red behaviour of these systems with infra-red catastrophies, which cannot be done by the Callan-Symansyk equations. The Callan-Symansyk equations can tell you how the strength of that interaction changes. But what they cannot tell you is if you have to have a whole new Hamiltonian to describe what's going on. It's not quite that bad in the Kondo problem, I mean you can start with the same Hamiltonian just with j = infinity, but if you're really going to get a complete description of its low energy behavior, you have to throw in some extra scattering terms and you only get that from a non perturbative renormalization group. The situation is clearly worse in the quark case where the spectrum itself has to change or presumable has to change from being free quarks to confined quarks and I can't conceive of the Callan-Symansyk equations telling you what the effective Hamiltonian for confined quarks is.

ANDERSON

Before Steve answers that if he wants to, I think I would have a couple of comments about that. Yes, I think I agree with you that the great thing about the Wilson renormalization group was the statement that the Hamiltonian wasn't necessarily qualitatively the same and - where universality is sufficiently adequate - one doesn't necessarily have to introduce new terms but it contains the general possibility of introducing new terms. The Callan-Symansyk equation on the other hand I think is useful so long as the coupling constant is small, it gets you to something with a very small coupling constant to a moderately small coupling constant and can be used so long as perturbation theory holds.

S. WEINBERG (Harvard)

I'm clearly out of my depth but I'll try anyway. In many cases where the Callan-Symansyk equation has been used to study critical phenomena, I presume that it is being used because the model has not qualitatively changed and it is not a case like the ones you've been talking about today Ken. But even in the case where it is and you're really talking about a model Hamiltonian, it doesn't seem to me that one is restricted, even though it is a model Hamiltonian, from imposing a cut off instead of by a lattice by the method of introducing a variable renormalization point.

K. WILSON (Cornell)

Now the problem is how do you know what terms are in the effective Hamiltonian?

S. WEINBERG (Harvard)

Well how do you know what terms there are in the effective Hamiltonian?

K. WILSON (Cornell)

Well that's my trade secret. That's what I get out of the computer.

VOICE FROM THE AUDIENCE

Could you tell me how difficult it is to go from something like the BCS Hamiltonian which involves self-interacting fermions to a phenomenological theory involving just the order parameter, and in particular can you use the BCS theory to calculate things like the scattering in collective excitations and such? Is that a well understood problem in solid state?

ANDERSON

Yes, fairly. It's essentially the problem of finding the appropriate generalized Ginzberg-Landau functional for the system and in many cases that doesn't describe all of the collective excitations of the system. One has a serious problem of time scales. One has this zero frequency but finite space scale problem, and then as one goes to higher frequencies one often has to introduce new time behaviors. One will have a collisionless regime, and a non collisionless regime depending on the time scale relative to the space scale.

VOICE FROM THE AUDIENCE

Do you use the BCS Hamiltonian to calculate the parameters in the phenomenological theory?

ANDERSON

One can yes. Well, first one can take the phonons and the electrons. That can then in many cases be transcribed into some effective BCS Hamiltonian and that finally in turn into a Landau-Ginzberg functional which tells you how the order parameter varies. Each of these are fairly complicated stages.

B. WARD (Purdue)

In your introduction, you mentioned both the Kondo problem and the results of Luther. You chose to describe to us only the Kondo problem, presumably because of the lack of time. Could you please describe how the result of Luther would have fitted into your discussion if you had had more time?

ANDERSON

Well what Luther essentially does is to take an asymptotically free case for instance and show that the renormalization group equations will carry you to a certain case with a finite coupling constant and then that case is mathematically soluble, and it is again a theory in which there are no phase transitions, no changes in symmetry. But in this case there is a gap in the triplet excitation spectrum, even though in the original Hamiltonian there appeared to be no gap in the excitation spectrum for triplet excitations as opposed to singlet excitations. So then a gap is appearing out of nowhere without symmetry change. Ken Wilson emphasized that there are a lot of caveats here but there are also a lot of opportunities. The standard dictionary says energy gap equals mass. That isn't necessarily always true in going from a many body analogy to a field theory analogy. Energy gap may be instead of that a certain distance off the mass shell rather than the mass. I think it may be in the quark gluon case that is the way that it goes, that when an energy gap appears in the theory one might interpret that as saying that there isn't any stable particle of that sort. I think that's not an impossibility.

MAGNETIC CHARGE

Julian Schwinger*

*Physics Department, University of California at Los Angeles, Los Angeles, Cal. 90024.

I am privileged to talk to you at a time when evidence for the existence of magnetic charge has at long last appeared, and before the strident voices raised in opposition have managed to effectively discredit that evidence. For the purposes of today's discussion we shall accept the interpretation of the unusual highly ionizing track discovered by P. B. Price et al., that it represents a particle carrying the magnetic charge

$$g_0 = 137e \tag{1}$$

and having a mass many times larger than the proton mass.

My main point is that this situation has two possible interpretations, only one of which was mentioned in the initial experimental report. The latter is the unsymmetrical viewpoint of Dirac (which is what is usually intended when people speak of monopoles) wherein electric and magnetic charges reside on different particles, with an absolute distinction between the two kinds of charge, and the charge quantization conditions reads ($\hbar = 1$, $c = 1$, unrationalized units)

$$e_1 g_2 = \frac{1}{2} n, \qquad n = 0, \pm 1, \pm 2, \ldots \tag{2}$$

Thus, relative to the known unit of electric charge e, $e^2 = 1/137$, the Dirac unit of magnetic charge is

$$g_D = \frac{1}{2e} = \frac{137}{2} e, \tag{3}$$

and the observed magnetic particle is doubly charged.

The second, symmetrical, interpretation takes seriously the dual rotational invariance possessed by the general Maxwell equations as expressed pictorially by the freedom to adopt new electric and magnetic axes:

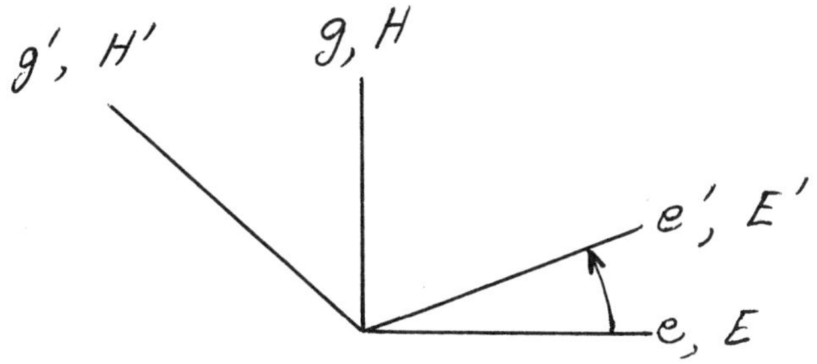

Now, the charge quantization condition has the rotationally invariant form

$$e_1 g_2 - e_2 g_1 = n = 0, \pm 1, \pm 2, \ldots, \tag{4}$$

[we shall review later how the shift from half-integer to integer quantization comes about], and the unit of magnetic charge, induced by the known unit of electric charge, e, is

$$g_0 = \frac{1}{e} = 137e \tag{5}$$

as observed.

But charge rotational invariance is far from obvious empirically and, indeed, seems to fly in the face of the charge asymmetry conventionally expressed by the remark (before August 1975) that no magnetically charged particle has ever been discovered! It is the charge quantization condition that reconciles the postulated, hidden, symmetry with the observed asymmetry between electric and magnetic charge. To see this, let us make an invariant distinction between small charges, such that

$$e_1^2 + g_1^2 < 1, \tag{6}$$

and large charges

$$e_1^2 + g_1^2 \geq 1, \tag{7}$$

and note that the familiar unit of electric charge is comfortably small ($e^2 = 1/137 \ll 1$). Having in mind the Euclidean geometry of the two-dimensional charge plane, we see that the individual charges of any pair of small charges obey

$$|e_1 g_2 - e_2 g_1| \leq (e_1^2 + g_1^2)^{1/2} (e_2^2 + g_2^2)^{1/2} < 1. \tag{8}$$

But the left side of this inequality is $|n| = 0, 1, \ldots$, and therefore can only be zero. The resulting statement,

$$e_1 g_2 - e_2 g_1 = 0, \tag{9}$$

is conveyed geometrically by the remark that all small charge points must lie on the same, absolute, charge line:

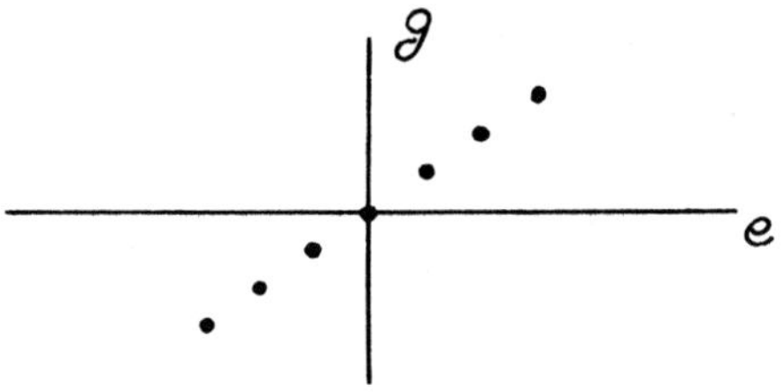

which, by a permissible rotation of the charge axes, appears as

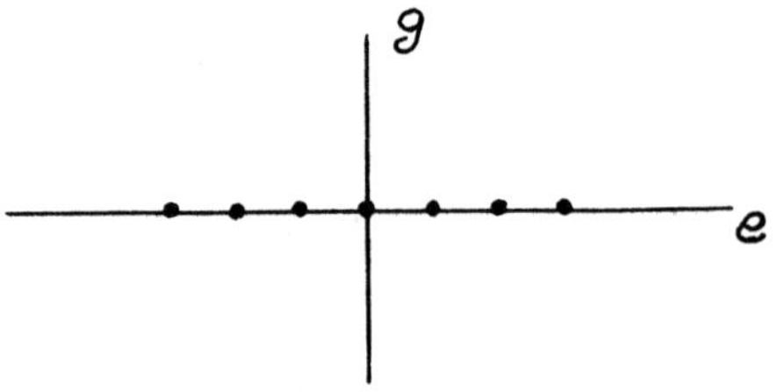

and we have thereby recovered the condition described by the statement that no magnetic charge exists. From this point of view, then, what was special about the situation before Aug. 1975 was simply that no large atomic charge had yet been detected. [Incidentally, the distinction between small and large charges used here for simplicity is not the best possible one since a charge that does not lie on the absolute line of small charges (and therefore, by convention, is magnetically charged) must be such that

$$e_1^2 + g_1^2 \geq 137; \tag{10}$$

any nucleus of charge Ze, $Z < 137$, is still a small charge.]

The unsymmetrically, Dirac, interpretation of the existence of the magnetic charge $g_0 = 137\,e$ would permit electric charges of value

$$e_1 = \frac{1}{2} ne, \quad n = 0, \pm 1, \ldots, \tag{11}$$

thereby excluding (as Price et al. remark) the odd fractional charges ($\pm\frac{1}{3}$, $\pm\frac{2}{3}$) that are required in the "quark" model. What is the situation with the symmetrical viewpoint?

We first note that the existence of pure electric charge, with the familiar unit e, implies that any magnetic charge is a multiple of g_0,

$$g_1 = mg_0, \quad m = 0, \pm 1, \ldots, \quad g_0 = 137e. \tag{12}$$

Now choose $g_2 = g_0$, and $e_2 = e_0$ (by definition the smallest electric charge that can accompany the magnetic charge unit g_0) and deduce that

$$e_1 = me_0 + ne, \tag{13}$$

which exhibits the possibility of two electric charge units. Since this relation is linear, it also holds for any composite structure, and we are assured that a magnetically neutral system (m = 0) has only e as its electric charge unit, as observed. Obviously, the only new possibility here is that $|e_o| < |e|$, since, otherwise, e_o can be reduced, modulo e. We now explore the hypothesis that the second unit of charge does exist, $e_o \neq 0$.

The counterpart of pure electric charge (unit e) is pure magnetic charge (a true monopole) with unit charge $g > g_o$. The charge quantization condition then asserts of any electric charge e_1 that

$$e_1 g = \ell, \quad \ell = 0, \pm 1, \ldots, \tag{14}$$

or

$$e_1 = \ell e_o, \quad e_o = 1/g, \tag{15}$$

where the latter is the analogue of Eq. (12) with $g_o = 1/e$. Accordingly, since $|e| > |e_o|$, there must exist an integer $N > 1$ such that (the minus sign is conventional)

$$e = -Ne_o, \tag{16}$$

from which it follows that $(g/g_o = e/e_o)$

$$g = -Ng_o \tag{17}$$

We conclude that the smallest electric and magnetic charges are a common fractional multiple of the corresponding units of pure charge:

$$e_o = -e/N, \quad g_o = -g/N. \tag{18}$$

It should also be recognized from Eq. (13) that, for a given magnetic charge, the possible electric charges differ by multiples of e, and the analogous relation

$$g_1 = l g_0 + n g \tag{19}$$

shows that, for a given electric charge, possible magnetic charges differ by a multiple of g. There are, then, only a small set of fractional charges, those less in magnitude than the appropriate unit of pure charge, namely

$$-\frac{1}{N} e, \quad \frac{N-1}{N} e, \tag{20}$$

for electric charge, and

$$-\frac{1}{N} g, \quad \frac{N-1}{N} g, \tag{21}$$

for magnetic charge. I introduced the term dyon for particles with these dyads of fractional charges. It is natural to regard them as members of one family which, for reasons soon to be evident, is of the spin $\frac{1}{2}$, Fermi-Dirac type. The anti-dyon family reverses the signs of both types of charge.

What is the value of N? That is answered by the dyon theory of hadronic matter, according to which all known hadrons, including the recently discovered ψ particles, are magnetically neutral composites of dyons. There are just two basic ways of producing magnetic neutrality. One combines a dyon with an anti-dyon, which composite can only be a meson. The other combines dyons [magnetic charge $\frac{N-1}{N} g$ and N-1 magnetic charges $-\frac{1}{N} g$] or antidyons [magnetic charge $-\frac{N-1}{N} g$ and N-1 magnetic charges $\frac{1}{N} g$]

which can only be baryons since particle and anti-particle are physically different. Accordingly the dyon constituents must be F-D particles and their number, N, must be odd. The smallest possible odd integer is

$$N = 3. \tag{22}$$

The conclusion that baryons are composites of dyons carrying the magnetic charges (unit g)

$$\frac{2}{3}, \quad -\frac{1}{3}, \quad -\frac{1}{3} \tag{23}$$

immediately raises the question whether the dyons with equal magnetic charge are duplicates, or physically different particles, owing to some other physical property than magnetic charge. Various arguments can be adduced here, but it is probably most important now to point out that the purely electromagnetic theory so far developed is incomplete since it makes no reference to the known weak interactions, which are mechanisms for the exchange of electric charge, nor to their hypothetical but plausible magnetic counterparts that (strongly) exchange magnetic charge. It is likely that, among other functions that we do not discuss here, the physical role of the magnetic charge exchange mechanism is to distinguish the two dyons of charge $-\frac{1}{3}g$ and, analogously, to distinguish the two options of charge $-\frac{1}{3}e$ that we also assign to the electric charges of the dyon family. In its latter role it provides the physical mechanism that defines hypercharge and isotopic spin for, as you have all long since recognized, we have made contact with the empirical "quark" model. But what a difference there is!

The search for free "quarks" as weakly ionizing particles is now seen to be misguided since no such particle can exist without an accompanying magnetic charge; the artificial problem of "quark" confinement is dismissed since free magnetically charged particles are found to be very massive, and, indeed, the "quark" model as it is usually applied is quite divorced from reality; the additional multiplicity of "quarks" conveyed by the deliberately meaningless and overly cute label of "color" appears automatically, and endowed with a physically significant name — magnetic charge; as for "gluons", "flavor", "charm", ...

We have obviously been saying that the particle observed by Price et al. is not a monopole but a dyon, and as such carries a fractional electric charge of magnitude $\frac{1}{3}e$ or $\frac{2}{3}e$. How could one verify that experimentally? The most immediate way would be through a study of ionization. I take this opportunity to review (and possibly improve) the theory of energy loss of electric and magnetic charged particles in dense media.

We first recall that, in the absence of matter, but with a prescribed electric or magnetic current J^{μ} (metric -+++), the vacuum persistence amplitude is $exp[iW]$;

$$W = \frac{1}{2}\int (dx)(dx')\, J^{\mu}(x)\, D(x-x')\, J_{\mu}(x'), \tag{24}$$

$$\frac{1}{4\pi}D(x-x') = \int \frac{(dk)}{(2\pi)^4}\, \frac{e^{ik(x-x')}}{k^2}\, .$$

The presence of a dispersive, absorptive medium, as represented by the parameters $\epsilon,\ \mu \simeq 1,$ is effectively introduced by the replacements [our initial choice of unit, $c = 1$, obscures the substitution $c \to c/\epsilon^{1/2}$]

$$k^0 \to \epsilon^{1/2} k^0, \quad \text{in} \quad k^2, \tag{25}$$

and

$$\begin{aligned} J^0 &\to \epsilon^{-1/2} J^0, \qquad \underset{\sim}{J} \to \underset{\sim}{J} \qquad &\text{electric charge} \\ J^0 &\to J^0, \qquad \underset{\sim}{J} \to \epsilon^{1/2} \underset{\sim}{J} \qquad &\text{magnetic charge.} \end{aligned} \tag{26}$$

Thus, for a point charge e_1, moving with velocity $\underset{\sim}{v}$,

$$W = e_1^2 \int dt\, dk^0 \int \frac{(d\underset{\sim}{k})}{(2\pi)^3} dt' \frac{v^2 - \frac{1}{\epsilon}}{\underset{\sim}{k}^2 - \epsilon (k^0)^2} e^{i(\underset{\sim}{k} \cdot \underset{\sim}{v} - k^0)(t - t')} \tag{27}$$

where, according to the interpretation of the 'vacuum' persistence probability, $\exp[-2\,\mathrm{Im}\,W]$, attention focusses on $2\,\mathrm{Im}\,W/T$, with T a microscopically long time interval during which the constancy of v is a valid approximation:

$$\begin{aligned} \frac{2\,\mathrm{Im}\,W}{T} &= \frac{e_1^2}{\pi} v\, \mathrm{Im} \int_0^\infty d\omega \left(1 - \frac{1}{v^2 \epsilon}\right) \int_{\omega^2/v^2}^{K^2} d\underset{\sim}{k}^2 \frac{1}{\underset{\sim}{k}^2 - \omega^2 \epsilon} \\ &= \frac{e_1^2}{\pi} v\, \mathrm{Im} \int_0^\infty d\omega \left(1 - \frac{1}{v^2 \epsilon}\right) \log \frac{K^2 v^2}{\omega^2 (1 - v^2 \epsilon)} . \end{aligned} \tag{28}$$

In the latter $k^0 > 0$ is denoted by ω , and K is a high-momentum cutoff, which could be set by the circumstances of a particular experiment, or be merely a device for isolating those processes in which the collective properties of the medium are significant. The inferred spatial rate of energy loss is then

$$-\frac{dE}{dz} = \frac{e_1^2}{\pi} \operatorname{Im} \int_0^\infty d\omega\, \omega \left(1 - \frac{1}{v^2\epsilon}\right) \log \frac{k^2 v^2}{\omega^2(1 - v^2\epsilon)} , \tag{29}$$

and the analogous result for a magnetic charge g_1 appears with the substitutions

$$e_1^2 \to g_1^2 , \quad 1 - \frac{1}{v^2\epsilon} \longrightarrow \epsilon - \frac{1}{v^2} . \tag{30}$$

The dielectric constant and its inverse have the following spectral representations,

$$\epsilon(\omega) = 1 + \frac{4\pi N e^2}{m} \int_0^\infty d\omega' \frac{p(\omega')}{\omega'^2 - \omega^2 - i0} ,$$

$$\frac{1}{\epsilon(\omega)} = 1 - \frac{4\pi N e^2}{m} \int_0^\infty d\omega' \frac{q(\omega')}{\omega'^2 - \omega^2 - i0} , \tag{31}$$

in which the real spectral weight functions are non-negative and obey

$$\int_0^\infty d\omega' p(\omega') = \int d\omega' q(\omega') = 1 ; \tag{32}$$

the symbols e and m refer to the atomic electrons which are of spatial density N. For the example of electric charge energy loss we find that

$$-\frac{dE}{dz} = \frac{2\pi N e^2 e_1^2}{m v^2} \left[\log \frac{k^2 v^2}{\omega_e^2 (1 - v^2)} - v^2 - \int_{1/\epsilon(0)}^{v^2} dv'^2 f(v'^2) \right] \tag{33}$$

where

$$\log \omega_e^2 = \int_0^\infty d\omega' q(\omega') \log \omega'^2 \tag{34}$$

while $f(v^2)$, which vanishes if $v^2 < 1/\epsilon(0)$, is defined implicitly by $[\omega_p^2 = 4\pi N e^2/m]$

$$\int_0^\infty d\omega' q(\omega') \frac{1}{(\omega'^2/\omega_p^2) + f(v^2)} = 1 - v^2 . \tag{35}$$

We also note the relation

$$\int_0^\infty d\omega' q(\omega') \log\left[\frac{\omega'^2}{\omega_p^2} + f(v^2)\right] = \log\frac{\omega_e^2}{\omega_p^2} + (1 - v^2) f(v^2) + \int_{1/\epsilon(0)}^{v^2} dv'^2 f(v'^2) \tag{36}$$

from which it directly follows that

$$v^2 \rightarrow 1: \quad -\frac{dE}{dz} = \frac{2\pi N e^2 e_1^2}{m} \log\frac{K^2}{\omega_p^2} \qquad (\text{Fermi}). \tag{37}$$

The magnetic analogue is

$$-\frac{dE}{dz} = \frac{2\pi N e^2 g_1^2}{m}\left[\log\frac{K^2 v^2}{\omega_g^2(1 - v^2)} - 1 - \int_{1/v^2}^{\epsilon(0)} d\left(\frac{1}{v'^2}\right)\hat{f}\left(\frac{1}{v'^2}\right)\right], \tag{38}$$

where

$$\log \omega_g^2 = \int_0^\infty d\omega' p(\omega') \log \omega'^2 < \log \omega_e^2 , \tag{39}$$

and

$$\hat{f}(1/v^2) = f(v^2) \tag{40}$$

is also defined directly by the implicit equation

$$\int_0^\infty d\omega' p(\omega') \frac{1}{(\omega'^2/\omega_p^2) + \hat{f}(1/v^2)} = \frac{1}{v^2} - 1 , \tag{41}$$

to which we add

$$\int d\omega' p(\omega') \log\left[\frac{\omega'^2}{\omega_p^2} + \hat{f}\left(\frac{1}{v^2}\right)\right] =$$

$$\log \frac{\omega_g^2}{\omega_p^2} + \left(\frac{1}{v^2} - 1\right)\hat{f}\left(\frac{1}{v^2}\right) + \int_{1/v^2}^{\epsilon(0)} d\left(\frac{1}{v'^2}\right)\hat{f}\left(\frac{1}{v'^2}\right), \tag{42}$$

with its consequence

$$v^2 \to 1: \quad -\frac{dE}{dz} = \frac{2\pi N e^2 g_1^2}{m} \log \frac{K^2}{\omega_p^2} \quad (\text{Magnetic Fermi}). \tag{43}$$

Although we have emphasized the differences between the two situations, it is approximately true that the ionization properties of electric and magnetic charges are connected by the substitution

$$g_1^2 \leftrightarrow e_1^2/v^2. \tag{44}$$

Accordingly, a dyon with charges $g_0 = 137\ e$, and $\frac{2}{3}\ e$, for example, ionizes at a rate proportional to

$$g_0^2 + \left(\frac{2e/3}{v}\right)^2 \sim 1 + \left(\frac{2/3}{137\ v}\right)^2, \tag{45}$$

and the roughly constant ionization, characteristic of magnetic charge, should switch over to increasing ionization as the particle slows, characteristic of electric charge, for

$$v \sim \frac{1}{137}. \tag{46}$$

This, however, is also the order of magnitude of atomic electron speeds so that the collisions begin to become adiabatic, with corresponding loss in effectiveness, and the rise of ionization near the end of the track (which is not seen in the Price et al. event) may not be very pronounced.

Let us then shift the focus from distant, low momentum transfer collisions with electrons ($|\underset{\sim}{k}| < K$) to impacts that transfer nearly the maximum possible energy, which produce the δ rays that accompany an ionizing track. A quick, but regrettably misleading way of suggesting the experimental possibilities in such collisions is to consider the classical theory of scattering, based on the non-relativistic equation of relative motion

$$\mu \frac{d^2\underset{\sim}{r}}{dt^2} = (e_1 e_2 + g_1 g_2)\frac{\underset{\sim}{r}}{r^3} + (e_1 g_2 - e_2 g_1)\underset{\sim}{v} \times \frac{\underset{\sim}{r}}{r^3}, \tag{47}$$

in which μ is the reduced mass of the two particles. For convenience, we introduce the symbols

$$n = e_1 g_2 - e_2 g_1, \qquad q = \mu(e_1 e_2 + g_1 g_2), \tag{48}$$

change the time variable

$$dt = \mu r^2 d\psi, \tag{49}$$

and, in the spherical coordinates indicated by

$$\underset{\sim}{r} = r\underset{\sim}{\nu}, \qquad |\underset{\sim}{\nu}| = 1, \tag{50}$$

invert the radial coordinate,

$$u = \frac{1}{r}. \tag{51}$$

The ensuing form of the equation of motion is

$$u\frac{d^2\underset{\sim}{\nu}}{d\psi^2} - \underset{\sim}{\nu}\frac{d^2 u}{d\psi^2} = q\underset{\sim}{\nu} + nu\frac{d\underset{\sim}{\nu}}{d\psi} \times \underset{\sim}{\nu}, \tag{52}$$

or, projecting parallel and perpendicular to $\underset{\sim}{\nu}$,

$$\frac{d^2 u}{d\psi^2} + \left(\frac{d\underset{\sim}{\nu}}{d\psi}\right)^2 u + q = 0, \tag{53}$$

$$\underset{\sim}{\nu} \times \frac{d^2 \underset{\sim}{\nu}}{d\psi^2} - n \frac{d\underset{\sim}{\nu}}{d\psi} = 0, \tag{54}$$

respectively. Immediately apparent are first integrals constituting the total energy, multiplied by μ , and the total angular momentum:

$$\frac{1}{2}\left(\frac{du}{d\psi}\right)^2 + \left(\frac{d\underset{\sim}{\nu}}{d\psi}\right)^2 \frac{1}{2} u^2 + qu = \frac{1}{2} p^2, \tag{55}$$

$$\underset{\sim}{\nu} \times \frac{d\underset{\sim}{\nu}}{d\psi} - n\underset{\sim}{\nu} = \underset{\sim}{J}, \quad \underset{\sim}{J}^2 - n^2 = \left(\frac{d\underset{\sim}{\nu}}{d\psi}\right)^2 = (p\rho)^2, \tag{56}$$

where p is the magnitude of the asymptotic relative momentum and ρ is the impact parameter of the collision. The angular equation of motion in (56) can also be presented as

$$\frac{d\underset{\sim}{\nu}}{d\psi} = \underset{\sim}{J} \times \underset{\sim}{\nu}, \quad \underset{\sim}{J} \cdot \underset{\sim}{\nu} = -n, \tag{57}$$

which describes the precession of $\underset{\sim}{\nu}$ about $\underset{\sim}{J}$ at the constant rate $|\underset{\sim}{J}|$, and with the constant projection $-n$ on the precession axis. Thus, the motion of $\underset{\sim}{\nu}$ defines a cone, which is such that $\frac{1}{2}\chi$, the magnitude of the angle between the conic surface and the plane perpendicular to $\underset{\sim}{J}$, is given by

$$\cot \tfrac{1}{2}\chi = \frac{p\rho}{|n|}. \tag{58}$$

The solution of the radial equation (53) is

$$u = -\frac{q}{(p\rho)^2}\left[1 - \epsilon \cos p\rho\psi\right], \tag{59}$$

where the eccentricity ϵ is evaluated [Eq. (55)] as

$$\epsilon = \left[1 + (p^2\rho/q)^2\right]^{1/2} = \left[1 + (np/q)^2 \cot^2 \tfrac{1}{2}\chi\right]^{1/2}. \tag{60}$$

During the course of the collision, which begins with $u = 0,\ \psi = \psi_2$ and ends with $u = 0, \psi = \psi_1$, the phase of the cosine in (59) changes by

$$\varphi = p\rho(\psi_1 - \psi_2) = \begin{cases} q > 0: & 2\cos^{-1}\frac{1}{\epsilon} \\ q < 0: & 2\pi - 2\cos^{-1}\frac{1}{\epsilon} \end{cases} \tag{61}$$

which possibilities are unified in

$$\frac{1}{2}\varphi = \tan^{-1}\left(\frac{p|n|}{q}\cot\frac{1}{2}\chi\right) \tag{62}$$

when this angle is restrained to the interval between 0 and π.

The angle between the initial and final radius vectors defines the scattering angle θ:

$$-\cos\theta = \underset{\sim}{\nu}_1 \cdot \underset{\sim}{\nu}_2 = \sin^2\frac{1}{2}\chi + \cos^2\frac{1}{2}\chi\,\cos|\underset{\sim}{J}|(\psi_1 - \psi_2), \tag{63}$$

where

$$|\underset{\sim}{J}|(\psi_1 - \psi_2) = \frac{|\underset{\sim}{J}|}{p\rho}\varphi = \frac{\varphi}{\cos\frac{1}{2}\chi}. \tag{64}$$

This relation is also presented as

$$\cos\frac{1}{2}\theta = \cos\frac{1}{2}\chi\left|\sin\frac{\frac{1}{2}\varphi}{\cos\frac{1}{2}\chi}\right|. \tag{65}$$

In the situation of monopole-electron scattering (which already appears in the literature), the parameter q is zero, so that $\frac{1}{2}\varphi = \frac{1}{2}\pi$ and

$$\text{monopole:}\quad \cos\frac{1}{2}\theta = \cos\frac{1}{2}\chi\left|\sin\frac{\frac{1}{2}\pi}{\cos\frac{1}{2}\chi}\right|. \tag{66}$$

These are implicit relations between the scattering angle θ and the impact parameter ρ, which we contrast with the pure Coulomb situation (obtained by placing $n=0$)

$$\cot\frac{1}{2}\theta = \frac{p^2\rho}{|q|}. \tag{67}$$

For the latter, as ρ varies from ∞ down to zero, the scattering angle θ moves monotonically from $\theta=0$ to $\theta=\pi$. In particular, back scattering ($\theta=\pi$) occurs only for head-on collisions ($\rho=0$). An inspection of Eq. (66) shows that, in contrast, $\theta=\pi$ occurs whenever

$$\cos\frac{1}{2}\chi = \frac{1}{2k}, \qquad k = 1, 2, \ldots . \tag{68}$$

The physical significance of this property follows from the formula for the differential scattering cross section per unit solid angle:

$$\frac{d\sigma}{d\Omega} = \frac{\rho d\rho}{\sin\theta d\theta} = \left(\frac{n}{2p}\right)^2 \frac{1}{\sin^4\frac{1}{2}\chi} \frac{\sin\chi d\chi}{\sin\theta d\theta}; \tag{69}$$

where $\theta=\pi$, but $\chi\neq\pi$, the differential cross section is infinite (this is the phenomenon known in classical optics as the glory). Between these $\theta=\pi$ points it must happen that $d\theta/d\chi=0$, where the cross section is also infinite (the rainbow). For monopole-electron scattering this occurs at

$$\cos\frac{1}{2}\chi = \frac{1}{2k+1} + \frac{4}{\pi^2}\frac{1}{(2k+1)^3} + \ldots, \tag{70}$$

$$k = 1, 2, \ldots,$$

Thus, for the large scattering angles that, in the physical process of electron knock-on, correspond to near maximum energy δ rays, the characteristics of a monopole are very different from those of an electric charge, providing this classical theory is applicable. Unfortunately, the numerical evaluations available for the quantum monopole-electron cross section suggest that n must be a large number for such interesting structure to develop; our concern is with $|n| = 1$.

We add a few remarks about the classical theory of dyon-electron scattering. For small scattering angles, (65) is approximated by

$$\left(\frac{1}{2}\theta\right)^2 \simeq \left[1+\left(\frac{q}{pn}\right)^2\right]\left(\frac{1}{2}\chi\right)^2 \tag{71}$$

and the differential cross section becomes

$$\frac{1}{2}\theta \ll 1: \quad \frac{d\sigma}{d\Omega} = \left(\frac{1}{2p}\right)^2\left[n^2+\left(\frac{q}{p}\right)^2\right]\frac{1}{(\theta/2)^4}, \tag{72}$$

which joins on to the combination of processes indicated in Eq. (45). The condition for $\theta = \pi$ to be realized according to (65) is (n and q positive, for simplicity)

$$\frac{1}{\cos\frac{1}{2}\chi}\tan^{-1}\left(\frac{pn}{q}\cot\frac{1}{2}\chi\right) = k\pi, \quad k = 1, 2, \ldots, \tag{73}$$

where the parameter pn/q assumes some value between ∞ , appropriate to the monopole, and 0, which is the Coulomb situation. It is helpful to notice that (73) has a solution at $\chi = \pi$ when

$$\frac{pn}{q} = \frac{137v}{2/3} = k\pi, \tag{74}$$

or

$$v = v_k = k\pi \frac{2/3}{137}, \tag{75}$$

in which we have inserted the particular realization associated with the dyon of charges $g_0 = 137\ e$, $\frac{2}{3}\ e$. Indeed, a numerical analysis shows that $\theta = \pi$, $\chi < \pi$ does not occur for $v < v_1$, but appears once for $v_1 < v < v_2$ and k times when $v_k < v < v_{k+1}$, with each additional possibility moving down from $\chi = \pi$ when v just exceeds v_k. Thus, the distinction between monopole and dyon should be evident for moderate values of k, corresponding to particle speeds that could be an order of magnitude larger than $v \sim \frac{1}{137}$ and therefore more accessible experimentally. But, although numerical results for the quantum cross section are not yet at hand, there is little basis for hope that this intricate structure will survive in the situation of interest.

Finally, I want to keep my promise to review the origin of the charge quantization conditions, and, specifically, contrast the unsymmetrical and symmetrical viewpoints. [More details can be found in a Physical Review paper that was submitted almost coincidentally with the public announcement by Price et al.]

An unsymmetrical but convenient action expression is [for the present purpose, rationalized units are employed]

$$W = \int (dx) \Big[J_e^{\mu} A_{\mu} + J_m^{\mu} B_{\mu} - \frac{1}{2} F^{\mu\nu} (\partial_{\mu} A_{\nu} - \partial_{\nu} A_{\mu}) + \frac{1}{4} F^{\mu\nu} F_{\mu\nu} \Big], \tag{76}$$

where A^{μ}, $F_{\mu\nu}$ are independent variables and B_{μ} is <u>defined</u> by

$$B_\mu(x) = \int (dx')\, {}^*F_{\mu\nu}(x') f^\nu(x' - x) + \partial_\mu \lambda_m(x). \tag{77}$$

Here $\lambda_m(x)$ is an arbitrary function, ${}^*F_{\mu\nu}$ is the tensor dual to $F_{\mu\nu}$, and $f^\nu(y)$ obeys

$$\partial_\nu f^\nu(y) = \delta(y), \tag{78}$$

but is otherwise arbitrary. We also introduce

$${}^*f^\nu(y) = -f^\nu(-y), \quad \partial_\nu {}^*f^\nu(y) = \delta(y), \tag{79}$$

so that (77) appears alternatively as

$$B_\mu(x) = -\int (dx')\, {}^*f^\nu(x - x')\, {}^*F_{\mu\nu}(x') + \partial_\mu \lambda_m(x). \tag{80}$$

The stationary property of W for A-variations directly yields the first Maxwell set

$$\partial_\nu F^{\mu\nu} = J_e^\mu . \tag{81}$$

The consequence of F-variations is

$$\begin{aligned} F_{\mu\nu}(x) &= (\partial_\mu A_\nu - \partial_\nu A_\mu)(x) \\ &\quad + \int (dx')^*\big(f_\mu(x-x') J_{m\nu}(x') - f_\nu(x-x') J_{m\mu}(x')\big), \end{aligned} \tag{82}$$

which has the dual form

$$\begin{aligned} {}^*F_{\mu\nu}(x) &= {}^*(\partial_\mu A_\nu - \partial_\nu A_\mu)(x) \\ &\quad - \int (dx')\big(f_\mu(x - x') J_{m\nu}(x') \\ &\quad - f_\nu(x - x') J_{m\mu}(x')\big), \end{aligned} \tag{83}$$

yielding the second Maxwell set

$$\partial_\nu {}^*F^{\mu\nu} = J_m^{\ \mu} . \tag{84}$$

Another implication of Eq. (82) is the construction

$$A_\mu(x) = -\int (dx')f^\nu(x - x')F_{\mu\nu}(x') + \partial_\mu \lambda_e(x), \tag{85}$$

which is analogous to (80). The analogy is not a symmetry in the sense of dual rotational invariance unless

$$^*f^\nu(y) = f^\nu(y) = -f^\nu(-y) . \tag{86}$$

Then the action is invariant under the infinitesimal rotation

$$\begin{aligned} \delta J_e &= \delta\varphi J_m, & \delta J_m &= -\delta\varphi J_e, \\ \delta F &= \delta\varphi {}^*F, & \delta {}^*F &= -\delta\varphi F, \\ \delta A &= \delta\varphi B, & \delta B &= -\delta\varphi A. \end{aligned} \tag{87}$$

The basic solution of the non-conservation equation of (78) is

$$f_\mu(y) = \int_0^\infty d\xi_\mu \delta(y - \xi), \tag{88}$$

where the integration follows some arbitrary path between the end points. A change of path is represented by

$$\begin{aligned} f_\mu^{(1)}(y) - f_\mu^{(2)}(y) &= \left(\int^{(1)} - \int^{(2)}\right) d\xi_\mu \delta(y - \xi) \\ &= -\int d\sigma_{\mu\nu}(\partial/\partial\xi_\nu)\delta(y - \xi) = \partial^\nu \int d\sigma_{\mu\nu}\delta(y - \xi), \end{aligned} \tag{89}$$

with the surface integration extending over the region bounded by the two paths. We shall write (89) as

$$\delta f_\mu(y) = \partial^\nu m_{\mu\nu}(y) \tag{90}$$

where, in this realization,

$$m_{\mu\nu}(y) = \int d\sigma_{\mu\nu}\,\delta(y-\xi). \tag{91}$$

If f is to obey the symmetry property (86) of the symmetrical formulation, one must change (88) to

$$f_\mu(y) = \int_0^\infty d\xi_\mu \frac{1}{2}\left[\delta(y-\xi) - \delta(y+\xi)\right] \tag{92}$$

and, for alterations within this class, (91) becomes

$$m_{\mu\nu}(y) = \int d\sigma_{\mu\nu}\frac{1}{2}\left[\delta(y-\xi) + \delta(y+\xi)\right] = m_{\mu\nu}(-y). \tag{93}$$

We now examine the change in W produced by another choice of f, as it enters in the vector potential $\mathcal{B}$:

$$\begin{aligned} \delta W &= \int (dx')\,\delta\mathcal{B}_\nu(x')\,J_m^{\ \nu}(x') \\ &= -\int (dx)(dx')\,{}^*F_{\mu\nu}(x)\,\delta f^\mu(x-x')\,J_m^{\ \nu}(x') \\ &= -\int (dx)(dx')\,J_e^{\ \mu}(x)\,{}^*m_{\mu\nu}(x-x')\,J_m^{\ \nu}(x'). \end{aligned} \tag{94}$$

Into this expression we insert a general point charge realization of the two currents:

$$J_{e,m}^{\ \mu}(x) = \sum_a (e_a, g_a)\int ds\,\frac{dx_a^{\ \mu}(s)}{ds}\,\delta(x - x_a(s)), \tag{95}$$

which gives

$$\delta W = -\sum_{ab} e_a g_b \int dx_a^{\mu} dx_b^{\nu} \, {}^*m_{\mu\nu}(x_a - x_b)$$
$$= -\sum_{ab} e_a g_b \int \tfrac{1}{2} {}^*d\sigma_{ab}^{\mu\nu} m_{\mu\nu}(x_a - x_b), \tag{96}$$

with

$$d\sigma_{ab}^{\mu\nu} = dx_a^{\mu} dx_b^{\nu} - dx_a^{\nu} dx_b^{\mu}$$
$$= -d\sigma_{ba}^{\mu\nu} = -d\sigma_{ab}^{\nu\mu}. \tag{97}$$

If the symmetrical formulation is being considered, where $m_{\mu\nu}(y) = m_{\mu\nu}(-y)$, according to Eq. (93), the antisymmetry of $d\sigma_{ab}^{\mu\nu}$ in a and b converts (96) into

$$\delta W = -\frac{1}{2}\sum_{ab}(e_a g_b - e_b g_a)\int \frac{1}{2} {}^*d\sigma_{ab}^{\mu\nu} m_{\mu\nu}(x_a - x_b). \tag{98}$$

To derive the charge quantization conditions we first consider the unsymmetrical viewpoint, where (91) and (96) are appropriate:

$$\delta W = -\sum_{ab} e_a g_b \int \frac{1}{2} {}^*d\sigma_{ab}^{\mu\nu} d\sigma_{\mu\nu} \delta(x_a - x_b - \xi). \tag{99}$$

We recognize that $\frac{1}{2} {}^*d\sigma_{ab}^{\mu\nu} d\sigma_{\mu\nu}$ is a four-dimensional volume element for the variable $x_a - x_b - \xi$, and that the basic values of the delta function integral are 0 and 1. Accordingly, a change in the choice of ξ can induce a jump in W that is given typically by the individual charge product $(-) e_1 g_2$. But the physical vacuum amplitude exp[iW] will remain unchanged if

$$e_1 g_2 = 2\pi n, \qquad n = 0, \pm 1, \pm 2, \ldots, \tag{100}$$

or

$$\frac{e_1 g_2}{4\pi} = \frac{1}{2} n, \tag{101}$$

which is (rationalized units!) the Dirac condition (2).

For the symmetrical formulation, we combine (93) and (98) to get

$$\delta W = -\frac{1}{2}\sum_{ab} (e_a g_b - e_b g_a) \int \frac{1}{2} {}^*d\sigma_{ab}^{\mu\nu} \, d\sigma_{\mu\nu} \frac{1}{2}\left[\delta(x_a - x_b - \xi) + \delta(x_a - x_b + \xi)\right]. \tag{102}$$

The domains of the two delta functions are quite distinct, so that the basic non-zero value of this integral is ½, leading to a jump in the value of W associated with a particular particle pair that is $(-)\frac{1}{2}(e_1 g_2 - e_2 g_1)$. The uniqueness of exp[iW] is then realized if

$$\frac{1}{2}(e_1 g_2 - e_2 g_1) = 2\pi n, \tag{103}$$

or

$$\frac{(e_1 g_2 - e_2 g_1)}{4\pi} = n, \tag{104}$$

which is the integer charge quantization condition (4).

DISCUSSION

B. ZUMINO (CERN)

Actually I have two questions. I don't know if either of these make sense. The first one is - I've always been puzzled about this quantization condition. Does it apply to the unrenormalized or to the renormalized charges? As I said, I'm not sure if this makes sense.

SCHWINGER

This whole discussion has been entirely carried out in the language of the physical particles. So I am talking about what you would call the renormalized charges. In this discussion there is no reference to unrenormalized charges so I can't answer the question.

B. ZUMINO (CERN)

When you try to make a theory of strong interactions based on the dyons, namely on particles which are carrying both electric and magnetic charge, then at least in the basic equations it seems to me there are violations of discrete symmetries like C or P or T or something is not valid. But in the strong interactions those symmetries are valid. How does one argue there?

SCHWINGER

Yes, well, this is of course an old story that the presence of magnetic charge or rather the presence of magnetic charge and electric charge in such a way that not all the possibilities of flipping the signs are present is of course a natural physical model, let's say for CP violation, to put it in those terms. That has always raised the question of why isn't the CP violation very strong instead of very weak, and long ago when I first began talking about this dyon theory I faced up to the problem by simply reminding myself and hopefully other people that this is not the whole story. That is to say, there is more to the strong interactions than simply electromagnetism in a larger context. The answer I gave was after all to look back at the world and recognize that there is more to electric charge too. There are the weak interactions, so called, which I viewed simply as mechanisms for transferring electric charges from one particle to another, and so the symmetry between electric and magnetic charges naturally suggests that there are also mechanisms for transferring magnetic charges. And so I took that up seriously and also of course the numerical asymmetry between electric and magnetic charges and said this magnetic charge transfer would in fact be very rapid inside, let's say, a baryon where the dyons are very close together, and could be arranged in such a way (since the spectrum of magnetic charges is on the average zero, the double charge is weighted half as much as the single charges) one could easily imagine this rapid exchange of magnetic charge occurs in such a way that over a very short time average the magnetic charge simply averages to zero. Which is to say that for ordinary experiments in which not very large energies or not too short time scales are relevant one would not see the presence of magnetic charge and one would not see the CP violations. On the other hand, for short time intervals, very high energies, one would have CP violations then of the kinds that are - sorry, I forget the standard term but you know what I mean - the formulation that says that they are present only under very energetic conditions. In short, one has to enlarge the theory to answer your question. One does not stop with the theory of pure electromagnetism. There must be the magnetic analog of the W - it must interact strongly and that must play also a significant part in the whole theory of the strong interactions based on this dyon theory. By that one has the possibility of incorporating CP violations into the theory in a natural way at the same time without overdoing it. That's of course the whole idea. I'm not sure if that's the sort of answer you had wanted.

D. LICHTENBERG (Indiana)

I recall that at one time you considered a quantization condition that eg is an even integer. Can you say something about that?

SCHWINGER

Yes sure, easily. I can but let me tell you a little bit of history. Of course this whole discussion of electric and magnetic charges has been going on for a very long time. Bruno Zumino has had a hand in some of these discussions, but as it happened I found myself last spring giving lectures on electric and magnetic charges in which I went over much the same ground that I did here trying to persuade myself that not only that the quantum number should be an integer but that it should be an even integer was correct. Of course that was my view as of June 1. However, in the process of giving those lectures, instead of persuading myself and everybody else, I dissuaded myself. And so, as a result, I wrote a paper in June - essentially espousing that - removing the requirement of even integers and as it happened that paper was in the process of being distributed just when the great announcement occurred. So I do believe I put myself on record as changing my mind just in time.

H. S. TSAO (IAS)

Could you elaborate more about why a "consistent theory can only be constructed for point particle"?

SCHWINGER

You notice how the consistency depends on the fact that we sweep over a δ function - a δ function means point particle. We sweep over a δ function and the changes depending on whether we do or don't are discreet values and so we can insist that the discreet values are integers. But if the charge is not a point particle, but is distributed then you will have here instead of the δ function, a general charge distribution and therefore you need not necessarily pick up the whole charge distribution or none of it or part of it. Then you see the integral will be a continuously variable quantity, instead of being zero or something. To avoid the ambiguities we must have discreteness of the change of W so that exp iW does not change. If you have a continuous distribution of charge then the change of W can be anything depending upon how much of a particular charge you're cutting. If you cut all of it or none of it then the possibility exists; if you cut part of it then it's impossible. You get a continuous change instead of a discreet change.

H. FRITZSCH (Caltech)

There is some evidence that the low-lying hadrons we observe consist of two sub-units or three sub-units - quarks or perhaps dyons or whatever. Do you see any reason in your scheme for two or three structures?

SCHWINGER

Well I don't understand it. Oh! I see what you mean two in the sense of the meson.

H. FRITZSCH (Caltech)

Two in the meson and three in the baryon.

SCHWINGER

Well I do believe I made some reference to that, didn't I? Obviously in this scheme I proposed to replace the quark by the dyon. Incidentally, there is the additional remark to all of this, namely that nobody should ever find a fractionally charged electrical particle without a magnetic charge accompanying it. That's the whole difference. This scheme I had in mind of course is precisely that the baryons are made out of three - I mentioned the reason for three - it has to do with the simplest way of balancing the magnetic charge and so on. And the symmetry between electric and magnetic charge tells you why 1/3 in one place would be 1/3 in the other and so on. I went over that very quickly, I'm sorry about that, but there was a specific reference to that picture in which the idea is that the dyons with their fractional electric and magnetic charges are the constituents and that the underlying physical reason for the particular combination is the necessity of having zero magnetic charge. That's the stablest situation. And there are two ways of doing it: a positive charge versus a negative charge, that's the two (that's the meson) or one double (in fractional units) magnetic charge against two oppositely signed. (That's the other way of doing it.) Of course that's a simple model. The real physics would be immensely more complicated because that is a picture not in terms of particles really but charges. Obviously we know very well inside the strongly interacting system you have no right to talk about particle numbers although everybody does it. But the charges as absolutely conserved quantities can be followed but what additional processes of pair creation and all the rest going on I have no way of describing. In addition, there is the complication I referred to in answering Bruno's question that the magnetic charges in fact will not sit there but will move around the particles. The charges will always be there but their association with particles will be elaborate and the possibility of other pairs existing is also very elaborate and so on. This is a strongly interacting system. I think, however, the saving grace of the picture is that it assigns the quantum numbers that we have learned have something phenomenologically to do with the families of particles - families of observed particles - not to underlying particles so much as charges, because that has a physical meaning and we can understand that but the particle picture will be complicated.

H. FRITZSCH (Caltech)

In the usual quark theory of hadrons one introduces a new non-abelian quantum number (color) and requires all physical states to be color singlets. This leads to the familiar SU_6 multiplets, solves the spin-statistics problem, gives the correct decay rate for $\pi_o \rightarrow 2\gamma$, etc. Your definition of a hadron as a composite system with magnetic charge zero leads to a wider class of hadron states, and the normal SU_6 problem is not reproduced. How do you resolve this problem?

SCHWINGER

Yes, may I point out that instead of color which is just simply a cute name for an unknown property, one has magnetic charge which is a familiar name for a physical property. That's what plays the role of color.

SUPERSYMMETRIC ANSATZ FOR SPONTANEOUSLY BROKEN GAUGE FIELD THEORIES

Feza Gürsey*

It is shown that a new type of Ansatz involving spinors in a gauge theory with Higgs fields based on SU(3) leads to 't Hooft type monopole solutions with finite energy which carry fractional charges of the quarks.

*Physics Department, Yale University, New Haven, Conn. 06520
This chapter is a transcript of Session Chairman Gürsey's comments.
Work (Yale Report COO-3075-129) supported in part by the Energy Research and Development Administration, Contract AT(11-1)-3075.

I. Introduction

In this talk I shall present a sketch of some recent work done in collaboration with Dr. J. Swank[1]. Although it is not directly related to the topic of this session which is devoted to supersymmetry, it concerns special solutions of an SU(3) gauge theory obtained by means of an Ansatz that has a supersymmetric structure. Indeed, the new Ansatz associates three SU(3) directions with a vector, one with a scalar, and the remaining four with the real components of a spinor in x-space. The resulting solutions describe 't Hooft monopoles with fractional charges and finite energy that are quark-like. Thus, at least in the static case, we are able to obtain from a Yang-Mills type SU(3) theory which involves only vector fields and Higgs scalars, special, rotationally covariant classical solutions that transform like spinors under the rotation group and carry fractional charges. If the procedure could be generalized to fully covariant solutions one might be able to imbed supersymmetry schemes involving fermions and bosons into purely vectorial Yang-Mills schemes with Higgs scalars, illustrating once more the relation between graded Lie algebras and supersymmetric structures.

II. The Supersymmetric Ansatz for SU(3)

We start from an SU(3) gauge field theory of 8 vector bosons in interaction with a Higgs scalar multiplet that is also an octet under SU(3). In the SU(2) case Wu and Yang[2] have obtained classical solutions in the pure Yang-Mills case by making the Ansatz

$$A^i_m(x) = \varepsilon^{imn}\,\hat{r}^n f(r), \quad (\hat{r}^n = x^n/r) \quad , \tag{2.1}$$

where i is the isospin index, m = 1,2,3 labels the three space directions. Furthermore, we have chosen a gauge in which

$$A^i_o(x) = 0 \quad . \tag{2.2}$$

The Wu-Yang Ansatz couples the three isospin directions with the three space directions. The corresponding solutions transform covariantly under the combined action of space rotations and isospin rotations[3]. In the case of a SU(3) gauge theory with Higgs scalars 't Hooft obtained[4] finite energy monopole solutions with quantized charges by using the same Ansatz (2.1) for the vector fields and identifying the electromagnetic direction in isospin space with the asymptotic direction of the Higgs multiplet. Subsequent work on the SU(2) case has dealt with the topological meaning of charge quantization[5], solutions with non vanishing V_o[6] and some exact solutions[7].

Many authors[8,9,10] have adapted the Wu-Yang and 't Hooft methods to SU(3) in the following way. SU(3) has a subgroup SO(3). In the 3x3 representation of SU(3) the three antisymmetrical λ matrices λ^a (namely λ^2, λ^5 and λ^7) transform like a vector (J=1) under SO(3) while the symmetrical ones λ^s (namely λ^1, λ^4, λ^6, λ^3 and λ^8) transform like the five components of a quadrupole (J=2). It is therefore natural to associate λ^a with $\hat{r}^a$ and λ^s with $\hat{r}^m\hat{r}^n - \frac{1}{3}\delta^{mn}$ in a generalization of the Wu-Yang Ansatz. The corresponding monopole solutions imply charge quantization[11], but these states, unlike quarks carry integer charges.

In our work[1] we select the SU(2) isospin subgroup of SU(3) and, in the spirit of Wu-Yang and 't Hooft, apply the Ansatz (2.1) to the isospin directions λ^1, λ^2 and λ^3. Now since the SU(3) octet decomposes in I=1, I=$\frac{1}{2}$ and I=0 pieces with respect to this subgroup we are then forced to associate the λ^8 direction with a scalar under rotations and the directions $\lambda^4+i\lambda^5$ and $\lambda^6+i\lambda^7$ respectively with the spin up and down components of a spinor under spatial rotations. Thus we start from the Lagrangian

$$\mathcal{L} = \frac{1}{2} D_\mu\phi^a D^\mu\phi^a - \frac{1}{4} G^a_{\mu\nu}G^{a\mu\nu} - V(\phi) \qquad (2.3)$$

where a=1,...,8 is the octet index, μ, ν are space-time vector indices, D_μ is the covariant derivative, such that, with f^{abc} denoting the structure constants of SU(3) we have

$$D_\mu\phi^a = \partial_\mu\phi^a + e\, f^{abc} A^b\phi^c \quad , \qquad (2.4)$$

$G^a_{\mu\nu}$ are the gauge invariant fields

$$G_{\mu\nu}{}^a = \partial_\mu A^a{}_\nu - \partial_\nu A_\mu{}^a + e\, f^{abc} A_\mu{}^b A_\nu{}^c \qquad (2.5)$$

and the Higgs potential

$$V(\phi) = -\frac{1}{2}\mu^2\phi^2 + \frac{1}{3}\kappa\phi^3 + \frac{1}{4}\lambda\phi^4 \qquad (2.6)$$

with

$$\phi^2 = \phi^a\,\phi^a \quad . \qquad (2.7)$$

$\mathcal{L}$ is then invariant under local SU(3) transformations. In principle SU(3) could be the unitary group or the color group. In our work we identify the electromagnetic charge direction with a direction in the octet space which leads us to consider the unitary symmetry group.

Restriction to static solutions with the gauge restriction (2.2) $A^a_o = 0$ puts the Lagrangian in the form

$$\mathcal{L}^{stat} = -\frac{1}{2} D_\ell\phi^a D_\ell\phi^a - \frac{1}{4} G^a_{\ell m}\, G^{a\ell m} - V(\phi) = -\mathcal{H} \quad , \qquad (2.8)$$

where $\mathcal{H}$ is the Hamiltonian density.

We now introduce the unit vector $\hat{r}$ and the unit spinor ξ in x-space defined by

$$\hat{r} = \vec{r}/r \quad , \qquad \xi = \begin{pmatrix} \xi_1 \\ \xi_2 \end{pmatrix} \tag{2.9}$$

with the properties

$$\hat{r}.\hat{r} = 1 \;, \quad \xi^{\dagger}\xi = 1 \tag{2.10}$$

and

$$\hat{r}^i = \xi^{\dagger}\sigma^i\xi \;, \qquad \xi\xi^{\dagger} = \tfrac{1}{2}(1 + \vec{\sigma}.\hat{r}) \quad , \tag{2.11}$$

A new solution is

$$\xi_1 = \frac{x_1 - ix_2}{\sqrt{2r(r-x_3)}} = \cos\frac{\theta}{2}\, e^{-i\phi} \tag{2.12a}$$

$$\xi_2 = \sqrt{\frac{r-x_3}{2r}} = \sin\frac{\theta}{2} \tag{2.12b}$$

where θ and ϕ are the polar angles. The general solution ξ' is obtained from this ξ by an arbitrary phase transformation

$$\xi' = e^{i\alpha(x)}\,\xi \,. \tag{2.13}$$

We make the following Ansatz for the special solutions we seek

$$\phi^i = \hat{r}_i \frac{Q(r)}{er} \;, \qquad \phi^8 = \frac{P(r)}{er} \quad , \tag{2.14}$$

$$\phi = \begin{pmatrix} \phi^7 - i\phi^6 \\ -\phi^5 + i\phi^4 \end{pmatrix} = \xi \frac{U(r)}{er} \quad , \tag{2.15}$$

$$A^i{}_{\ell} = \epsilon_{\ell ik}\, \hat{r}_k \frac{K(r)-1}{er} \tag{2.16}$$

$$A^8{}_{\ell} = \frac{2i}{e\sqrt{3}}\, \xi^{\dagger}\partial_{\ell}\, \xi \tag{2.17}$$

$$A_{\ell} = \begin{pmatrix} A^7_{\ell} - iA^6_{\ell} \\ -A^5_{\ell} + iA^4_{\ell} \end{pmatrix} = \sigma_{\ell}\, \xi \frac{F(r)}{er} \quad . \tag{2.18}$$

The form we have chosen for A_ℓ^8 ensures that under the phase transformation (2.13), the term $D_\ell \phi^a$ behaves like $\partial_\ell \xi - \xi^\dagger(\partial_\ell \xi)\xi$, making the choice of α immaterial as (2.13) becomes a local hypercharge transformation.

III. Quark-like Solutions

The Higgs potential is minimized by the vacuum expectation value of the Higgs vector ϕ_0 which satisfies the relations

$$\sqrt{3}\, d_{abc}\, \phi_0{}^b \phi_0{}^c = -\, \varepsilon(\phi_0)_a \quad , \tag{3.1}$$

where d_{abc} are the familiar SU(3) coefficients in the symmetric algebra of Gell-Mann's λ matrices. Here ε can take two values ε_0 or ε_1 given by

$$\begin{pmatrix}\varepsilon_0\\ \varepsilon_1\end{pmatrix} = \frac{\kappa}{2\lambda} \pm \sqrt{\left(\frac{\kappa}{2\lambda}\right)^2 + \frac{\mu^2}{\lambda}} \quad . \tag{3.2}$$

Then ε_0 (the absolute minimum) specifies the mass scale, so that we can introduce the dimensionless variable

$$x = e\, \varepsilon_0\, r \quad . \tag{3.3}$$

Note that eq. (3.1) is the equation of the q-vectors of Michel and Radicati[12] which is satisfied by the hypercharge and electric charge directions. Hence the ϕ_0 is a natural direction chosen by spontaneous symmetry breaking for the electric charge Q which is an SU(3) generator and satisfies

$$\sqrt{3}\, d_{abc}\, Q^b Q^c = \frac{1}{\sqrt{3}}\, Q_a \quad . \tag{3.4}$$

If we normalize ϕ_0 to $\hat{\phi}$ which satisfies the same equation (3.4) as Q, then $\hat{\phi}$ which in general is not diagonal is related to the charge

$$Q = Q^a \lambda^a = \frac{1}{2}\left(\lambda_3 + \frac{1}{\sqrt{3}}\lambda_8\right) \tag{3.5}$$

by a SU(3) transformation. Q and $\hat{\phi}$ have the normalization

$$\frac{1}{2}\,\mathrm{Tr}\, Q^2 = Q^a Q^a = \frac{1}{3}, \quad \mathrm{Det}\, Q = \frac{2}{27} \quad . \tag{3.6}$$

The monopole solutions are those which converge asymptotically ($x \to \infty$) to their vacuum expectation values which are q-vectors. This in turn determines the asymptotic behavior of the gauge potentials. The boundary conditions at x=0 are determined by the finiteness of the energy.

Using the $\hat{\phi}$ direction and following 't Hooft[4] we now construct the Maxwell field tensor

$$F_{\mu\nu} = \hat{\phi}^a G_{\mu\nu} - \frac{4}{e} f^{abc} \hat{\phi}^a D_\mu \hat{\phi}^b D_\nu \hat{\phi}^c \quad . \tag{3.7}$$

The corresponding magnetic field is

$$B_k = - \frac{1}{2} \varepsilon_{k\ell m} F_{\ell m} \quad . \tag{3.8}$$

In the special case of G = U = 0, the Hamiltonian takes the form

$$\begin{aligned} H = \frac{4\pi}{e^2} (\varepsilon_0 e) \int_0^\infty \frac{dx}{x^2} \{ & \tfrac{1}{2}(xQ'-Q)^2 + \tfrac{1}{2}(xP'-P)^2 \\ & + (xK'+\tfrac{1}{2}F^2)^2 + (xF'-\tfrac{1}{2}KF)^2 + Q^2K^2 \\ & + \tfrac{1}{8} [(\sqrt{3}Q+P)^2+P^2]F^2 + \tfrac{1}{2}(K^2+F^2-1)^2+V \} \end{aligned} \tag{3.9}$$

where V is the Higgs potential (2.6).

We find 3 solutions with the following asymptotic behavior

$$\text{I.} \quad Q_0/x = \sqrt{3}/2, \qquad P_0/x = 1/2, \qquad U_0 = 0 \quad , \tag{3.10a}$$

$$\text{II.} \quad Q_0/x = -\sqrt{3}/2, \qquad P_0/x = 1/2, \qquad U_0 = 0 \quad , \tag{3.10b}$$

$$\text{III.} \quad Q_0/x = 0 \; , \qquad P_0/x = 1 \; , \qquad U_0 = 0 \quad . \tag{3.10c}$$

The corresponding magnetic fields are

$$\vec{B}_I = \frac{2}{3e} \frac{\hat{r}}{r^2} \; , \quad \vec{B}_{II} = - \frac{1}{3e} \frac{\hat{r}}{r^2} \; , \quad \vec{B}_{III} = - \frac{1}{3e} \frac{\hat{r}}{r^2} \tag{3.11}$$

All these solutions have finite energy of the order of the mass of the 't Hooft monopole. The quantization of magnetic flux then gives for possible values of the electric charge multiples of $\frac{2}{3}e$, $-\frac{1}{3}e$, and $-\frac{1}{3}e$ which are the basic quark charges.

Note that solutions obtained through $\hat{\phi} \to -\hat{\phi}$ would carry antiquark charges.

Our method generalizes to higher gauge theories with an SU(2) subgroup.

I am indebted to J. Swank for a critical reading of the manuscript.

References

1. F. Gürsey and J. Swank, Yale report (1975) to be published.
2. T.T. Wu and C.N. Yang, in "Properties of Matter under Unusual Conditions", ed. H. Mark and S. Fernbach, p. 349 (Interscience, New York, 1969).
3. I. Bars, Proc. of the SLAC Summer Institute on Particle Physics, August 1974, vol. II, p. 237 (1974).
4. G. 't Hooft, Nucl. Phys. B79, 276 (1974).
5. J. Arafune, P.G.O. Freund and C.J. Goebel, J. Math. Phys. 16, 433 (1975).
6. B. Julia and A. Zee, Phys. Rev. D11, 2227 (1975); T. Dereli, J.H. Swank and L.J. Swank, Phys. Rev. D11, 3541 (1975).
7. M.K. Prasad and C.M. Sommerfield, Phys. Rev. Lett. 35, 760 (1975).
8. A.C.T. Wu and T.T. Wu, J. Math. Phys. 15, 53 (1974); A. Chakrabarti, Centre de Physique Theorique, Ecole Polytechnique Report No. A200, 1075 (1975).
9. W.J. Marciano and H. Pagels, Phys. Rev. D12, 1093 (1975).
10. T. Dereli, Ph.D. Thesis, Middle East Technical University (1975).
11. P.A.M. Dirac, Proc. Roy. Soc. A133, 60 (1934), Phys. Rev. 74, 817 (1948); J. Schwinger, Phys. Rev. 144, 1087 (1966).
12. L. Michel and L.A. Radicati, Annals of Phys. 66, 758 (1971).

CHARGE AND MASS SPECTRUM OF QUANTUM SOLITONS

Roman Jackiw*

A perturbative method for soliton sectors in quantum field theory is reviewed and the soliton mass spectrum is discussed. It is shown that quantizing the Yang-Mills monopole forces the presence of charge-bearing monopoles, which are almost degenerate in mass.

*Laboratory for Nuclear Science and Department of Physics, Massachusetts Institute of Technology, Cambridge, Mass. 02139.
Work supported in part by the Energy Research and Development Administration, Contract AT(11-1)-3069.

I. INTRODUCTION

Exact solutions for most quantum field theories are unavailable; one extracts the physical content by an expansion in powers of a coupling constant. When this constant is small, a sufficiently accurate description is provided by a few terms of the perturbative series. But how should the first term be computed? In the familiar Born expansion, one begins with a free, non-interacting theory - the coupling constant is set to zero. Yet certainly there exist phenomena which can not be described by non-interacting models, even for weak coupling. I shall describe an alternative perturbation series developed in collaboration with Goldstone[1], in which the first approximation uses solutions for the classical field equations, and is proportional to an inverse power of the coupling constant. Our approach, applicable to some models, when the coupling is weak, has exposed new sectors in the space of quantum states, populated by heavy stable particles, which are collective bound states of an indefinite [infinite] number of the usual particles. The new particles, called variously solitons, baryons, monopoles, to distinguish them from the usual particles which we call mesons, have strong interactions with each other; while the interaction strength with the mesons is of order unity. [The mesons interact weakly with each other, since the strength of interaction is proportional to the coupling constant.] Stability is assured by the emergence of new conserved quantum numbers, not directly associated with symmetries of the theory, but rather with topological properties of the classical solution. I shall first example the method by describing simple models in one spatial dimension, and then I shall explore the quantum theory associated with the Yang-Mills monopole.[2]

The program of expanding a quantum theory around a non-trivial solution of classical field equations is already familiar from studies of spontaneous symmetry breaking. One knows that a constant solution is the lowest order approximation to the vacuum expectation value of the quantum field; typically it is proportional to an inverse power of the coupling. Once this large effect is isolated, the smallness of the coupling constant allows for a systematic calculation of corrections. Our new approach is similar, except that the starting point is a non-constant solution of the classical equations.

II. MODELS IN ONE SPATIAL DIMENSION

Field theories in one spatial dimension illustrate adequately all aspects of our method; also they are simple to describe, hence I discuss them first. Consider a scalar field Φ with dynamics governed by a Lagrangian depending on a coupling constant g, in a way which can be scaled according to the following rule.

$$\mathcal{L}(\Phi;g) = \frac{1}{g^2}\mathcal{L}(g\Phi;1) \tag{2.1}$$

The reason for making this Ansatz is that the classical field equation for $g\Phi$ is independent of g, while the quantum theory, which involves $\frac{1}{\hbar}\mathcal{L}(\Phi;g) = \frac{1}{\hbar g^2}\mathcal{L}(g\Phi;1)$, quantum corrections, proportional to powers of $\hbar$, are also proportional to powers of g, which is taken to be small. [Henceforth $\hbar$ is set to unity.] A Lagrangian of the form

$$\mathcal{L}(\Phi) = \frac{1}{2}\partial^\mu\Phi\,\partial_\mu\Phi - U(\Phi) \tag{2.2}$$

leads to an equation of motion

$$\ddot{\Phi} - \Phi'' = -U'(\Phi) \tag{2.3}$$

I shall be concerned with time-independent classical solutions.

$$\begin{gathered} \Phi'' = U'(\Phi) \\ \frac{1}{2}(\Phi')^2 = U(\Phi) \end{gathered} \tag{2.4}$$

The models of interest are further delimited by two requirements. First we demand that the energy of a classical static field ϕ

$$E_C(\varphi) = \int dx \left[\tfrac{1}{2}(\varphi')^2 + U(\varphi)\right] \tag{2.5}$$

be finite for ϕ solving (2.4). [A classical field satisfying (2.4) shall be denoted by ϕ_C.]

$$E_C(\varphi_C) = \int dx\, (\varphi_C')^2 < \infty \tag{2.6}$$

The reason for this restriction is that in the quantum theory $E_C(\phi_C)$ is the lowest approximation to the mass of the soliton, a quantity which should be finite. Second we demand that ϕ_C be classically stable. Stability can be defined in two equivalent ways: ϕ_C, which obviously renders $E_C(\phi)$ stationary, should also minimize it; alternatively a time dependent perturbation to ϕ_C, $\phi(t,x) = \phi_C(x) + e^{i\omega t}\delta\phi(x)$ should not grow exponentially with time. Either definition has the consequence that the Schrödinger-like equation

$$\left[-\frac{d^2}{dx^2} + U''(\varphi_C)\right]\psi_k(x) = \omega_k^2\, \psi_k(x) \tag{2.7}$$

must have a non-negative spectrum: $\omega_k^2 \geqslant 0$. Classical stability is required so that the corresponding quantum states be stable.

One example [of many] which meets our requirements is the Φ^4 theory with spontaneous breaking of the $\Phi \leftrightarrow -\Phi$ symmetry.

$$U(\Phi) = \frac{m^4}{2g^2} - m^2 \Phi^2 + \frac{g^2 \Phi^4}{2} \tag{2.8}$$

There exists a family of static solutions with finite energy.

$$\varphi_c(x - x_0) = \frac{m}{g} \tanh m(x - x_0) \tag{2.9}$$

$$E_c(\varphi_c) = \frac{4}{3} \frac{m^3}{g^2} \tag{2.10}$$

The solutions are stable, as is verified by solving the associated Schrödinger-like equation (2.7): all eigenvalues are non-negative. There is a discrete zero-frequency mode; this is a consequence of translational invariance. The normalized eigen-function is

$$\psi_0(x - x_0) = N^{-1/2} \varphi_c'(x - x_0) \tag{2.11a}$$

$$N = \int dx \, (\varphi_c')^2 = E_c(\varphi_c) \tag{2.11b}$$

This mode, which will play a central role in the quantum theory, is called the translation mode.

The quantum theory, which is build around the classical solution is the following. With the class (2.9), we associate a class of quantum states |p> which describe a heavy particle, the soliton, with momentum p. Also we make the Ansatz that to lowest order

$$\langle p|\Phi(0,x)|p'\rangle = \int dx_0\, e^{i(p-p')x_0}\varphi_c(x-x_0) = O(g^{-1}) \tag{2.12}$$

In lowest order all states are degenerate in energy; this is because the mass, identified with the classical energy, is very large and dominates the kinematical momentum dependence.

$$E(p) \approx M \approx E_c(\varphi_c) = O(g^{-2}) \tag{2.13}$$

The apparent degeneracy of the states is an artifact of the approximation and disappears when higher orders are taken into account. To this end we identify all normalized solutions of (2.7), except the translation mode, with matrix elements of the quantum field between one soliton and one soliton - one meson states: |p; k> where p is the total momentum and k is meson momentum.

$$\langle p|\Phi(0,x)|p';k\rangle = \frac{1}{\sqrt{2\omega_k}}\int dx_0\, e^{i(p-p')x_0}\,\psi_k(x-x_0) \tag{2.14}$$

Also ω_k is interpreted as the lowest order energy difference between $|p\rangle$ and $|p';k\rangle$. However no physical state is associated with the translation mode, since there is no degeneracy in the quantum problem.

The translation mode reappears in a calculation of E(p) to higher order. This can be done by considering a sum rule, derived by taking matrix elements of the canonical commutator.

$$\langle p|[\Phi(0,x),\dot{\Phi}(0,y)]|p'\rangle = i\delta(x-y)(2\pi)\delta(p-p') \tag{2.15a}$$

The product of the two field operators is saturated with intermediate soliton states; only the no meson and one meson terms are retained. The one meson terms give

$$\frac{i}{2}\sum_k{}' \int dz\, \psi_k^*(x-z)\,\psi_k(y-z)\,e^{i(p'-p)z} + x \leftrightarrow y \tag{2.15b}$$

where the prime on the summation recalls that the zero frequency mode is excluded - it does not correspond to a physical state. Since the sum is not ove a complete set of functions, we get for (2.15b)

$$i\delta(x-y)(2\pi)\delta(p-p') - i\int dz\, e^{i(p'-p)z}\, \frac{1}{N}\, \varphi_c'(x-z)\, \varphi_c'(y-z) \tag{2.15c}$$

The no meson intermediate states involve expressions of the form

$$\int \frac{dq}{(2\pi)} \langle p|\Phi(0,x)|q\rangle\langle q|\dot{\Phi}(0,y)|p'\rangle =$$

$$i\int \frac{dq}{(2\pi)} \left[E(q) - E(p')\right] \int dz\,dz'\, e^{i(p-q)z}\, e^{i(q-p')z'}\, \varphi_c(x-z)\, \varphi_c(y-z') \tag{2.15d}$$

These cancel the extra term in (2.15c), provided $E(q)-E(p') = \frac{q^2 - p'^2}{2N}$. Thus we find that to order $M^{-1} = O(g^2)$ the degeneracy is removed and $E(p) = M + \frac{p^2}{2N} = M + \frac{p^2}{2M}$.

By following powers of g, one can compute all quantities of interest in the one-soliton sector. For example matrix elements of Φ between soliton–n meson and soliton–n' meson states are $O(g^{n+n'-1})$. Thus the soliton-multimeson S matrix is determined, specifically soliton–one meson scattering is described by $\psi_k(x)$ which is $O(g^0)$. These calculations are by now well known and need not be further reviewed.[3]

Let us instead focus on a general aspect of the theory which must be appreciated for further applications. The procedure of associating the degeneracy of the classical solution ϕ_C with a set of quantum states is entirely general. Because of this degeneracy, the small oscillation equation (2.7) will have zero frequency modes, with eigenfunctions $\delta\phi_C$, which are the infinitesimal transforms of the classical solution under the symmetry transformation which is responsible for the classical degeneracy. If $\int dx(\delta\phi_C)^2 < \infty$, the zero frequency mode is an isolated eigensolution of the small-oscillations equation, and does not correspond to a quantum state - it is not a Goldstone particle since no symmetry is spontaneously broken in the quantum theory. In lowest order the energy of all the soliton states is degenerate, but a higher order calculation removes the degeneracy with the help of the zero frequency mode.

We have already seen this scheme operating for the case of classical degeneracy associated with translations. Another example is given by charge rotations. Suppose we have a doublet of fields Φ^a, $a = 1,2$, in a theory with charge conservation, i.e. with invariance under infinitesimal rotations.

$$\delta\Phi^a = \varepsilon^{ab}\Phi^b \tag{2.16}$$

If ϕ_C^a solves the classical static field equations, then $I^{-1/2}\epsilon^{ab}\phi_C^b$, $I = \int dx(\phi_C^a\phi_C^a) < \infty$, will be a normalized zero frequency solution of the small oscillation equation. In the corresponding quantum theory we must allow the soliton states to carry a label additional to momentum, viz. charge. The generalization of (2.12) is

$$\langle pn|\frac{1}{\sqrt{2}}\left(\Phi^1 + i\,\Phi^2\right)|p'n'\rangle =$$

$$\delta_{n,n'-1}\int dx\, e^{i(p'-p)x}\frac{1}{\sqrt{2}}\left(\varphi_c^1(x) + i\,\varphi_c^2(x)\right) \qquad (2.17)$$

In lowest order, the energy of all states is independent of p and n; however a calculation entirely analogous to that for translation invariance, shows that the charge degeneracy is broken in higher order.[4]

$$E_n(p) = M + \frac{p^2}{2M} + \frac{n^2}{2I} \qquad (2.18)$$

Thus when the classical theory supports large, $0(g^{-1})$, classical solutions for charged fields, which are localized in space , then the corresponding quantum theory requires charged quantum soliton states with a closely spaced, $O(g^2)$, mass spectrum.

We conclude the discussion of one dimensional models with several additional observations. The trivially conserved current

$$J^\mu = \varepsilon^{\mu\nu}\partial_\nu \Phi \tag{2.19}$$

leads to an associated constant of motion

$$\int dx\, J^0 = \int dx\, \Phi' = \Phi\Big|_{x=\infty} - \Phi\Big|_{x=-\infty} \tag{2.20}$$

In the soliton sector, this constant is non-zero since the field tends to different values as $x\to\pm\infty$. This is an example of a topological conservation law which assures that the solitons are stable.[1]

When classical time-dependent scattering solutions are available, the phase shift for soliton-soliton scattering can be obtained, to leading order, from the classical time delay. The results are $O(g^{-2})$ indicating strong forces for weak coupling.[5]

In the Φ^4 example we may consider the $\langle p|\Phi|p'\rangle$ matrix element to be analytic continuation of $\langle pp'|\Phi|0\rangle$. Since the former is anti-symmetric in p↔p', so is the latter. This shows that the solitons are Fermions, though in one dimension in the absence of spin, this need not be a startling result.[1]

III. A MODEL IN THREE SPATIAL DIMENSIONS

Models in three spatial dimensions that possess static, stable solutions of finite energy require something more complicated than a set of self-interacting spinless fields.[1] A well known example is the gauge theory of a spinless iso-triplet, with spontaneous breaking of the isotopic symmetry.[2] The Lagrangian is

$$\mathcal{L} = \frac{1}{2}(D^{\mu}\underline{\Phi})^{a}(D_{\mu}\underline{\Phi})^{a} - \frac{1}{4}F^{a\mu\nu}F^{a}_{\mu\nu} - U(\underline{\Phi}^{2})$$

$$(D_{\mu}\underline{\Phi})^{a} = \partial_{\mu}\underline{\Phi}^{a} + g\,\varepsilon^{abc}A^{b}_{\mu}\underline{\Phi}^{c}$$

$$F^{a}_{\mu\nu} = \partial_{\mu}A^{a}_{\nu} - \partial_{\nu}A^{a}_{\mu} + g\,\varepsilon^{abc}A^{b}_{\mu}A^{c}_{\nu}$$

$$U(\underline{\Phi}^{2}) = \frac{m^{4}}{2\lambda}\left[1 - \frac{\lambda^{2}}{m^{2}}\underline{\Phi}^{2}\right]^{2}$$

$$\underline{\Phi}^{2} = \underline{\Phi}^{a}\underline{\Phi}^{a}, \quad a = 1,2,3 \tag{3.1}$$

The form of U indicates that the Φ field acquires a vacuum expectation value; while the attendant Goldstone bosons are removed by the Higgs mechanism. In order to discuss the quantum mechanics in terms of only physical states we pass to the unitary gauge, that is we fix the scalar field's direction in isospace, and allow only its magnitude to be a dynamical variable.

$$\Phi^a = \hat{\varphi}^a \Phi \qquad \hat{\Phi}^2 = 1 \tag{3.2}$$

Correspondingly the gauge field is decomposed into parts parallel and perpendicular to $\hat{\phi}_a$

$$\begin{aligned} A_\mu^a &= \hat{\varphi}^a A_\mu + \varepsilon^{abc} \hat{\varphi}^b \left(W_\mu^c - \frac{1}{g} \partial_\mu \hat{\varphi}^c \right) \\ A_\mu &= \hat{\varphi}^a A_\mu^a \\ W_\mu^a &= \varepsilon^{abc} A_\mu^b \hat{\varphi}^c + \frac{1}{g} \partial_\mu \hat{\varphi}^a \end{aligned} \tag{3.3}$$

The gauge of A^μ must be still specified; we take $\nabla \cdot \vec{A} = 0$.

The known static solution is most frequently exhibited by

choosing $\hat{\phi}^a = \hat{r}^a$, which is convenient in discussions of the mathematical structure.[2] For quantization, the more familiar choice is made $\phi^a = \begin{pmatrix} 0 \\ 0 \\ 1 \end{pmatrix}$, whence it is realized that the theory describes a neutral Higgs particle Φ, the photon A_μ, and two massive, charged mesons $W_\mu^{1,2}$. The static solution in this gauge has the following properties: $\phi_C(r)$ is radially symmetric, vanishes at $r = 0$ and approaches $\frac{m}{\lambda}$ at $r = \infty$. The time components of all vector fields are zero. The vector potential $\vec{A}_C$ is that of a magnetic, Dirac monopole[6] of strength $1/g$. The $\vec{W}_C$ fields are localized in space; they take the form

$$\vec{W}_C^a = \hat{e}^a w(r), \quad a = 1, 2 \tag{3.4}$$

where $w(r)$ has the singularity $-1/gr$ at the origin and vanishes exponentially at infinity. The unit vectors $\hat{e}^a$ are orthonormal and orthogonal to $\hat{r}$. There are no unique formulas for $\vec{A}_C$ and $\hat{e}^a$; they depend on the remaining gauge freedom: the possibility of performing a purely electromagnetic gauge transformation with a static, harmonic gauge function, which shifts $\vec{A}_C$ by $-\frac{1}{g}\vec{\nabla}\theta$, $\nabla^2\theta = 0$, and rotates $\hat{e}^a$ by the angle θ in the two dimensional charge space. [One explicit expression is

$$\hat{e}^2 = \frac{\hat{\eta} \times \hat{r}}{|\hat{\eta} \times \hat{r}|}, \quad \hat{e}^1 = \hat{e}^2 \times \hat{r},$$

$$\vec{A}_C = \frac{1}{2g}\, \hat{\eta} \times \vec{\nabla} \ln \frac{1 + \hat{\eta}\cdot\hat{r}}{1 - \hat{\eta}\cdot\hat{r}}$$

where $\hat{\eta}$ is any constant unit vector; changes in $\hat{\eta}$ are equivalent to gauge transformations.]

Provided the constant λ in (3.1) is chosen to be βg^2 where $\beta = O(1)$, so that (2.1) is satisfied, the quantum theory can be analyzed by our methods. Detailed calculations are in progress, but the general picture can be inferred by recalling the one-dimensional examples.[7] All classical functions are $O(g^{-1})$, hence large. The quantum-monopole states carry not only momentum, but also electric charge since the theory admits static solutions for the charge bearing field W_μ^a. Therefore the Ansatz is

$$\langle \underset{\sim}{p} n | \Phi | \underset{\sim}{p}' n' \rangle = \delta_{nn'} \int d\underset{\sim}{r}\, e^{i(\underset{\sim}{p}' - \underset{\sim}{p})\cdot \underset{\sim}{r}}\, \varphi_c(r)$$

$$\langle \underset{\sim}{p} n | \vec{A} | \underset{\sim}{p}' n' \rangle = \delta_{nn'} \int d\underset{\sim}{r}\, e^{i(\underset{\sim}{p}' - \underset{\sim}{p})\cdot \underset{\sim}{r}}\, \vec{A}_c(\underset{\sim}{r})$$

$$\langle \underset{\sim}{p} n | \frac{1}{\sqrt{2}} \left(\vec{W}^1 + i \vec{W}^2 \right) | \underset{\sim}{p}' n' \rangle =$$

$$\delta_{n,n'-1} \int d\underset{\sim}{r}\, e^{i(\underset{\sim}{p}' - \underset{\sim}{p})\cdot \underset{\sim}{r}}\, \frac{1}{\sqrt{2}} \left(\vec{W}_c^1(\underset{\sim}{r}) + i \vec{W}_c^2(\underset{\sim}{r}) \right) \tag{3.5}$$

Matrix elements of gauge dependent operators $[\vec{A}, \vec{W}^a]$ are gauge dependent, $[\hat{\eta}$ dependent] but the mass of a monopole with electric charge n is gauge invariant.

$$M_n = M + \frac{n^2}{2} \Delta M \tag{3.6}$$

Contributions to ΔM include a part analogous to (2.17): $1/I$, $I = \int d\underset{\sim}{r} \rho(r)$, $\rho(r) = 2w^2(r) = O(g^{-2})$. Also there is a further term arising from the Coulomb interactions energy:

$$\frac{g^2}{I^2} \frac{1}{4\pi} \int d\underset{\sim}{r}\, d\underset{\sim}{r}' \frac{\rho(r)\rho(r')}{|\underset{\sim}{r} - \underset{\sim}{r}'|} = O(g^2).$$

The lowest order approximation to M is of course the classical energy, which is $O(g^{-2})$.

Thus the single monopole of the classical theory leads in the quantum theory to a whole family of dyons[8] carrying electric and magnetic charge; they are heavy, with $O(g^2)$ mass splittings. The charged monopoles are stable, since they cannot decay by emitting a charged vector meson whose mass is O(1).[9]

IV. CONCLUSION

An assessment of this work leads to two questions: Is it correct? Is it useful? We certainly believe that our method has exposed true aspects of some field theories for weak coupling. We are encouraged by the analysis of the non-linear

Schrödinger equation - a non-relativistic field theory which can be solved exactly by conventional procedures and also approximately by the new methods - results agree when expected to do so.[10] Furthermore soliton calculations for the Sine-Gordon theory are in agreement with conventional analysis of the equivalent Thirring model.[11] Nevertheless it is also clear that certain effects will never be found in any order: creation and annihilation of solitons does not appear in an obvious way in our perturbation series. Whether this is a defect of the perturbation series or a hint of new phenomena is an open question.

The utility of these ideas for physical theory is unclear. The most conservative viewpoint is that the, as yet undiscovered, gauge theory of nature possess coherent solutions which correspond to some heavy particles, whose properties we have explored but which have little significance for phenomenological applications. More venturesome is the notion that some of the observed "fundamental" particles are not bound states of a few elementary quarks, rather they are coherent bound states of the soliton variety, and that some of the absolute conservation laws are topological. Finally the most speculative idea makes reference to the quarks themselves. Their number has steadily increased with time - two would have been sufficient to account for the low lying isotopic multiplets of the fifties; in the sixties SU(3) and strangeness required a third quark; today evidence is mounting for a fourth degree of freedom. It is most natural to suppose that the progression does not end here - we are dealing with an infinite number which is not to be represented by an infinite component field, but rather by the quantization of a continuous classical degeneracy, as in the Yang-Mills example.

REFERENCES

1. J. Goldstone and R. Jackiw, Phys. Rev. D 11, 1486 (1975).

2. G. 't Hooft, Nuclear Physics B79, 276 (1974);
A. M. Polyakov, Zh ETF Pis. Red. 20, 430 (1974); [English translation: JETP Lett. 20, 194 (1974)].

3. Some higher order calculations are discussed in Ref. 1; others have been performed by the following authors:
R. Dashen, B. Hasslacher and A. Neveu, Phys. Rev. D 10, 4130 (1974);
J. L. Gervais and B. Sakita, Phys. Rev. D 11, 2943 (1975);
C. G. Callan and D. J. Gross, Nuclear Physics B93, 29 (1975);
E. Tomboulis, Phys. Rev. D 12, 1678 (1975);
J. L. Gervais, A. Jevicki, B. Sakita, Phys. Rev. D 12, 1038 (1975);
N. Christ and T. D. Lee, Phys. Rev. D 12, 1606 (1975);
E. Tomboulis and G. Woo, Annals of Physics (N.Y.)(in press);
A. Klein and F. Krejs, Phys. Rev. D (in press);
L. Jacobs, MIT preprint.
For general reviews see R. Jackiw, "Quantum Mechanical Approximations in Quantum Field Theory", in Theories and Experiments in Higher Energy Physics, B. Kursunoglu et al. Plenum Press (New York, 1975);
R. Rajaraman, Physics Reports 21C (1975);
R. Jackiw, Acta Physica Polonica B (in press).

4. The removal of degeneracy due to charge rotations has been analyzed with functional integration techniques by R. Rajaraman and E. Weinberg, Phys. Rev. D 11, 2950 (1975).

5. R. Jackiw and G. Woo, Phys. Rev. D 12, 1643 (1975);
S. Coleman, Phys. Rev. D 12, 1650 (1975);
V. Korepin and L. Faddeev, (unpublished);
L. Dolan, Phys. Rev. D (in press).

6. P.A.M. Dirac Proc. Royal Soc. A133, 60 (1931).

7. The general aspects of the quantum theory associated with the Yang-Mills monopole were analyzed by J. Goldstone and R. Jackiw; detailed calculations are performed by E. Tomboulis and G. Woo, MIT preprint.

8. J. Schwinger, Science 165, 757 (1969).

9. Classical charge bearing solutions have been obtained by B. Julia and A. Zee, Phys. Rev. D 11, 2227 (1975). Their role in the quantum theory is explained by Tomboulis and Woo, Ref. 7.

10. P. P. Kulish, S. V. Manakov and L. D. Faddeev, Landau Institute Preprint;
L. Dolan, Phys. Rev. D (in press);
C. Nohl, Annals of Physics (N.Y.)(in press).
A. Klein and F. Krejs, MIT preprint.

11. R. Dashen, B. Hasslacher and A. Neveu, Phys. Rev. D 11, 3424 (1975).

DISCUSSION

S. COLEMAN (Harvard)

There is a point that may be too technical for discussion but which puzzled me in your discussion of the monopole. You established a gauge and of course you did not consider the removal of degeneracies associated with gauge transformations. On the other hand if you have decided to also fix your electromagnetic gauge in a parallel way by fixing the direction of $\hat{\eta}$, it would seem that you would also remove the degenracy which you subsequently quantized and therefore I'm confused. Why is one degeneracy different from another degeneracy?

JACKIW

No you see I don't remove that degeneracy because the degeneracy that I quantize is a global degeneracy. It's the global rotation that gets quantized, and I cannot remove the possibility of performing global rotations by fixing the gauge locally. That's what happens in the Higg's phenomenon. In the Higg's phenomenon you have a vector - let's take this example - it's got three components and you can make three rotations. Now in the vacuum sector you eliminate the possibility of making two of the rotations by choosing the unitary gauge and the third rotation which you can still make, doesn't lead to anything new because the charge bearing fields don't have a non-zero solution in the vacuum sector. In the one soliton sector, after removing the possibility of making two rotations, there is a third rotation that I can still perform and even if I perform it only globally and not locally, I am rotating one charge state into the other and I'm quantizing that.

COLEMAN (Harvard)

I do not fully understand what you are saying. We'll talk about it privately.

JACKIW

I should mention incidentally that the quantization result that I got was independent of that unit vector. It was gauge invariant. At least that test was passed.

J. SUCHER (Maryland)

You have defined the field for your solitons states by giving a number of its matrix elements. Do you know whether this procedure will lead to a local field?

JACKIW

I don't. But my colleagues in mathematical physics who have gotten interested in some of these questions are beginning to address themselves - and I see the author raising his hand so maybe he'll answer the question.

A. JAFFEE (Harvard)

In the vacuum sector, we now have established the existence of a convergent expansion about the classical weak coupling limit of the $g^2\phi^4 - \phi^2$ problem. This expansion gives a local field and phase transitions. I believe that these methods can also be applied to the soliton ϕ^4 problem, and will show that everything you said in that case is qualitatively correct.

H.S. TSAO (IAS)

Is the $\lambda = \beta g^2$ condition modified in quantum corrections, i.e., is this condition consistent with renormalization?

JACKIW

That's an ansatz. That doesn't get modified. I have a parameter, a free parameter λ and in order to develop my quantum theory I say that it is of order g^2 so I can compare the effects of the Bose self-interaction with the gauge interaction. And if I were to say it's order is g^4 then I wouldn't do the kind of perturbation theory that I'm doing. That doesn't get modified in any way.

H. S. TSAO (IAS)

Is that consistent with the renormalization in higher order?

JACKIW

Yes. We just say that λ is of order g^2. You do that very frequently. Typically you say that λ is in fact of order g^4 because in scalar electrodynamics for example when you remove the infinity associated with charge particle self-scattering in fourth order, you remove it by a counter term $\lambda\phi^4$ and the infinity is of order e^4 and therefore you say λ is of order e^4 to remove that infinity. That's a familiar procedure. Only I don't say that it is of order e^4. I say it's of order e^2, which incidentally also works for removing infinities in the vacuum sector.

J. HSU (Texas)

You said that the magnetic monopole is absolutely stable. Is this a model-dependent statement or a general statement, and how do you show it?

JACKIW

Oh yes. This is a statement about this model. Now, I should say that unlike the one dimensional model in which there are explicit calculations of everything, and one can see that all those little eigen frequencies are positive, and one can see that various functions have various convergence properties that you want them to have, we don't have such an explicit construction for the Yang-Mills case. Now there exists general arguments which assure you, or they can assure you if you want to be assured, that, that object is stable in this model. However there possibly is some loophole and unless one has really an explicit numerical calculation of everything one is always uncertain. In fact 't Hooft's original paper did numerical computations to verify at least that the energy was finite. He did not do numerical computations to verify that it was classically stable.

S. COLEMAN (Harvard)

As far as I know, the situation is that nobody knows whether 't Hooft's solution is stable. 't Hooft's solution has a form of spherical symmetry and we do not know when you perturb the solution slightly it might not want to develop an asymmetric bump. What we do know from general arguments and very simple general arguments is that even if it is unstable it certainly cannot decompose into systems that carry no magnetic charge. That is simply a consequence of Gauss's theorem that the magnetic charge must be conserved. So whether or not this is the stable monopole, there certainly must be a stable monopole around someplace.

JACKIW

Except that if it is not spherically symmetric then there is yet another degeneracy.

S. COLEMAN (Harvard)

Yes then it becomes like an asymmetric nucleus and therefore you get rotational bands as well as the translational spectrum and the charge spectrum which you displayed. You get rigid body type rotational bands.

SEMI-CLASSICAL METHODS IN FIELD THEORY

Roger Dashen*

A brief review of the application of semi-classical WKB techniques to model field theories is presented.

*Institute for Advanced Study, Princeton University, Princeton, N.J. 08540.

Work supported by the Energy Research and Development Administration, Contract E(11-1)-2220-56.

I. Introduction

Interacting classical field theories often have solutions which look like particles. They are little lumps of energy that may pulsate and move around but retain a finite size as time progresses. Over the last few years, it has become apparent that these classical objects correspond to particle states in the quantized theory. This subject has now become rather large and I cannot hope to review everything. Jackiw,[1] in his talk at this conference, has already covered a number of topics. I will try to make my talk complimentary to his.

For a theory with a single scalar field a particle like solution is a function φ which

(i) satisfies the classical field equations and

(ii) has the property that

$$\lim_{|\vec{x}|\to\infty} |\varphi(\vec{x},t)| = |<\varphi>_{vac}| \tag{1}$$

Given such a solution one can define a rest frame in which the classical three momentum vanishes. In the rest frame all known particle like solutions are either time independent or periodic; i.e.,

$$\varphi(\underset{\sim}{x}, t+T) = \varphi(\underset{\sim}{x}, t) \tag{2}$$

for some T. I will concentrate on periodic solutions. There may be non-periodic particle like solutions. They would be hard to find either analytically or numerically.

The generalization of these ideas to theories with many boson fields is straightforward, but Fermi fields require some special devices. I will mention fermions briefly at the end of the talk.

The mass of one of these classical "particles" is equal to its energy in the rest frame. There has been a great deal of work on the center of mass motion of these objects. It is known that when properly quantized they become particles with the correct energy-momentum relation $E = \sqrt{M^2 + \vec{p}^{\,2}}$ and that the uncertainty principle is not violated. Jackiw[1)] has discussed this point in his talk. I will assume that the center of mass motion is understood and work in the rest frame.

II. Semi-Classical Quantization

Typically, the periodic particle-like solutions exist for a continuum of periods T. Classically, the mass spectrum is therefore continuous. This is not surprising. The energy spectrum of classical hydrogen is continuous. Quantum mechanics makes the energy levels of hydrogen discrete and will also make our mass spectrum discrete.

In simple potential theory problems the WKB method tells us that the discrete quantum levels are approximately determined by

$$\oint p\,dq = (2n+1)\pi \tag{3}$$

For hydrogen the WKB energy levels are exact. This is an accident, but as we will see, not an isolated one.

For general periodic motion the analog of Eq. (3) is[2)]

$$\sum_k \oint p_k dq_k = S + ET = 2(n + \xi)\pi \tag{4}$$

where S is the classical action and E is the classical energy. The quantity ξ is analog of the $\frac{1}{2}$ in the $n + \frac{1}{2}$ which appears in WKB formulas. It can be calculated by examining small oscillations around the classical solution. In calculating ξ one encounters divergences which are cancelled by the usual counter terms in the Lagrangian. This is how renormalization enters into a semi-classical calculation. There is no simple general formula for ξ. It has to be computed separately in each case. For a zeroth order approximation ξ can be neglected and Eq. (4) becomes the usual Bohr-Sommerfeld quantization rule.

Plugging a periodic particle like solution in Eq. (4) will fix the period leading to a discrete mass spectrum labeled by an integer n.

III. Examples in Two Dimensional Space Time

I now come to the main part of the talk, namely the results of calculations in model field theories in 1 + 1 dimensions.

A. The Non-Linear Schroedinger Equation.[3,4)]

Consider a set of non-relativistic bosons in one space dimension interacting through an attractive contact potential $\frac{\gamma}{8}\,\delta(x)$. Quantum mechanically this is an exactly solvable problem. For n bosons there is a single bound state with binding energy

$$|E_B(n)| = m\left(\frac{\gamma}{16}\right)^2 \left(\frac{n^3 - n}{6}\right) \tag{5}$$

where m is the boson mass.

Writing this theory in its second quantized form one finds a Lagrangian density

$$\mathcal{L} = i\Psi^* \frac{\partial}{\partial t}\Psi - \frac{1}{2m}\left|\frac{\partial\Psi}{\partial x}\right|^2 + \frac{\gamma}{16}|\Psi|^4 + C_1|\Psi|^2 + C_2 \tag{6}$$

where C_1 and C_2 are counter terms to take care of the normal ordering The usual normal ordering prescription is equivalent to choosing C_1 and C_2 such that the energy of the state with no particles is zero and the energy of the single boson state, is just $p^2/2m$.

We can think of Eq. (6) as defining a non-relativistic field theory. The field equation is

$$i\frac{\partial}{\partial t}\Psi + \frac{1}{m}\frac{\partial^2}{\partial x^2}\Psi + \frac{\gamma}{8}\Psi|\Psi|^2 + C_1\Psi = 0 \tag{7}$$

the classical version of which is called the nonlinear Schroedinger equation. It can be solved exactly. That is, given $\Psi(x, t = 0)$ there is a linear algorithm which gives $\Psi(x,t)$ for all times. In particular one can ask for particle like solutions. It is known that there is a unique such solution which in its rest frame is

$$\Psi_\epsilon = \left(\frac{16}{\gamma}\right)^{\frac{1}{2}} e^{-iC_1 t} \left(\frac{e^{i\frac{\epsilon^2 mt}{2}}}{\cosh \epsilon mx}\right) \tag{8}$$

where ϵ is an arbitrary parameter related to an arbitrary period by $T = 4\pi/\epsilon^2 m$. To quantize ϵ one uses Eq. (4). After renormalization ξ vanishes and the quantization condition reduces to

$$\int |\Psi|^2 dx = n \tag{9}$$

which means that the quantum number n is just the number of bosons in the state. The energy of the n-th state is then found to be[3,4]

$$E_n(p) = \frac{p^2}{2nm} - m\left(\frac{\gamma}{16}\right)^2 \left(\frac{n^3 - n}{6}\right) \tag{10}$$

when center of mass motion has been taken into account. It is evident that the n-th state is an n boson bound state with a binding energy which agrees with the exact result in Eq. (5). This is another case where the WKB method is accidentally exact. Note that the $n = 1$ state is the single boson or "elementary particle" of the theory.

Let $|n\,X\rangle$ be the n body bound state with center of mass located at X. One can show[3] that for large n

$$\Psi^{c\ell}_{\epsilon(n)}(x - X, t)\ \delta\left(X - \frac{x + nX'}{n+1}\right)$$

$$= \langle n+1,\ X|\ \Psi^{+}(x,t)\,|\ nX'\rangle \qquad (10)$$

where $\Psi^{c\ell}_{\epsilon(n)}$ is given by Eqs. (8) and (9). Note that $\Psi^{c\ell}$ is not the wave function of anything. It is also not an expectation value in a single state ($\langle n|\Psi^{+}|n\rangle$ vanishes). What $\Psi^{c\ell}$ does is to interpolate between the different bound states. One may think of this as analogous to the way the classical orbits of hydrogen interpolate between the quantized Bohr orbits. If we interpret the classical solutions as objects which interpolate between quantum mechanical bound states then it is clear that a semi-classical method can only work if there are many bound states. This is consistent with what one knows about the WKB approximation in potential theory and with our experience in model field theories.

B. The Sine-Gordon Equation.[2,4,5]

The Lagrangian density

$$\mathcal{L} = \tfrac{1}{2}(\partial_\mu\varphi)^2 + \frac{m^2}{\gamma}(\cos\sqrt{\gamma}\,\varphi - 1)$$

$$- \frac{\delta m^2}{\gamma}(\cos\sqrt{\gamma}\,\varphi - 1) - \Delta E \qquad (12)$$

where δm^2 and ΔE are counter terms defines a most interesting model. In our 1 + 1 dimensional world the coupling constant γ is dimensionless and is not renormalized. Like the nonlinear Schroedinger equation the Sine-Gordon equation

$$\Box^2 \varphi - \frac{m^2 - \delta m^2}{\sqrt{\gamma}} \sin\sqrt{\gamma}\varphi = 0 \tag{13}$$

is exactly solvable. In particular one can compute ξ for any classical solution.[2)] It turns out that ξ is proportional to the action S. Defining

$$\gamma' = \frac{\gamma}{1 - \frac{\gamma}{8\pi}} \tag{14}$$

the result is that one can use the usual Bohr-Sommerfeld quantization rule with γ replaced by γ' .

The Sine-Gordon equation has two types of particle like solutions. First there is the soliton and antisoliton which in the rest frame are

$$\varphi = \left(\frac{16}{\gamma}\right)^{\frac{1}{2}} \tan^{-1}\left(e^{\pm mx}\right) \tag{15}$$

There is nothing to quantize here and one finds a soliton mass[2,4]

$$M(\text{soliton}) = \frac{8m}{\gamma'} \tag{16}$$

The other kind of particle like solution is the breather

$$\varphi = \left(\frac{16}{\gamma}\right)^{\frac{1}{2}} \tan^{-1} \epsilon \left[\frac{\sin\left(\frac{mt}{\sqrt{1+\epsilon^2}}\right)}{\cosh\left(\frac{\epsilon mx}{\sqrt{1+\epsilon^2}}\right)}\right] \tag{17}$$

which depends on an arbitrary parameter ϵ. Upon quantization the breather produces a series of particles at masses[2)]

$$M_n = \frac{16m}{\gamma'} \sin\frac{n\gamma'}{16}$$

$$n = 1, 2 \ldots < \frac{8\pi}{\gamma'} \tag{18}$$

There are a finite number of these states all of which lie below the soliton-antisoliton threshold at $2M(\text{soliton}) = 16m/\gamma'$. As the coupling γ' increases they move up one by one to the soliton-antisoliton threshold and disappear from the spectrum. At $\gamma' = 8\pi$, the $n = 1$ state moves through threshold and disappears. For $\gamma' > 8\pi$ there are no breather states.

For weak coupling the breather mass formula can be approximated by

$$M_n = nM_1 - M_1 \left(\frac{\gamma}{16}\right)^2 \left(\frac{n^3 - n}{6}\right) + 0(\gamma^4)$$

$$M_1 = m + 0(\gamma^2) \tag{19}$$

Evidently the $n = 1$ state is the original "elementary particle" of the theory. It disappears for strong coupling $\gamma' > 8\pi$. For $n \geq 2$ we can interpret the states as loosely bound composites of n of the "elementary" bosons.

In its range of validity, $\gamma' n \ll 1$, Eq. (19) agrees with the exact quantum mechanical result. For the $n = 2$ state further checks are available . The formula

$$\frac{2 M_1 - M_2}{M_1} = 2 - 2 \cos \frac{\gamma'}{16} \tag{20}$$

has been checked against the exact (perturbation theory) quantum mechanical answer[2)] in orders γ^2, γ^3 and γ^4. Remarkably it agrees. To compute the quantum mechanical γ^4 term one has to keep two loop diagrams in the kernel of the Bethe-Salpeter equation.

Coleman[5)] has shown that there is a correspondence between the Sine-Gordon Lagrangian and the massive Thirring model

$$\bar{\Psi} i \not\partial \Psi - M \bar{\Psi} \Psi - \frac{g}{2} (\bar{\Psi} \gamma^\mu \Psi)^2 \tag{21}$$

with coupling constant

$$g = \frac{\pi}{2} \left[\frac{8\pi - \gamma'}{8\pi} \right] \tag{22}$$

The fermions are the solitons. For γ' near 8π, g is small and one can sum diagrams in the Thirring model. For small positive g there is one bound state in agreement with Eq. (18). The formula

$$\frac{2M(\text{soliton}) - M_1}{M(\text{soliton})} = 2 - 2\cos\left[\frac{g}{1 + \frac{2g}{\pi}}\right] \tag{23}$$

has been checked[2)] in order g^2 and g^3.

For $\gamma' > 8\pi$, the soliton-antisoliton interaction is repulsive so that there are no bound states.

It is almost certainly true that the WKB mass formulas are exact for the Sine-Gordon equation. This is another accident which I would not expect to occur in more realistic theories. I do want to stress one thing however. It could have turned out that a WKB method is nonsense for a relativistic system with an infinite number of degrees of freedom. The Sine-Gordon equation shows that this is not the case.

C. The φ^4 Theory.[2)]

The Lagrangian

$$\tfrac{1}{2}(\partial\mu\,\varphi)^2 + \frac{m^2}{2}\varphi^2 - \frac{\lambda}{4}\varphi^4 \tag{24}$$

does not lead to equations which are exactly solvable. One does know that the theory has two degenerate vacua at $\varphi = \pm \frac{m}{\sqrt{\lambda}}$ and that there is a kink state which connects the two. To go further, it is interesting to look for the analog of the breather, which can be found only approximately.[2)]

One introduces a small arbitrary parameter ϵ and scaled variables

$$\tau = \frac{\sqrt{2}\, m\, t}{\sqrt{1+\epsilon^2}}$$

$$\xi = \frac{\sqrt{2}\, m\, \epsilon\, x}{\sqrt{1+\epsilon^2}} \tag{25}$$

A periodic solution of the form

$$\frac{\varphi - \langle\varphi\rangle_{vac}}{\langle\varphi\rangle_{vac}} = \epsilon f_1(\xi) \sin \tau + \epsilon^2 g_1(\xi) + \epsilon^2 g_2(\xi) \cos 2\tau$$

$$+ \epsilon^3 f_3(\xi) \sin 3\tau + \ldots\ldots \tag{26}$$

is then assumed. For small ϵ the solution is

$$f_1 = \frac{2}{\sqrt{3} \cosh \xi}$$

$$g_1 = g_2 = -\frac{3}{4} f_1^2$$

$$f_3 = -\frac{1}{16} f_1^3 \tag{27}$$

The quantization rule forces ϵ to be of order λ so that weak coupling is implied. The particle spectrum turns out to be

$$M_n = nM_1 - \frac{9}{16}\left(\frac{\lambda}{m^2}\right)^2 M_1 \left(\frac{n^3 - n}{6}\right) + 0(\lambda^4) \tag{28}$$

$$M_1 = m\sqrt{2}\left[1 - \frac{3}{32}\left(\frac{\lambda}{m^2}\right)^2\right] + 0(\lambda^4)$$

Again the n = 1 state is the "elementary" particle and the higher states are the loosely bound composites of n elementary bosons. To the indicated order the mass formula is actually exact. It will cease to be exact in higher orders. In this calculation M_n has a maximum at $n_{max} \approx \frac{4\sqrt{2}m^2}{3\lambda}$. It is probable that only those states with $n < n_{max}$ exist. The kink-antikink threshold is very close to $M_{n\,max}$ and as λ increases the higher states probably break up into kink-antikink pairs.

The real point of this φ^4 calculation is that semi-classical methods can be used even when the classical equations are not exactly solvable.

D. The Gross-Neveu Model.[6)]

The Gross-Neveu model is defined by the Lagrangian density

$$\mathcal{L} = \bar{\Psi}\, i\not{\partial}\, \Psi + \frac{g^2}{2}\left(\bar{\Psi}\Psi\right)^2 \tag{29}$$

where Ψ is a column vector made out of N two component fermion fields. An equivalent Lagrangian density is

$$\mathcal{L} = \bar{\Psi} i \not{\partial} \Psi - g \sigma \bar{\Psi} \Psi - \frac{\sigma^2}{2} \tag{30}$$

To do a semi-classical calculation it is convenient to integrate out the fermions obtaining an effective action for σ

$$S_{eff} = - \int \frac{\sigma^2}{2} d^2x + \mathrm{tr}\, \ell n(i \not{\partial} + \sigma) \tag{31}$$

where the trace is defined with boundary conditions appropriate to periodic σ' s.

Surprisingly enough it is possible to find analytic solutions to the functional equation[6)]

$$\frac{\delta}{\delta\sigma(x,t)} S_{eff} = 0 \tag{32}$$

They are of three types: kinks and breathers analagous to the solitons and breathers in the Sine-Gordon equation and static "bags." The "bags" are time independent solutions with the property that $\sigma(x = +\infty) = \sigma(x = -\infty) = \langle \sigma \rangle$ vac. and are stable because they contain trapped fermions.

It has not been possible to evaluate ξ for the Gross-Neveu model. Thus we have to content ourselves with a zeroth order approximation, valid for large N, where ξ is neglected. In this approximation the states arrange themselves into supermultiplets labelled by an integer n. The common mass of the members of a supermultiplet is[6)]

$$M_n = g < \sigma >_{vac} \frac{2N}{\pi} \sin \frac{n\pi}{2N} \tag{33}$$

$$n = 1, 2 \ldots . N - 1$$

Within each supermultiplet the states are arranged into ordinary symmetry multiplets labeled by an internal symmetry index n_0. The total number of states is very large. Counting all degeneracies, for N = 10 there are roughly twenty thousand states.

The spectrum for N = 7 is shown in Fig. 1. The lowest state at n = 1 is the elementary fermion. The states above it are fermion resonances and, classically, are σ breathers containing a trapped fermion. The $n_0 = 0$ states, those furthest to the left, are mesons with the quantum numbers of the vacuum. Along the diagonal $n = n_0$ are nuclei and above them are excited nuclei. The spectrum ends at the kink-antikink threshold $g < \sigma >_{vac} 2N/\pi$.

From Eq. (33) one sees that coupling g cancels out of mass ratios which depend only on N. This is a general consequence of the renormalization group applied to this model.

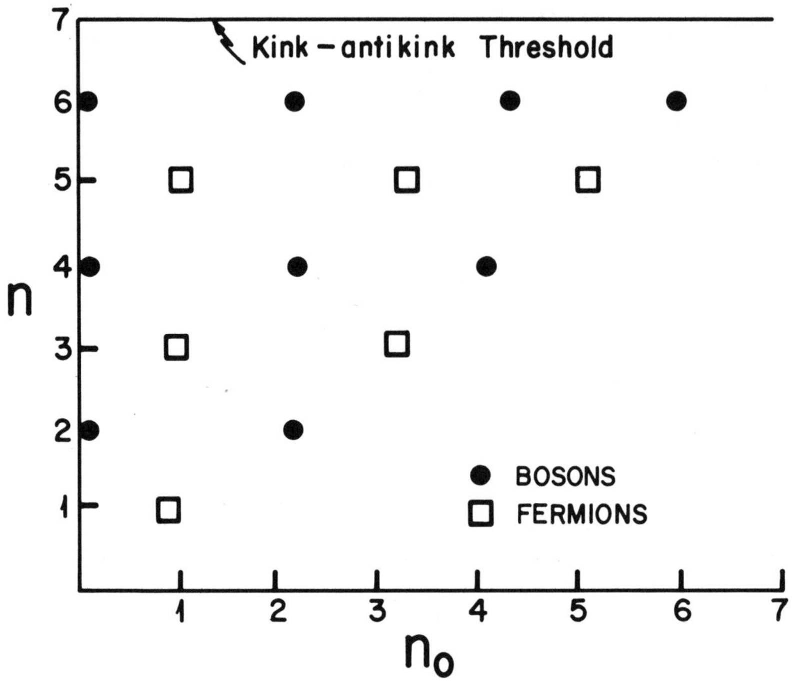

FIG. 1

The particle spectrum in the Gross-Neveu model for N = 7. The meaning of the quantum numbers n and n_0 is explained in the text.

REFERENCES

There are now many papers on this subject. A bibliography complete through spring 1975 can be found in R. Rajaraman, IAS preprint COO 2220-47, to be published in Physics Reports. I will list only a few papers which are directly relevant to this talk.

1) R. Jackiw, Talk at the conference on Gauge Theories and Modern Field Theory, Northeastern, 1975. See also J. Goldstone and R. Jackiw, Phys. Rev. D11, 1486 (1975).

2) R. Dashen, B. Hasslacher and A Neveu, Phys. Rev. D 11, 3424 (1975).

3) C. Nohl, Princeton preprint.

4) L. Faddeev, IAS preprint.

5) S. Coleman, Phys. Rev. D 11, 2038 (1975).

6) R. Dashen, B. Hasslacher and A. Neveu, IAS preprint, COO 2220-46, to be published in Phys. Rev.

DISCUSSION

P. MARTIN (Harvard)

Luther has pointed out that the Thirring and Sine-Gordon models are also isomorphic to the ground state of the X-Y model which has been exactly solved. The bound states of the spin waves corresond to the bound states you obtained by WKB. This seems to prove that the classical approximation is not necessary.

DASHEN

I got a letter from him telling me just that, but thank you. Let me make one remark about that. It's really too bad that the electron is spin ½. Because in solid state physics you can essentially study one dimensional systems for fermions by expanding something that's sufficiently isotropic and in the Gross-Neveu model, the number of fermions in there in the one dimensional electron gas is of course just the number of spin directions of electron because in one dimension the spin doesn't really mean anything. Now for n=2 that spectrum is not interesting because there's really only one state sitting there. If we only had spin 3/2 electrons so we can look in a natural crystal and see whether or not that thing was there, it would be a lot of fun but unfortunately we don't.

R. JACKIW (MIT)

I would like to understand better your discussion of the corrections to the non-linear Schroedinger equation. The question I have is the following: Let me distinguish the Bohr-Sommerfeld quantization from the WKB. The WKB includes the Gaussian and the Bohr-Sommerfeld does not. Incidentally, even in potential theory in general it's not one but it's a complicated thing that depends on the potential. Now, if you do just Bohr-Sommerfeld for that problem then that phase integral is identical to the number operator which has to be n. If you say that you're doing WKB then that phase integral is equal to n plus something, but the number integral is still equal to n and how do you reconcile those two things?

DASHEN

That gets a little bit technical. It's a renormalization phenomena.

JACKIW (MIT)

n is the correct quantization?

DASHEN

Yes, n is the correct quantization but there's a renormalization thing as you would say that comes in there.

VOICE FROM THE AUDIENCE

Can you say anything in the one dimensional attractive δ function model, about the relationship between the classical field solution and the bound state wave functions in the quantum mechanical problem.

DASHEN

These classical field solutions haven't got anything to do with the wave functions because if you think of the classical solution in hydrogen as an orbit that's running around in time, what that means is that the bound state wave function is concentrated around the orbit. So I have an orbit going around and the wave function is sort of at it's largest around the orbit. But other than that there isn't any direct connection.

LIST OF GTMFT CONFERENCE ATTENDEES

A

Aaron, R. - Northeastern University
Abbott, L. - Brandeis University
Adler, S. L. - Institute for Advanced Study
Alles, W. - Istituto di Fisica, Bologna, Italy
Anderson, P. W. - Bell Laboratories
Argyres, P. N. - Northeastern University
Arnowitt, R. - Northeastern University

B

Baker, H. C. - Berea College
Baluni, V. - Massachusetts Institute of Technology
Barnes, K. J. - University of Southampton
Barnett, M. - Harvard University
Bars, I. - Yale University
Baym, G. - University of Illinois
Bender, C. - Massachusetts Institute of Technology
Bernstein, J. - Stevens Institute of Technology
Bicerano, J. - Harvard University
Bincer, A. M. - University of Wisconsin
Blanar, G. - Northeastern University
Bludman, S. A. - University of Pennsylvania
Bose, S. K. - University of Notre Dame
Boyer, C. - Northeastern University

C

Campbell, B. A. - McGill University
Cardarelli, D. - Northeastern University
Casella, R. - National Bureau of Standards
Chahine, C. - Brandeis University
Chang, L. N. - University of Pennsylvania
Chang, S. S. - Syracuse University
Chase, N. - Northeastern University
Chiaverini, D. - Northeastern University
Chodos, A. - Massachusetts Institute of Technology
Choudhury, A. L. - Elizabeth City State University
Coleman, S. - Harvard University
Collins, J. C. - Princeton University
Cooper, F. - Los Alamos Scientific Laboratory
Cornwall, J. M. - University of California at Los Angeles
Cung, V. K. - Johns Hopkins University
Cvitanovic, P. - Institute for Advanced Study

D

Dashen, R. - Institute for Advanced Study
DeGrand, T. A. - Massachusetts Institute of Technology
De Rújula, A. - Harvard University
Deser, S. - Brandeis University
Detar, C. - Massachusetts Institute of Technology
Diamond, P. - Massachusetts Institute of Technology
Dokos, C. - Harvard University
Dolan, L. - Massachusetts Institute of Technology
Donoghue, J. F. - University of Massachusetts at Amherst
Dookovskoy, A. - Massachusetts Institute of Technology

E

Elias, V. - University of Massachusetts at Amherst

F

Federbush, P. - University of Michigan
Feldman, D. - Brown University
Fleming, G. N. - Pennsylvania State University
Freedman, D. Z. - SUNY at Stony Brook
Freund, P. G. O. - Enrico Fermi Institute, Chicago
Friedman, M. - Northeastern University
Fritzsch, H. - California Institute of Technology
Fulton, T. - Johns Hopkins University

G

Gauthier, L. - Universite de Montreal
Gavrielides, A. - Purdue University
Gettner, M. W. - Northeastern University
Glashow, S. L. - Harvard University
Glaubman, M. J. - Northeastern University
Goldberg, H. - Northeastern University
Goldstone, J. - Massachusetts Institute of Technology
Golowich, E. - University of Massachusetts at Amherst
Greenberg, O. W. - University of Maryland
Grisaru, M. T. - Brandeis University
Gross, E. P. - Brandeis University
Gulshan, A. - Columbia University
Guralnik, G. - Brown University
Gürsey, F. - Yale University

H

Haavisto, J. R. - Boston University
Haller, K. - University of Connecticut
Harnad, J. - Dawson College
Hellman, W. S. - Boston University
Hongoh, M. - Universite de Montreal
Hsu, J. P. - University of Texas at Austin

J

Jackiw, R. - Massachusetts Institute of Technology
Jacobs, L. - Massachusetts Institute of Technology
Jaffe, A. - Harvard University
James, P. B. - University of Missouri
Joglekar, S. - Institute for Advanced Study
Johnson, K. - Massachusetts Institute of Technology
Jose, J. - Brown University
Ju, I. - University of Michigan

K

Kaku, M. - City College of CUNY
Kanavi, S. - Northeastern University
Kang, J. S. - Brandeis University
Kang, K. - Brown University
Kannenberg, L. C. - University of Lowell
Kayser, B. - National Science Foundation
Kellogg, S. - Brandeis University
Kephart, T. W. - Northeastern University
Kerman, A. - Massachusetts Institute of Technology
Kim, J. E. - Brown University
Kiskis, J. - Massachusetts Institute of Technology
Klein, A. - Massachusetts Institute of Technology
Krapchev, V. - Massachusetts Institute of Technology
Krejs, F. - University of Pennsylvania
Kummer, W. - Princeton University
Kwiecien, J. A. - Northeastern University

L

Labonte, G. - Universite de Montreal
Langacker, P. G. - University of Pennsylvania
Lee, B. W. - Fermi-National Accelerator Laboratory
Lee, S. Y. - Purdue University
Li, L. F. - Carnegie-Mellon University
Lichtenberg, D. B. - Indiana University

M

Mallett, R. L. - University of Connecticut
Mandula, J. E. - Massachusetts Institute of Technology
Mansouri, F. - Yale University
Marciano, W. J. - Rockefeller University
Martin, P. - Harvard University
McLerran, L. - Massachusetts Institute of Technology
Mello, P. A. - Instituto de Fisica, Universidad Nacional Autonema de Mexico
Migneron, R. - University of Western Ontario
Minkowski, P. - California Institute of Technology
Monsay, E. - Princeton University
Muzinich, I. J. - Brookhaven National Laboratory

N

Nath, P. - Northeastern University
Ng, J. N. - University of Pennsylvania
Ng, W. C. - University of Maryland

O

O'Donnell, P. J. - University of Toronto
Oseguera, U. - Massachusetts Institute of Technology

P

Pagels, H. - Rockefeller University
Pais, A. - Rockefeller University
Parke, S. - Harvard University
Parsa, Z. - Polytechnic Institute of New York
Paschos, E. A. - Brookhaven National Laboratory
Patera, J. - Universite de Montreal
Pati, J. C. - University of Maryland
Peaslee, D. C. - Australian National University
Poggio, E. R. - Harvard University

Q

Quinn, H. R. - Harvard University

R

Roman, P. - Boston University
Ronan, M. T. - Northeastern University
Russo, M. J. - Westfield State University

S

Scalettar, R. - California State University
Schechter, J. - Syracuse University
Schnitzer, H. J. - Brandeis University
Schroer, B. - Freie Universitat, Germany
Schwartz, L. - Brandeis University
Schwinger, J. - University of California at Los Angeles
Shankar, R. - Harvard University
Sharp, R. T. - Universite de Montreal
Shei, S. S. - Institute for Advanced Study
Sobelman, G. - Harvard University
Soh, K. S. - Cornell University
Sohn, R. - University of Connecticut
Srikanth, M. L. - Boston University
Srivastava, G. - Northeastern University
Srivastava, Y. - Northeastern University
Steinhardt, P. - Harvard University
Sucher, J. - University of Maryland
Swank, L. J. - Yale University

T

Teplitz, V. - Virginia Polytechnic Institute
Thacker, H. - SUNY at Stony Brook
Thorn, C. - Massachusetts Institute of Technology
Tomboulis, E. - Massachusetts Institute of Technology
Tomozawa, Y. - University of Michigan
Townsend, P. - Brandeis University
Treiman, S. - Princeton University
Tsao, H. S. - Institute for Advanced Study
Tudron, T. N. - Princeton University

U

Ukawa, A. - Cornell University
Ulehla, M. - Massachusetts Institute of Technology

V

Van Nieuwenhuizen, P. - SUNY at Stony Brook
Vaughn, M. - Texas A & M University
von Goeler, E. - Northeastern University

W

Ward, B. F. L. - Purdue University
Weinberg, E. J. - Columbia University
Weinberg, S. - Harvard University
Weingarten, D. - University of Rochester
Wick, G. C. - Columbia University
Widom, A. - Northeastern University
Willemsen, J. - Massachusetts Institute of Technology
Willey, R. S. - University of Pittsburgh
Wilson, K. - Cornell University
Winternitz, P. - Universite de Montreal
Woo, G. - Massachusetts Institute of Technology
Wu, T. T. - Harvard University

Y

Yalin, P. - Harvard University
Yao, P. Y. - University of Michigan
Yeung, P. - Massachusetts Institute of Technology
Yildiz, A. - Harvard University
Yilmaz, H. - Perception Technology Corporation

Z

Zepeda, A. - Centro de Investigacion del I.P.N., Mexico
Zumino, B. - CERN